Lecture Notes in Physics

The Editorial Policy for Proceedings

The series Lecture Notes in Physics reports new developments in physical research and teaching – quickly, informally, and at a high level. The proceedings to be considered for publication in this series should be limited to only a few areas of research, and these should be closely related to each other. The contributions should be of a high standard and should avoid lengthy redraftings of papers already published or about to be published elsewhere. As a whole, the proceedings should aim for a balanced presentation of the theme of the conference including a description of the techniques used and enough motivation for a broad readership. It should not be assumed that the published proceedings must reflect the conference in its entirety. (A listing or abstracts of papers presented at the meeting but not included in the proceedings could be added as an appendix.)
When applying for publication in the series Lecture Notes in Physics the volume's editor(s) should submit sufficient material to enable the series editors and their referees to make a fairly accurate evaluation (e.g. a complete list of speakers and titles of papers to be presented and abstracts). If, based on this information, the proceedings are (tentatively) accepted, the volume's editor(s), whose name(s) will appear on the title pages, should select the papers suitable for publication and have them refereed (as for a journal) when appropriate. As a rule discussions will not be accepted. The series editors and Springer-Verlag will normally not interfere with the detailed editing except in fairly obvious cases or on technical matters.
Final acceptance is expressed by the series editor in charge, in consultation with Springer-Verlag only after receiving the complete manuscript. It might help to send a copy of the authors' manuscripts in advance to the editor in charge to discuss possible revisions with him. As a general rule, the series editor will confirm his tentative acceptance if the final manuscript corresponds to the original concept discussed, if the quality of the contribution meets the requirements of the series, and if the final size of the manuscript does not greatly exceed the number of pages originally agreed upon. The manuscript should be forwarded to Springer-Verlag shortly after the meeting. In cases of extreme delay (more than six months after the conference) the series editors will check once more the timeliness of the papers. Therefore, the volume's editor(s) should establish strict deadlines, or collect the articles during the conference and have them revised on the spot. If a delay is unavoidable, one should encourage the authors to update their contributions if appropriate. The editors of proceedings are strongly advised to inform contributors about these points at an early stage.
The final manuscript should contain a table of contents and an informative introduction accessible also to readers not particularly familiar with the topic of the conference. The contributions should be in English. The volume's editor(s) should check the contributions for the correct use of language. At Springer-Verlag only the prefaces will be checked by a copy-editor for language and style. Grave linguistic or technical shortcomings may lead to the rejection of contributions by the series editors. A conference report should not exceed a total of 500 pages. Keeping the size within this bound should be achieved by a stricter selection of articles and not by imposing an upper limit to the length of the individual papers. Editors receive jointly 30 complimentary copies of their book. They are entitled to purchase further copies of their book at a reduced rate. As a rule no reprints of individual contributions can be supplied. No royalty is paid on Lecture Notes in Physics volumes. Commitment to publish is made by letter of interest rather than by signing a formal contract. Springer-Verlag secures the copyright for each volume.

The Production Process

The books are hardbound, and the publisher will select quality paper appropriate to the needs of the author(s). Publication time is about ten weeks. More than twenty years of experience guarantee authors the best possible service. To reach the goal of rapid publication at a low price the technique of photographic reproduction from a camera-ready manuscript was chosen. This process shifts the main responsibility for the technical quality considerably from the publisher to the authors. We therefore urge all authors and editors of proceedings to observe very carefully the essentials for the preparation of camera-ready manuscripts, which we will supply on request. This applies especially to the quality of figures and halftones submitted for publication. In addition, it might be useful to look at some of the volumes already published. As a special service, we offer free of charge LATEX and TEX macro packages to format the text according to Springer-Verlag's quality requirements. We strongly recommend that you make use of this offer, since the result will be a book of considerably improved technical quality. To avoid mistakes and time-consuming correspondence during the production period the conference editors should request special instructions from the publisher well before the beginning of the conference. Manuscripts not meeting the technical standard of the series will have to be returned for improvement.

For further information please contact Springer-Verlag, Physics Editorial Department V, Tiergartenstrasse 17, W-6900 Heidelberg, FRG

K. Goeke P. Kroll H.-R. Petry (Eds.)

Quark Cluster Dynamics

Proceedings of the 99th WE-Heraeus Seminar
Held at the Physikzentrum Bad Honnef, Germany
29 June - 1 July 1992

Springer-Verlag Berlin Heidelberg GmbH

Editors

Klaus Goeke
Institut für Theoretische Physik, Ruhr-Universität Bochum
Postfach 10 21 48, W-4630 Bochum 1, Germany

Peter Kroll
Universität/Gesamthochschule Wuppertal
W-5600 Wuppertal, Germany

Herbert-Rainer Petry
Institut für Theoretische Kernphysik, Universität Bonn
Nussallee 14, W-5300 Bonn 1, Germany

ISBN 978-3-662-13942-4   ISBN 978-3-540-47560-6 (eBook)
DOI 10.1007/978-3-540-47560-6

Originally published by Springer-Verlag Berlin Heidelberg New York in 1993
Softcover reprint of the hardcover 1st edition 1993

Typesetting: Camera ready by author/editor
58/3140-543210 - Printed on acid-free paper

PREFACE

The 99th W.E. Heraeus-Seminar on "Quark Cluster Dynamics" was held in Bad Honnef from 29 June to 1 July 1992, in order to bring together theoretical physicists who are interested in quark correlations of hadronic matter. The contributions to these proceedings focus on the following subjects: dynamical symmetries of heavy quarks, diquarks, chiral-invariant quark interactions, models of confinement and quark aspects of hadronic interactions. The theoretical aspects of each of these fields are supplemented by a review of the most recent experimental data.

"Quark Cluster Dynamics" was financially supported by and profited from the help of the Physikzentrum Bad Honnef, the BMFT and the Heraeus foundation. We want to express our deep gratitude to these organizations.

This volume is dedicated to the late Konrad Bleuler, who died on 1 January 1992. Together with the whole physics community we share the loss of an outstanding physicist who was particularly associated with our area of research.

Physikzentrum Bad Honnef
01.12.1992

K. Goeke
P. Kroll
H.R. Petry

Konrad Bleuler
23.09.1912   01.01.1992
Orbituary

on the opening of the Symposium
"Quark Cluster-Dynamics"

by Max Georg Huber

Originally it had been the intention of the organizers of this conference to have this symposium dedicated to Prof. K. Bleuler - who was to have his 80th birthday later this summer. Unfortunately a year ago Prof. Bleuler became seriously ill and on January 1st 1992 he died as a consequence of an ill-natured brain tumor. The organizers of this conference - and I think all of us, who participate in this meeting - very strongly feel that we should honour Prof. Bleuler on this occasion - in particular since the agenda of this conference deals with problems he was most interested in during the last decade of his life.

Konrad Bleuler was born in Herzogenbuchsee (Kanton Bern/Switzerland) on September 23, 1912. He grew up in a liberal Swiss family which was particularly connected with the cultural life of Zurich. His grandfather was president of the Eidgenössische Hochschule (ETH) in Zurich.

Konrad Bleuler first studied engineering and physics at the ETH from 1931-1936 with the intention to pursue a career in industry. At this time physics in Zurich was dominated by the personalities of Pauli, Wenzel and Scherrer; at the same time Heinz Hopf was teaching mathematics. The tradition, founded by Hermann Weyl, to understand these two sciences as a deep spiritual unity, influenced Konrad Bleuler for the rest of his life. He, therefore, abandoned his industrial work, obtained a PhD in mathematics in 1942 at the ETH and started a career as a theoretical physicist, first as assistent of Stückelberg in Geneva, then of Wenzel at the ETH Zurich where he became Privatdozent and titular professor in 1945. A longer term of study in Rome founded his lifelong love for Italy and its culture. Guest visits in Birmingham, Stockholm, Helsinki and Genova followed. During these years he became an internationally renowned scientist through his work on the indefinite metric in quantum electrodynamics.

In 1957 he accepted an offer as full professor at the university of Neuchâtel (Switzerland), where he started his work on the structure of atomic nuclei within the framework of meson-theoretical field theories. This work was successfully continued after he accepted a professorship at the University of Bonn in 1960; it is still regarded as one of the pioneering contributions to the understanding of the nucleon-nucleon interaction. In his last years he regarded the quark structure of hadrons as a particular challenge and tried to derive a nuclear model directly from the quark picture.

His love for the mathematical, and that meant for him, geometrical foundations of physics, however, never stopped. He founded and inspired a series of international conferences on this subject, which remained a world-wide attractive forum for the most recent developments in this field.

Konrad Bleuler was a remarkable scholar with broad interests - in the fields of science as well as in music and literature. He has had many friends - not only among scientists but also among poets and artists. He was particulary fascinated by the esthetical aspects of physical laws and he had a genuine gift to reveal them in his talks and lectures. His possibility to immediately understand new and complex phemomena was impressive, and most of us who have had the opportunity to see him act during conferences, seminars or lectures never will forget his particular way to ask questions, the many comments he made spontaneously and unprepared, and his polite interventions which made the original contributions not only more transparent, but most often put the details in proper proportions and - sometimes - he most politely applied corrections to the speaker's results.

Basically he was a very sensitive person. He could become quite nervous and, in fact, often quite angry whenever he thought the physical issue was diluted by stupid comments or even distorted by ignorance or incompetence. He was devoted to the eduction of young students, and he considered it a crime when professors did not care sufficiently for their courses and for the education of their students and collaborators.

I am sure he would have enjoyed being here with us at this symposium. We all, who have had the pleasure to know him, do miss him. The best way to keep his legacy is to remember his principles and to act accordingly.

Contents

Weak Decays of Light and Heavy Hadrons [1]

Berthold Stech

Institut für Theoretische Physik, Universität Heidelberg,
Philosophenweg 16, D-6900 Heidelberg

Abstract: Nonperturbative QCD strongly influences weak decays. No exact
treatment is available. Nevertheless, using simple models, a semi-quantitative
understanding of exclusive decays has recently been achieved, from strange particle decays up to the decay of b-flavored hadrons. An outline of the physics
involved is presented.

1   Introduction

Weak decays have always been a rich source of information about the nature of
elementary particle interactions. Many years ago, β- and μ-decay experiments
revealed the form of the effective flavor-changing interaction at low momentum
transfer. At the same time, the selection rules and other consequences of the
weak interaction were instrumental in exploring the properties of nuclei. Today
we are in a similar situation: weak decays of heavy quarks are employed for tests
of the standard model and measurements of its parameters. Quite obviously,
they offer the most direct way to determine the weak mixing angles and to
test the unitarity of the Kobayashi-Maskawa matrix. On the other hand, weak
transitions are also of great help in studying that part of strong interaction
physics which is least understood, the nonperturbative confinement forces. Both
tasks complement each other: an understanding of long-range QCD — of the
connection between quark and hadron properties — is a necessary prerequisite
for a precise determination of the weak mixing angles and the CP-violating
phase.

The simplest processes are those involving a minimum number of hadrons,
ie. a single hadron (or hadron resonance) in the final state of a semi-leptonic
decay, or two hadrons in a nonleptonic decay. In recent years, much progress
has been achieved towards an understanding of these transitions. Simple bound-
state models are able to describe, in a semi-quantitative way, the current matrix

---

[1]Supported in part by Deutsche Forschungsgemeinschaft, Bonn

elements occurring in semi-leptonic decay amplitudes [1],[2]. A factorization prescription for reducing the hadronic matrix elements of four-quark operators to a product of matrix elements shed light onto the dynamics of nonleptonic processes, where even drastic effects had been lacking an explanation before [3],[4]. In this talk I will shortly describe the tools useful in dealing with the influence of nonperturbative QCD on weak decay processes. I will mention a few of the problems one encounters and also report some of the successes achieved. Since this is a meeting on quark clusters, I will make some remarks on $q - \bar{q}$-wave functions and emphasize particularly the effect of diquark states in low energy transitions. I then can only just mention the usefulness of the factorization approximation for the more energetic nonleptonic D- and B-decays. Nothing will be said about the interesting possibility to study CP-violating amplitudes in B-meson decays.

The theoretical basis is the Standard Model of particle physics. Possible deviations from this well established model are expected to be very small and will not be considered here.

2 Semi-Leptonic Decays

Semi-leptonic decays involve the leptonic and hadronic currents of the standard model. Simplest examples are the τ-decay processes $\tau^- \rightarrow \nu_\tau \pi^- (\rho^-, a_1^-, ...)$. Here the nonperturbative part concerns the production out of the vacuum of the $q - \bar{q}$-bound states $\pi, \rho, a_1, ...$ by the strangeness-conserving quark current. The corresponding matrix elements define so-called meson-decay constants. Their experimental determination can be used to check our very limited ways to deal with nonperturbative QCD: quark model estimates, QCD sum rule calculations and lattice gauge theory results.

The decay $K^- \rightarrow \pi^0 e^- \bar{\nu}_e$ is an example of a decay which involves a transition formfactor between light mesons. This decay amplitude is also proportional to the Cabibbo-Kobayashi-Maskawa matrix element V_{us}. The $K - \pi$ transition formfactor is again a nonperturbative quantity depending on the internal structure of the mesons. Thus, to obtain V_{us} from experiment, one had to find a point where this formfactor is known in spite of its nonperturbative nature. Fortunately, because of the small current quark masses involved, π-mesons and K-mesons seen at high velocity have almost the same structure. Thus, for $q^2 = 0$ which corresponds to mesons with equal momenta and high energy the $K - \pi$ formfactor differs very little from 1. This way V_{us} could be determined [5]. For $q^2 \neq 0$, on the other hand, there is no strict way of calculating the formfactor. But one learns much by confronting model calculations with the experimental data.

A somewhat different situation is present in the decay of heavy quarks, in so-called "heavy to heavy" transitions. Let us consider, for instance, the decay $\bar{B}^0 \rightarrow D^* e^- \bar{\nu}$. Again, we need a reliable value at least for one of the $B \rightarrow D^*$ transition formfactors in order to extract the Kobayashi-Maskawa matrix element V_{cb} from experiment. Here, the spin-flavor symmetry [6], strictly valid for infinitely heavy b and c quarks ($m_Q \to \infty$) may be used: For $q^2 = q^2_{max}$, (i.e. both mesons have the same velocity), the special formfactor $\xi_{A_1}(q^2)$ has the value [7]

$$\xi_{A_1}\left(q^2 = q^2_{max}\right) = 1 + O(\alpha_s) + O\left(\left(\frac{\Lambda_{QCD}}{m_Q}\right)^2\right) \qquad (1)$$

The QCD correction $O(\alpha_s)$ is known to next-to-leading order and very small [8]. The mass correction term of order $(\Lambda_{QCD}/m_Q)^2$ should be reasonably small as estimated from model calculations. For q^2-values different from $q^2 = q^2_{max}$, however, we are again in the fully nonperturbative regime.

I like to say now a few words about $q\bar{q}$ bound state models because most formfactor calculations have been performed using them. Here, hadrons are described by constituent quarks as the only internal degrees of freedom.

The simplest possibility is a nonrelativistic quark model (with some relativistic corrections included) using harmonic oscillator wave functions [2]. This model is suited for formfactor calculations near $q^2 = q^2_{max}$, where initial and final hadrons have small relative velocities. This region of validity covers only a small part of the phase space, however. Also, in some instances, the formfactor has a pole not very far from $q^2 = q^2_{max}$ making the calculation uncertain.

To avoid these problems, a relativistic quark model has been applied using relativistic oscillator wave function on the light cone [1],[3]. This model is best suited for formfactor calculations near $q^2 = 0$ where initial and final mesons have equal and large momenta. An extension to q^2 values away from zero is not straightforward, however, since the wave functions used are not solutions of the Bethe-Salpeter equation. Probability and (partial) current conservation is not guaranteed. Therefore, pole formulae have been used to extrapolate from $q^2 = 0$ to $q^2 = q^2_{max}$. In spite of its simplicity this model and related ones [9] turned out to be remarkably successful. They describe numerous transitions in fair agreement with experiment. But surely, they will no more be good enough as soon as very precise experimental data are available. What is really wanted — in the general framework of $q - \bar{q}$ bound state models — are solutions of the Bethe-Salpeter equation. A B-meson wave function, for instance, can then be written as a 4×4 matrix $\phi_B(x_1, x_2)$ and the vector formfactor to a meson X could be calculated for all q^2 using the Mandelstam formula

$$\langle X|j_\mu|B\rangle = \int tr\{\phi_X^+ \gamma_\mu \phi_B(i\overleftrightarrow{\partial}_2 + m_2)\}d^4x_2 \qquad (2)$$

x_2 : coordinate of the spectator quark.

Unfortunately, such a calculation requires fully covariant wave functions for space and time-like separations of the particle coordinates. They are not available at present.

For the special case of heavy to heavy transitions, a trick was found to obtain the complete q^2-dependence of formfactors from any model which gives the formfactors at $q^2 = 0$ [10]. In this case (to zero and first order in $1/m_Q$) the dependence of the formfactors on the particle masses at $q^2 = 0$ can be converted into the q^2-dependence for fixed masses. This method allowed to improve the BSW model [1], [3] for heavy to heavy transitions such that the heavy quark symmetries and the normalization conditions at $q^2 = q^2_{max}$ are fully respected [11].

The present status of formfactor calculations is as follows: For the semileptonic heavy to light transitions $D \rightarrow K, D \rightarrow K^*, \bar{B}^0 \rightarrow \pi^+$ the best determinations are from detailed QCD sum rule calculations [12]. The much simpler to perform relativistic quark model calculations are in good agreement with these results except for $D \rightarrow K^*$. For this decay all quark models give a transition rate too large by about a factor 2. For heavy to heavy transitions QCD sum rule results and quark model results agree and are in accord with the experimental decay spectra. Because of eq. (1) it is possible to extract a reliable value for V_{cb} (for $\tau_B = 1.3 \ ps$) [13]

$$|V_{cb}| = 0.040 \pm 0.005 \tag{3}$$

for which the error is mainly experimental.

3   Nonleptonic Decays

The dynamics of nonleptonic weak decays is strongly influenced by the confining color forces among the quarks. In contrast to semi- leptonic transitions, where all long-distance QCD effects are described by few hadronic formfactors appearing in hadronic matrix elements of color-singlet quark currents, nonleptonic processes are complicated by the phenomenon of quark rearrangement due to the exchange of soft and hard gluons. The theoretical description involves matrix elements of local four-quark operators, which are much harder to deal with than current operators. Up to very recently, these strong interaction effects prevented a coherent understanding of nonleptonic decays. The famous $|\Delta I| = 1/2$ rule is a prominent example: Already shortly after the discovery of strange particles it was observed that hyperon as well as K-decay processes associated with isospin change $|\Delta I| = 1/2$ are strikingly dominant over other transitions. For instance, in K-decays one found the decay ratio

$$\frac{\Gamma(K_s \rightarrow \pi^+\pi^-)}{\Gamma(K^- \rightarrow \pi^0\pi^-)} \simeq 450 \tag{4}$$

instead of an expected ratio close to one. With the discovery of charmed mesons a further puzzle appeared: the nonleptonic decay width of the D^0- and the D^+-mesons are not equal, but differ by the factor $\simeq 3.2$. Furthermore, the mass difference between the long- and short-lived K-mesons is about a factor of 3 larger than what is expected from the usual box diagram calculation.

To lowest order in the standard model, nonleptonic weak decays are governed by a single W-exchange diagram. Strong interactions affect this simple picture in a two-fold way. Hard gluon corrections can be accounted for by perturbative methods and renormalization group techniques. They give rise to new effective weak vertices. Long-distance confinement forces are responsible for the binding of quarks inside the asymptotic hadron states. The basic assumption in the calculation of nonleptonic amplitudes is that it is possible to separate the two regimes by means of the operator product expansion incorporating all long-range QCD effects in the hadronic matrix elements of local four-quark operators. This treatment appears well justified because of the long-time scale involved in the formation of the final hadrons.

3.1 The decay of strange particles

Integrating out the heavy W-boson and top quark, one derives an effective Hamiltonian consisting of local 4-quark operators multiplied by Wilson coefficients. For the case of $\Delta S = 1$ strange particle decays one has

$$
\begin{aligned}
\mathcal{H}_{eff} \;=\; & \frac{G}{\sqrt{2}} V_{ud}^* V_{us} \{ c_1 (\bar{d}_i \gamma_\mu (1 - \gamma_5) u_i)(\bar{u}_j \gamma^\mu (1 - \gamma_5) s_j) \\
+ \; & c_2 (\bar{u}_i \gamma_\mu (1 - \gamma_5) u_i)(\bar{d}_j \gamma^\mu (1 - \gamma_5) s_j) \quad + \quad h.c.\} \\
+ \; & \text{Penguin operators} \\
& i, j \;:\; \text{color indices}
\end{aligned}
\tag{5}
$$

The coefficients c_1 and c_2 are combinations of Wilson coefficients. These are scale-dependent numbers compensating the scale dependence of the operators. At the scale where $\alpha_s(\mu) \simeq 0.5$ one has [14]

$$
c_1(\mu) \simeq 1.5 \quad c_2(\mu) \simeq -0.88
\tag{6}
$$

At this scale the coefficients for the Penguin operators are still very small: $c_6 \simeq -0.02$. It is seen that the effect of hard gluon exchange enhances the $I = 1/2$ part of the Hamiltonian and suppresses the $I = 3/2$ part. However, the enhancement is not sufficient for an explanation of the huge ratio in eq. (4). Strong nonperturbative forces must be at work in addition.

The simplest way to deal with the Hamiltonian (5) is the use of the factorization approximation [15]. For the process $K \rightarrow 2\pi$ this gives rise to simple graphs in which the strangeness-changing currents turn the K-meson into a π-meson (as in semi-leptonic decays) while the strangeness-conserving currents generate the second π-meson. As observed in ref. [16], factorization is exact in leading order in the $1/N_c$ expansion (N_c: number of quark colors). But even when keeping c_1 and c_2 fixed in this limit, the calculated amplitude for $K \rightarrow (2\pi)_{I=0}$ is still a factor $\simeq 3.6$ too small. From the observed large deviation from the lowest order result we have to conclude that long-range QCD leads to a breakdown of the $1/N_c$ expansion in the $I = 1/2$ channel!

The matrix elements occurring in the factorization approach account for $q - \bar{q}$ interactions only and miss nonperturbative effects from the forces between two quarks. From the very existence of baryons we know, however, that the QCD forces between quarks are of similar strength as those between quarks and antiquarks. Since the effective weak Hamiltonian contains Lorentz scalar $q-q$ combinations as well as $q-\bar{q}$ currents, strongly correlated spin zero diquark states can be created and annihilated in a very effective way. To see the appearance of scalar diquark currents it is instructive to view the weak interaction as a scattering process:

$$s + u \rightarrow u + d$$

According to the decomposition $3 \times 3 = 3^* \oplus 6$ the initial and final states can be in an antitriplet or sextet state with respect to color. With the help of a Fierz transformation one can bring the effective weak Hamiltonian into the form [17],[4]

$$\mathcal{H}_{eff} = \frac{G_F}{\sqrt{2}} V_{ud}^* V_{us}$$
$$\times \{ c_-(\mu)(ud)_{3^*}^+ (su)_{3^*} + c_+(\mu)(ud)_6^+ (su)_6 + \text{h.c.} \}$$
$$+ \text{ Penguin contributions} \tag{7}$$

where $(su)_{3^*,k} = \epsilon_{kij} s_i^T (C(1-\gamma_5) u_j$ etc. are scalar and pseudoscalar diquark currents, and C is the charge conjugation matrix. The coefficients

$$c_- = c_1 - c_2 \simeq 2.4 \quad \text{and} \quad c_+ \simeq c_1 + c_2 \simeq 0.65 \tag{8}$$

give directly the perturbative enhancement and suppression factors mentioned before.

Color antitriplet diquarks exist in baryons as bound or at least as strongly correlated pairs of quarks. They have dynamical significance in hadron reactions and baryon spectroscopy [18]. The status of these diquarks is similar to the status of constituent quarks: they are extended particles which dissolve into smaller partons with increasing q^2. The three $0^\pm$ states $(sd), (su), (du)$ form an SU(3) antitriplet with the (du) diquark being an isospin zero state. For low q^2 processes we can introduce local diquark field operators with these quantum numbers. The c_- part of (7) can then be rewritten:

$$\mathcal{H}_{eff}^D = \frac{G_F}{\sqrt{2}} V_{ud}^* V_{us} \frac{4}{3} c_- g_{(ud)} g_{(us)} (\phi_L^+ \lambda_6 \phi_L). \tag{9}$$

In this expression ϕ_L describes a normalized left-handed (scalar-pseudoscalar) diquark field. Obviously, $\mathcal{H}_{eff}^D$ is a pure $I = 1/2$ operator. The replacement of the first term in (7) by (9) is useful only if the couplings $g_{(ud)}, g_{(us)}$, i.e., the diquark decay constants are sufficiently large. They are defined by

$$\langle 0|(su)_{3^*,k}|(su)_l\rangle = \sqrt{\frac{2}{3}} g_{(su)}(\mu)\delta_{kl} \text{ etc.} \tag{10}$$

and have been evaluated by various methods, e. g. in a careful investigation using QCD sum rules for the color gauge invariant time-ordered product of two diquark currents [19]. One finds large coupling constants. In (9) only the combination

$$c_-(\mu) g_{(ud)}(\mu) g_{(su)}(\mu) = (0.075 \pm 0.015) \text{ GeV}^4 \tag{11}$$

enters. We note the important fact that this combination is scale-independent, the anomalous dimension factors cancel [19]. The quoted numerical result referes to a scalar (ud) diquark mass of 0.5 GeV and increases slightly for larger mass values. It is obtained for the parity-conserving part of the Hamiltonian. The value for the combination of couplings which enter the parity-violating part of the Hamiltonian is (0.090 ± 0.015) GeV^4, in agreement with little explicit chiral symmetry breaking.

Nonleptonic hyperon decays [20]

In the description of hyperon decays, baryon pole formulae and current algebra results can be applied. For instance, in the decay $\Lambda \rightarrow P\pi^-$ the parity-conserving part of the weak interaction changes the Λ into a virtual neutron which then emits a π-meson, or the π-meson is emitted first leading to a Σ^+ which then turns into a proton through the action of the weak Hamiltonian. The determination of the main contribution to P-wave decays is thus reduced to calculating the baryon matrix elements of (7). Since in a constituent quark model for baryons[2] there cannot be color sextet diquarks, eq. (9) applies and simply replaces an $(su)_{0^+}$ diquark state in the initial baryon by a $(ud)_{0^+}$ diquark leaving the remaining quark unaffected. This calculation depends very little on details of the baryon wave function and is thus reliable. Using SU(3) symmetry, the d/f ratio turns out to be precisely -1, and the combination of couplings in (7) fixes the magnitude of these baryon matrix elements.

The soft meson limit of the S-wave amplitudes is determined by the same matrix elements. To obtain the decay amplitudes for on-mass-shell π-mesons, the necessary correction terms can be found by using pole formulae consistent with the soft limit and observing that the relevant $1/2^-$ baryon intermediate states, which contain 0^- diquarks, are heavier by about 700 MeV than the corresponding $1/2^+$ baryons. With this input, the diquark contributions to both S- and P-wave amplitudes are determined in magnitude and sign. They reflect the effect of the strong QCD binding forces between quarks.

To these amplitudes we have to add the amplitudes for the direct emission of π-mesons as obtained by factorizing the color singlet current products of the full effective Hamiltonian, which has $|\Delta I| = 1/2$ and $|\Delta I| = 3/2$ parts and now also include the Penguin operators. The factorization contribution accounts for the effect of the strong QCD attraction between quark and *antiquark* and not between two quarks. It is not a large contribution since matrix elements of Lorentz vectors are small for small particle momenta. Nevertheless, the addition of this amplitude is necessary, in particular in those decays where the diquark contribution is small or vanishing. Magnitude and sign of the factorization amplitudes are easily obtained using the known semi-leptonic matrix elements of baryons and the π-meson decay constant.

---

[2]At low momentum transfers a baryon is well described as a composite of three constituent quarks. This picture is not in conflict with recent knowledge about matrix elements of the flavor singlet axial current from the EMC experiment [21] if meson-quark couplings are properly taken into account [22].

Table 1. S-wave decay amplitude ($\times 10^7$)

decay	theory [20]	experiment
Λ^0_0	-2.11	-2.36 ± 0.03
Λ^0_-	2.74	3.25 ± 0.02
Ξ^0_0	3.46	3.43 ± 0.06
Ξ^-_-	-4.58	-4.49 ± 0.02
Σ^+_+	0.03	0.14 ± 0.03
Σ^+_0	-3.30	-3.26 ± 0.11
Σ^-_-	4.29	4.27 ± 0.01

It is quite remarkable that in spite of the approximations used, the theoretical amplitudes calculated in the way described here are in very satisfactory agreement with the experimentally determined amplitudes [20]. No unknown parameters had to be introduced. The problem of the P-waves and their connection with the S-waves which was a headache for many years is now solved. In Tables 1 and 2, I show the theoretical and experimental results for comparison. The calculation of the S-wave amplitudes for decays to $\pi - N$ states includes a correction for final state interaction suggested by the measured phase shifts.

In Ω^- decays a clear separation of diquark and factorization amplitudes occurs: The decay $\Omega^- \to K^-\Lambda$ proceeds by the steps $\Omega^- \to K^-\Xi^0 \to K^-\Lambda$. Its strength is determined by the diquark transition in the $\Xi^0 \to \Lambda$ matrix element and there is no factorization contribution. The decays $\Omega^- \to \Xi\pi$, on the other hand, cannot proceed by a diquark transition but allow a factorization calculation. The $\Omega^- \to \Xi\pi$ amplitudes are much smaller than the $\Omega^- \to K\Lambda$ amplitude and approximately given by factorization, even though these decays can be $|\Delta I| = 1/2$ decays! This is precisely what we expected and gives further support to our treatment. In the low energy regime weak decays are large if they can proceed by a diquark transition and small if not.

The $|\Delta I| = 1/2$ rule in K-decays [4]

The strong attractive quark-quark interaction in color triplet states which we take into account by the use of diquark fields will also influence K-decays. In order to calculate the diquark contribution to K-decay amplitudes we again use the effective weak Hamiltonian (9). To apply it we need to know how pseu-

Table 2. *P*-wave decay amplitudes ($\times 10^7$)

decay	theory [20]	experiment
Λ^0_0	-15.59	-15.61 ± 1.40
Λ^0_-	23.20	22.40 ± 0.54
Ξ^0_0	-14.43	-12.13 ± 0.71
Ξ^-_-	19.78	17.45 ± 0.58
Σ^+_+	44.22	41.83 ± 0.17
Σ^+_0	30.74	26.74 ± 1.32
Σ^-_-	-1.18	-1.44 ± 0.17
$\Omega^-_{K^-}$	7.70	5.37 ± 0.13
$\Omega^-_{\pi^-}$	1.46	1.80 ± 0.08
$\Omega^-_{\pi^0}$	-0.75	-1.10 ± 0.07

doscalar mesons interact with diquarks. Here we can make use of the fact that the pseudoscalar mesons are the Goldstone bosons associated with the spontaneously broken chiral symmetry of QCD. As usual we introduce a unitary matrix $\Sigma(x)$ which contains the pseudoscalar fields and transforms linearly under $SU(3)_L \times SU(3)_R$. In addition to the octet of pseudscalar mesons, we consider now the normalized diquark fields ϕ_L, ϕ_R as effective degrees of freedom and add to the well-known chiral Lagrangian for mesons the chiral Lagrangian for the diquark fields. It consists of a free part, which includes a chiral invariant mass term m, and an interaction part. For the interaction we take the most general non-derivative coupling terms. Derivative couplings — if not removable by field redefinitions — would imply Noether currents not separable in meson and diquark parts and would signify an unwanted internal structure of mesons and diquarks. One has, therefore,

$$\mathcal{L}_{int} = \frac{\Delta m^2}{4}(\phi_L^\dagger \Sigma^* \phi_R + \phi_R^\dagger \Sigma^T \phi_L) + ..., \qquad (12)$$

where the dots stand for terms proportional to the current quark masses m_q. The parameter $\Delta m^2 = (m^2_- - m^2_+)$ fixes the mass difference between states of different parity, as can be seen by setting $\phi_{L,R} = 1/\sqrt{2}\,(\phi_+ \mp \phi_-)$. We expect a mass splitting of about 900 MeV, as it is the case for mesons. This is nicely confirmed by the QCD sum rule analysis which indicates that $J^P = 0^-$ diquark states are considerably heavier than 0^+ states [19]. To processes like $K \to 2\pi$ many graphs contribute [4]. Each graph involves a diquark loop and diverges individually. The sum of all graphs is *finite*, however.

The diquark loop graphs or, more generally, the effective meson Lagrangian obtained by integrating out the diquark fields [23], allow now a quantitative calculation of the amplitude caused by quark-quark correlations. This pure

$|\Delta I| = 1/2$ contribution appears already at leading order of the chiral expansion (order p^2). It is large even though it is nonleading in the $1/N_c$ expansion. The factorization contribution mentioned above, on the other hand, is small but survives in the $N_c \to \infty$ limit and should thus be added. Its $|\Delta I| = 1/2$ part has the same sign as the diquark amplitude. It then turns out that to order p^2 of the chiral expansion the calculated amplitude $A_0^{th} = A^{th}(K \to (2\pi)_{I=0})$ reaches already $\simeq 75\%$ of the experimental one. By going up to order p^4 in the chiral expansion the diquark amplitude increases further. At this order we also have to include meson loop contributions [24], [4]. By their absorptive parts they give rise to non-zero phase values. The full calculation to order p^4 is quite tedious. It gives a $\simeq 25\%$ increase of the A_0^{th} amplitude bringing it into line with the experimental value. For the phases one finds $\delta_2 = -\delta_0/2 \simeq -13°$. These values are consistent with the analysis based on S-wave $\pi\pi$ scattering data [25]. The amplitude $A_2^{th} = A^{th}(K^0 \to (2\pi)_{I=2})$ is obtained by factorization (slightly corrected using the measured $I = 2$ $\pi\pi$ phase shifts). Diquarks cannot contribute to it.

Thus, our final result is [4]:

$$
\begin{aligned}
|A_0^{th}/A_0^{exp}| &= 1.0 \pm 0.2, \\
|A_2^{th}/A_2^{exp}| &= 1.2 \pm 0.3.
\end{aligned}
\tag{13}
$$

The quoted errors reflect the uncertainties in the diquark masses and couplings and of the factorization scale. The $|\Delta I| = 1/2$ rule is no more a mystery!

The diquark model achieves even more [23]: it fixes all constants [26] arising in higher order chiral perturbation theory involving Goldstone mesons, and thus provides for many testable predictions. It also led to an understanding of the long-range part of the $K_L - K_S$ mass difference since the corresponding $\Delta S = 2$ transition amplitude is closely related to the $|\Delta I| = 1/2$ enhancement mechanism [4], [27]. Finally, I should mention that scalar diquarks do not contribute to the CP-violating quantity ϵ_K' even though the corresponding operator is an $I = 1/2$ operator [17], [28]. This is in agreement with the observed smallness of this quantity ruling out earlier expectations of a *general* $|\Delta I| = 1/2$ enhancement in $\Delta S = 1$ decays.

3.2 D- and B-meson decays and factorization

The discovery of heavy charm and bottom quarks opened up the possibility to study a great variety of new nonleptonic decay channels. Thanks to the efforts of many experimental groups, there is already an impressive amount of experimental data available. For a theoretical calculation the factorization approximation can again be considered [3],[11]. It allows to express a decay amplitude as a product of two matrix elements of colour singlet currents: a current matrix element as occurring in semi-leptonic decays and a vacuum-meson transition matrix element. Factorization accounts in this way for nonperturbative QCD effects between quarks and antiquarks. But is factorization justified? We just saw that factorization applied in this manner provides only for a small part of the amplitude in strange particle decays. However, this fact is understood: In low energy $\Delta S = 1$ transitions, matrix elements of scalar currents dominate over those of vector or axial vector currents. In fact, f_π times the energy of the outgoing π-meson is here small compared to the diquark decay constant $g_{(ud)}$.

For energetic two-body (or quasi two-body) decays of D- and B-mesons, on the other hand, the situation is different. In these processes a much larger amount of energy is converted into kinetic energy favoring the direct current induced production of a fast meson over a more complicated process in which the final quarks would have to rearrange themselves. The energetic meson arises from a fast-moving quark-antiquark pair created in a point interaction with both quarks having high and almost equal velocities. This quark-antiquark pair will behave like a colorless point particle and thus have little interaction with the remaining quarks [29]. Moreover, the hadronization to the spatial extended meson will occur only after travelling a distance given by the γ-factor times the typical hadronization time of $\simeq 1\ fm/c$. For a process with large energy release, $\bar{B}^0 \rightarrow D^+\pi^-$, for instance, hadronization occurs far outside the interaction region excluding a recombination of quarks.

In the applications we adopted the factorization ansatz also for processes where the γ-factor is not very large. Nevertheless, the other properties mentioned will still hold for most two-body decays and could be important enough. But only the comparison with experiment can tell whether factorization is a useful concept also for these processes and, if so, how large the corrections to it are.

The special way how to factorize needs specification. The effective Hamiltonian can be written in a variety of ways which are — before approximation — fully equivalent, but not when factorization has been applied. In order to account for $q - \bar{q}$ correlations it appears best to write all operators occurring in the effective Hamiltonian in terms of products of colour singlet currents (by using a Fierz transformation if necessary). Factorization then singles out the prominent and hopefully most important part of the amplitude from the contributions of the full set of physical (color singlet) intermediate states.

Naturally, a problem arises if — for a given process — one of the important operators in the effective Hamiltonian gives in this way only a very small or vanishing amplitude. For instance, the color singlet currents in the operator

$$c_2(m_b)(\bar{c}_i\gamma_\mu(1 - \gamma_5)u_i)(\bar{d}_j\gamma^\mu(1 - \gamma_5)b_j) \tag{14}$$

cannot directly generate a π^- or D^+ for the decay $\bar{B}^0 \rightarrow D^+\pi^-$. On the other hand, (14) can be rewritten in the form

$$c_2(m_b)\frac{1}{3}(\bar{d}_i\gamma_\mu(1 - \gamma_5)u_i)(\bar{c}_j\gamma^\mu(1 - \gamma_5)b_j)$$

$$+ \text{ product of color octet currents} \tag{15}$$

In this form the first term contributes a $1/N_c$ suppressed but nevertheless finite factorization amplitude to the decay $\bar{B}^0 \rightarrow D^+\pi^-$. However, there is no reason to neglect the matrix element of the product of color octet currents in (15) which could be of the same magnitude. In fact, the phenomenological analysis of several exclusive D-decays strongly indicated a cancellation of the $1/N_c$

factorization term with the nonfactorizeable part [3]. Theoretical arguments for explaining this cancellation are based on the $1/N_c$ expansion [16] and on the direct evaluation of matrix elements involving color octet currents [30]; they are not fully conclusive yet. From a phenomenological point of view it is best to stick to operators in the form of products of color singlet currents as in (14) and to replace these currents by hadronic currents which can generate mesons of the same quantum numbers. To account for errors — for instance due to an incomplete cancellation of the $1/N_c$ term for $B^- \rightarrow D^+ \pi^-$ transitions in (15) — we replaced in our investigations the Wilson coefficients $c_1(\mu)$ and $c_2(\mu)$ by the two free parameters a_1 and a_2, respectively. For a detailed discussion I refer to ref. (3) and (11) and experimental reports [31].

For the dominant exclusive D-decays we obtained the very satisfactory result that within error limits a_1 and a_2 are little process-dependent and numerically very close to $c_1(m_c)$ and $c_2(m_c)$, respectively. In particular, the sign of a_2 is negative. The destructive interference of amplitudes resulting from the relative signs of a_1 and a_2 turned out to be a decisive factor for an understanding of the lifetime ratio of D^+ an D^0 [28].

In B-decays we found again a_1 to be very close to its expected value $a_1 \simeq c_1(m_b)$. The value for $|a_2|$ also appears to be close to $|c_2(m_b)|$, but because of the lack of precise data, process independence and the sign of a_2/a_1 are not yet established.

For special tests of factorization, for the determination of unknown decay constants using factorization, for the interesting case of B-meson decays to baryon-antibaryon states where the diquark mechanism may play an important role, as well as for the decays of b-flavored baryons I have again to refer to the literature.

4   Summary

Semi-leptonic Decays: QCD sum rules and quark models together with the study of the heavy quark mass limit allow for good semi-quantitative and in part even quantitative calculations of formfactors in the physical region. A very nearly model-independent determination of V_{cb} and V_{ub} is now possible.

Nonleptonic Decays: In recent years great progress has been achieved in understanding the physics involved in nonleptonic decays. Puzzling mysteries are now resolved and most exclusive two-body decays are under control.

For $\Delta S = 1$ transitions, the relations between S- and P-wave hyperon decays as well as between hyperon and K-decays are clarified. Most important, the $|\Delta \bar{I}| = 1/2$ rule and the exceptions to it found a coherent and selfconsistent explanation.

In D- and B-decays many decay channels are open. Quite remarkably, the decay rates of energetic two-body decays agree well with the predictions obtained from factorized amplitudes. Obviously, an important part of the confining forces acting in nonleptonic decays can be taken care of by the use of decay constants and formfactors of single currents.

Clearly, many problems remain open. Examples are the validity range of factorization, the understanding of multiparticle decays and Penguin-induced CP-conserving and CP-violating decays.

Acknowledgements: It is a pleasure to thank P. Kroll and H. R. Petry for the invitation to this well-organized and stimulating conference.

References

[1] M. Wirbel, B. Stech, and M. Bauer, Z. Phys. **C29** (1985) 637.

[2] B. Grinstein, M. B. Wise, and N. Isgur, Phys. Rev. Lett. **56** (1986) 298; N. Isgur, D. Scora, B. Grinstein, and M. B. Wise, Phys. Rev. **D39** (1989) 799; T. Altomari and L. Wolfenstein, Phys. Rev. Lett. **58** (1987) 1563.

[3] M. Bauer and B. Stech, Phys. lett. **152B** (1985) 380; M. Bauer, B. Stech, and M. Wirbel, Z. Phys. **C34** (1987) 103.

[4] M. Neubert and B. Stech, Phys. Rev. **D44** (1991) 775; B. Stech, Mod. Phys. Lett. A, Vol. 6, No. 34 (1991) 3113.

[5] H. Leutwyler and M. Roos, Z. Phys. **C25** (1984) 91.

[6] S. Nussinov and W. Wetzel, Phys. Rev. **D36** (1987) 130; M. B. Voloshin and M. A. Shifman, Yad. Fiz. **45** (1987) 463 [Sov. J. Nucl. Phys. **45** (1987) 292], Yad. Fiz. **47** (1988) 511 [Sov. J. Nucl. Phys. **47** (1988) 511]; N. Isgur and M. B. Wise, Phys. Lett. **B232** (1989) 113, Phys. Lett. **B237** (1990) 527.

[7] M. E. Luke, Phys. Lett. **B252** (1990) 447.

[8] M. Neubert, SLAC-PUB 5770 (1992), and references quoted therein.

[9] J. G. Körner and G. A. Schuler, Z. Phys. **C38** (1988) 511, Z. Phys. **C46** (1990) 93.

[10] M. Neubert and V. Rieckert, Heidelberg preprint HD-THEP-91-6 (1991), to be published in Nucl. Phys. B.

[11] M. Neubert, V. Rieckert, B. Stech and Q. P. Xu, Heidelberg preprint HD-THEP-91-28 (1991), to appear in "Heavy Flavors", ed. by A. J. Buras and M. Lindner, Advanced Series on Directions in High Energz Physics, World Scientific Publishing Co.

[12] P. Ball, V. M. Braun, and H. G. Dosch, Phys. Lett. **B273** (1991) 316, Phys. Rev. **D44** (1991) 3567, and literature quoted therein; P. Ball, HD-THEP-92-25 (1992).

[13] M. Neubert, SLAC-PUB-5842 (1992), and references cited therein.

[14] A. J. Buras et al., MPI/PTH 56/91, to be published in "Heavy Flavors", ed. A. J. Buras and M. Lindner, World Scientific Co.

[15] R. P. Feynman, in *Symmetries in Particle Physics*, ed. by A. Zichichi, Acad. Press 1965, p. 167; O. Haan and B. Stech, Nucl. Phys. **B22** (1970) 448.

[16] E. Witten, Nucl. Phys. **B160** (1979) 57; A. J. Buras, J. M. Gerard, and R. Rückl, Nucl. Phys. **B268** (1986) 16.

[17] B. Stech, Phys. Rev. **D36** (1987) 975, Nucl. Phys. (Proc. Suppl) **7a** (1989) 106; *Diquarks*, ed. M. Anselmino and E. Predazzi (World Scientific, 1989), p. 277.

[18] See D. B. Lichtenberg in *Diquarks*, ed. M. Anselmino and E. Predazzi (World Scientific, 1989) p. 1, and further articles in this book.

[19] G. Dosch, M. Jamin, and B. Stech, Z. Phys. **C42** (1989) 167; M. Jamin and M. Neubert, Phys. Lett. **B238** (1990) 387.

[20] B. Stech and Q. P. Xu, Z. Phys. **C49** (1991) 491.

[21] J. Ashman et al., Phys. Lett. **B206** (1988) 364, Nucl. Phys. **B328** (1990) 1.

[22] U. Ellwanger and B. Stech, Phys. Lett. **B241** (1990) 409, Z. Phys. **C49** (1991) 683.

[23] M. Neubert, Ph.D. thesis, Heidelberg 1990.

[24] W. A. Bardeen, A. J. Buras, and J.-M. Gérard, Phys. Lett. **B192** (1987) 138, Nucl. Phys. **B293** (1987) 787; A. J. Buras, Nucl. Phys. (Proc. Suppl.) **10a** (1989) 199.

[25] T. J. Devlin and J. O. Dickey, Rev. Mod. Phys. **51** (1979) 237.

[26] J. Gasser and H. Leutwyler, Nucl. Phys. **B250** (1985) 465, 517, 539; J. Kambor, J. Missimer, and D. Wyler, PSI preprint ETH-TH/89-42 (1989).

[27] M. Neubert, Z. Phys. **C50** (1991) 243.

[28] B. Stech, in *CP Violation*, Advanced Series on Directions in High Energy Physics, Vol. 3, ed. by C. Jarlskog, (World Scientific, 1989) p. 680.

[29] J. D. Bjorken, Nucl. Phys. (Proc. Suppl.) **B11** (1989) 325.

[30] B. Blok and M. Shifman, NSF-ITP-92/76, TPI-MINN-92/33/T.

[31] K. R. Schubert, *Recent experimental results on heavy quark decays*, these Proceedings;
B-Decays, ed. by S. Stone, (World Scientific, 1992);
Heavy Flavor Physics, ed. by M. Davier and G. Wormser, Editions Frontieres, 1992.

Recent Experimental Results
on Heavy Quark Decays

K. R. Schubert

Institut für Experimentelle Kernphysik, Universität Karlsruhe,
Postfach 6980, D-W-7500 Karlsruhe 1

Abstract. This contribution reviews recent results from experiments on weak decays of the four charmed baryons Λ_c^+, Ξ_c^0, Ξ_c^+, and Ω_c^0, on evidences for the b-flavoured baryon Λ_b^0, on decays of B mesons into baryons, and on exclusive semileptonic B meson decays.

1 Weak Decays of Charmed Baryons

Fig. 1 shows the dodecaeder of the 20 ground state baryons ($J = 1/2$) which can be formed out of d, u, s, and c quarks. The eight with $C = 0$ are well known. Out of the nine with $C = 1$, five are well established (Λ_c^+, Σ_c^0, Σ_c^{++}, Ξ_c^0, Ξ_c^+), two have not yet been seen ($\Xi_c'^0$ and $\Xi_c'^+$), and two (Σ_c^+ and Ω_c^0) need confirmation. The three with $C = 2$ have not been seen and their production is likely to be very difficult. The four $C = 1$ baryons Λ_c^+, Ξ_c^0, Ξ_c^+, and Ω_c^0 decay only by the weak interaction, their decay properties will be discussed in this section.

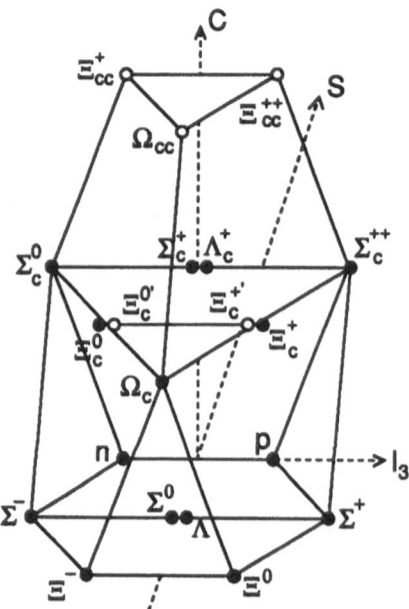

Fig. 1: The 20 baryons with spin 1/2 in the threedimensional space of charm, stangeness, and 3rd isospin component. The solid dots indicate observed states. The states with open dots have still to be found.

1.1 The Λ_c^+ Baryon

The first weak decay of a charmed baryon, $\Lambda_c^+ \rightarrow \Lambda\pi^+\pi^+\pi^-$, was observed 1975 in a bubble chamber at Brookhaven [1]. The first Λ_c^+ decays from e^+e^- annihilation followed in 1980 and gave a mass of $m(\Lambda_c^+) = (2285 \pm 6)$ MeV [2]. The best mass value of today [3] is

$$m(\Lambda_c^+) = (2284.9 \pm 0.6) \text{ MeV}; \quad \tau(\Lambda_c^+) = (1.91\,^{+0.15}_{-0.12}) \cdot 10^{-13} \text{ s}$$

is the best life time value [3]. 13 exclusive decay channels are well established; Table 1 gives their decay ratios relative to $pK^+\pi^-$ and Fig. 2 shows the two recently found channels $\Lambda\pi^+$ and $\Sigma^0\pi^+$ [4]. The Σ^0 is reconstructed from its $\Lambda\gamma$ decay. The lower satellite peak in the $\Lambda\pi$ spectrum is from $\Lambda_c \rightarrow \Sigma\pi$, $\Sigma \rightarrow \Lambda\gamma$ decays with missing γ, and the upper satellite in the $\Sigma\pi$ spectrum is from $\Lambda_c \rightarrow \Lambda\pi$ with inclusion of a random γ.

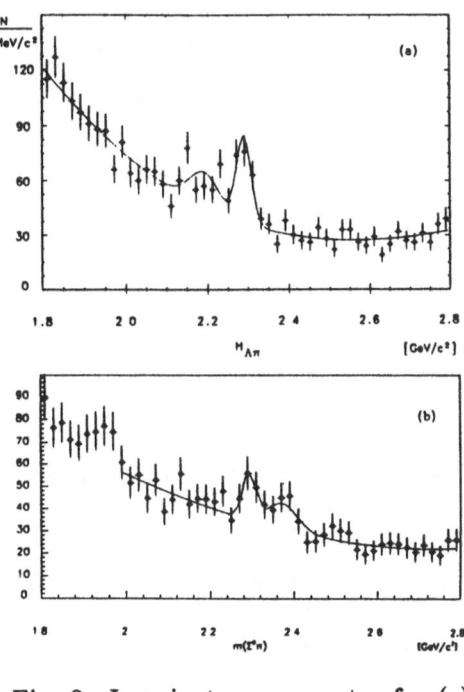

Table 1: Decay ratios of exclusive hadronic Λ_c^+ decays.

i	$\dfrac{B(\Lambda_c^+ \rightarrow i)}{B(\Lambda_c^+ \rightarrow pK^-\pi^+)}$
$pK^-\pi^+$	1
$p\overline{K}^0$	0.49 ± 0.07 [3]
$p\overline{K}^0\pi^+\pi^-$	0.54 ± 0.17 [3]
$pK^-\pi^+\pi^+\pi^-$	0.022 ± 0.015 [3]
$p\pi^+\pi^-$	0.069 ± 0.036 [3]
$p\pi^+\pi^-\pi^+\pi^-$	0.036 ± 0.023 [3]
pK^+K^-	0.048 ± 0.027 [3]
$\Lambda\pi^+$	0.180 ± 0.032 [3]
$\Lambda\pi^+\pi^+\pi^-$	0.64 ± 0.10
$\Sigma^0\pi^+$	0.17 ± 0.07 [3]
$\Xi^-K^+\pi^+$	0.15 ± 0.05 [3]
$\Sigma^+\pi^+\pi^-$	$0.54\,^{+0.18}_{-0.15}$ [28]
$\Sigma^+K^+\pi^-$	$0.13\,^{+0.12}_{-0.07}$ [28]

Fig. 2: Invariant mass spectra for (a) $\Lambda\pi^+$ and (b) $\Sigma^0\pi^+$ combinations in events $e^+e^- \rightarrow$ hadrons at $E_{\text{CM}} \approx 10$ GeV as observed by ARGUS [4].

Absolute decay fractions $B_i = B(\Lambda_c^+ \rightarrow i)$ cannot be determined so far; experiments with e^+e^- annihilation as well as with hadronic Λ_c^+ production determine only the products $\sigma \cdot B_i$ where σ are the pertinent Λ_c^+ production cross sections. There is, however, an indirect way to obtain $B(\Lambda_c^+ \rightarrow pK^-\pi^+)$, and one B_i is sufficient, from B meson decays as will be discussed in section 3.2.

1.2 Parity Violation in $\Lambda_c^+ \to \Lambda\pi^+$

Fig. 3 sketches the decay kinematics. In e^+e^- annihilation at high energies, Λ_c^+ baryons are produced in the fragmentation of c-quark jets. This mechanism produces unpolarized Λ_c^+. The only observable polarization quantity is, therefore, the longitudinal Λ polarization $P_\Lambda = \alpha(\Lambda_c)$. The subsequent Λ decay into $p\pi^-$ shows a forward-backward asymmetry if $\alpha(\Lambda_c) \neq 0$; the proton angular distribution is

$$dN/d\cos\vartheta = \text{const} \cdot [1 + \alpha(\Lambda_c)\alpha(\Lambda)\cos\vartheta],$$

where $\alpha(\Lambda)$ is the known decay parameter of the Λ hyperon.

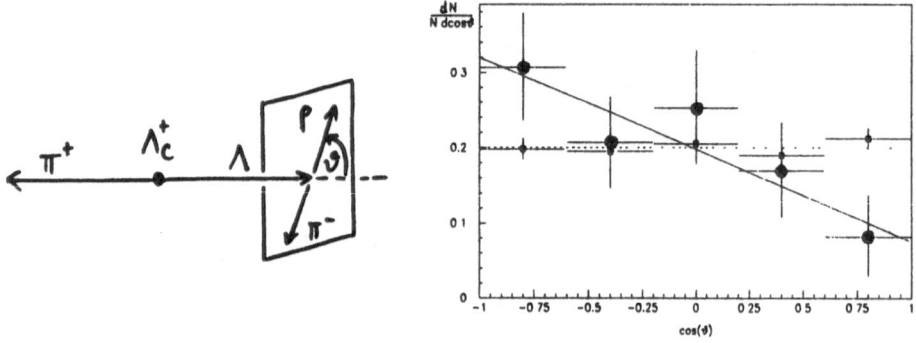

Fig. 3: The distribution of ϑ in the Λ rest frame measures parity violation in $\Lambda_c \to \Lambda\pi$.

Fig. 4: ARGUS results [4] on $dN/d\cos\vartheta$. The dots are from Λ_c decay, the small squares from background.

Fig. 4 shows the $\cos\vartheta$ distribution as observed by ARGUS [4] in e^+e^- annihilation at $\sqrt{s} \approx 10$ GeV. Using $\alpha(\Lambda) = 0.642 \pm 0.013$, the group obtains

$$\alpha(\Lambda_c) = -0.96 \pm 0.42,$$

in good agreement with a previous CLEO result [5], $\alpha(\Lambda_c) = -1.0 \pm 0.4$. The results show that there is maximal parity violation in $\Lambda_c^+ \to \Lambda\pi^+$ decay, i. e. maximal destructive interference between S and P wave decay.

1.3 Semileptonic Λ_c^+ Decays

ARGUS [6] has looked for Λe^+ and $\Lambda\mu^+$ correlations in e^+e^- annihilation at $\sqrt{s} \approx 10$ GeV. A momentum cut $p(\Lambda\ell^+) > 2.5$ GeV/c excludes events from $B\bar{B}$ pairs. After background subtraction, ARGUS obtains invariant $\Lambda\ell^+$ mass distributions as shown in Fig. 5. Their shape is compatible with Λ_c^+ decays, and the decay ratios are found to be

$$B(\Lambda_c \to \Lambda e^+ X)/B(\Lambda_c \to pK^-\pi^+) = 0.37 \pm 0.14,$$

$$B(\Lambda_c \to \Lambda\mu^+ X)/B(\Lambda_c \to pK^-\pi^+) = 0.35 \pm 0.20.$$

The decays are probably dominated by the exclusive channels $\Lambda_c \to \Lambda\ell^+\nu$ [6].

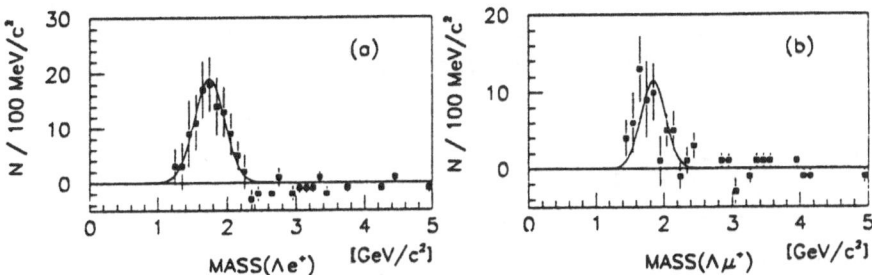

Fig. 5: ARGUS results [6] on $\Lambda\ell^+$ combinations with $p(\ell) > 0.5 \cdot p_{max}(\ell)$. The curves are Monte Carlo expectations from $\Lambda_c \to \Lambda\ell^+\nu$.

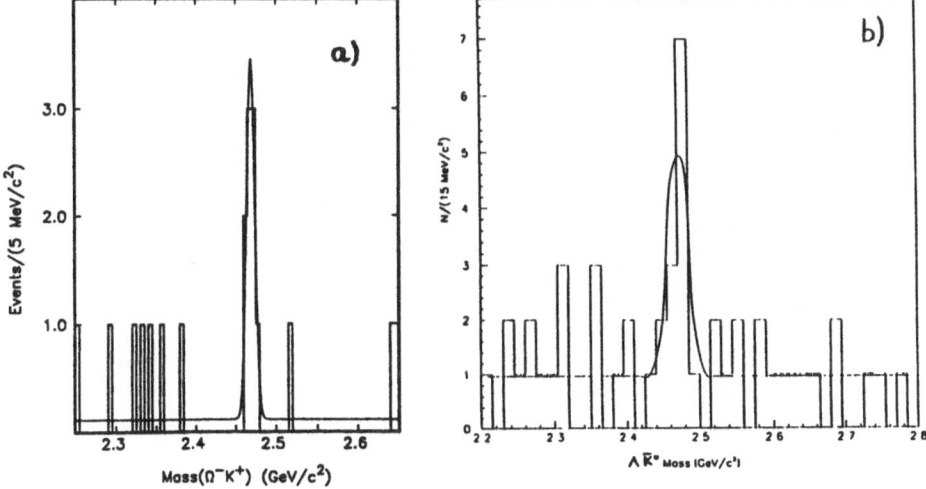

Fig. 6: CLEO results [9] on $\Xi_c^0 \to \Omega^-K^+$ (a) and ARGUS results [8] on $\Xi_c^0 \to \Lambda\overline{K}^0$ (b).

1.4 The Ξ_c^0 Baryon

The Ξ_c^0 has been observed by ACCMOR, ARGUS, and CLEO in five decay channels. Mass values agree with each other; the mean value [3] is

$$m(\Xi_c^0) = (2472.7 \pm 1.7) \text{ MeV}.$$

ACCMOR [7] determined also the mean life and found

$$\tau(\Xi_c^0) = (0.8 \, {}^{+0.6}_{-0.3}) \cdot 10^{-13} \text{ s}.$$

Fig. 6 shows some of the signals [8, 9], and Table 2 summarizes the known decay channels. It is worth noting that all three dominant decay mechanisms are present, $\Xi^-\pi^+$ is mediated by the colour-favoured spectator, $\Lambda\overline{K}^0$ by the colour-suppressed spectator, and Ω^-K^+ by the annihilation mechanism as sketched in Fig. 7. In contrast to charmed meson decays, annihilation is not helicity-suppressed in charmed baryon decays, $\Xi_c^0 \rightarrow \Omega^-K^+$ is a nice example.

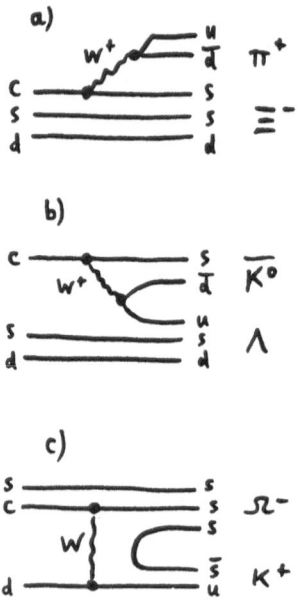

Table 2: Decay ratios of exclusive hadronic Ξ_c^0 decays.

i	$B_i/B_{\Xi^-\pi^+}$
$\Xi^-\pi^+$	1
$\Xi^-\pi^+\pi^+\pi^-$	3.3 ± 1.0 [3]
$\Lambda\overline{K}^0$	2.7 ± 1.5 [8]
Ω^-K^+	0.50 ± 0.22 [9]
$pK^-K^-\pi^+$	?

Fig. 7: The three tree diagrams for charmed baryon decays.

1.5 The Ξ_c^+ Baryon

The Ξ_c^+ has been seen by ACCMOR, ARGUS, CLEO, E400 (FNAL), and WA62 (CERN) in three decay channels, $\Xi^-\pi^+\pi^+$, $\Lambda K^-\pi^+\pi^+$, and $\Sigma^+K^-\pi^+$. Best values for mass and mean life are [3]

$$m(\Xi_c^+) = (2466.4 \pm 2.1) \text{ MeV},$$

$$\tau(\Xi_c^+) = (3.0 \ ^{+1.0}_{-0.6}) \cdot 10^{-13} \text{ s}.$$

1.6 The Ω_c^0 Baryon

The ssc baryon Ω_c^0, "the neutron of the second family", had been seen by the CERN hyperon experiment WA62 [10] with three $\Xi^-K^-\pi^+\pi^+$ events in 1985.

The mass was (2740 ± 20) MeV. This year, ARGUS [11] reported first evidence for Ω_c^0 production in e^+e^- annihilation. Their signal, also in the $\Xi^- K^- \pi^+ \pi^+$ decay mode with $\Xi^- \rightarrow \Lambda \pi^-$, is shown in Fig. 8. At a mass of (2719.0 ± 7.4) MeV, there are 14 events with a backgrond expectation of 4.4 ± 0.9. This could still be a statistical fluctuation; the significance of the signal [1] is 2.5 σ (standard deviations).

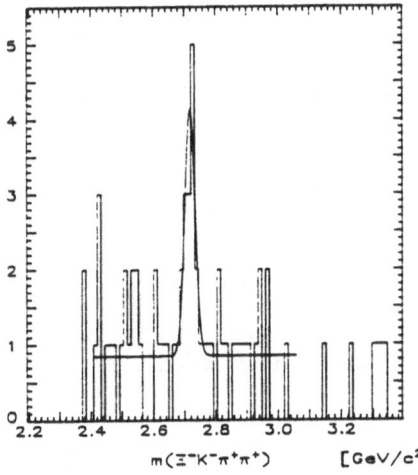

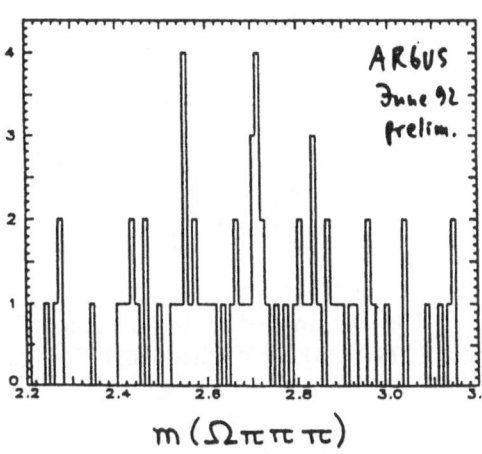

Fig. 8: ARGUS results [11] on $\Omega_c \rightarrow \Xi^- K^- \pi^+ \pi^+$.

Fig. 9: Preliminary ARGUS results [12] on $\Omega_c \rightarrow \Omega^- \pi^- \pi^+ \pi^+$.

Very recently, ARGUS could increase the significance of its observation by adding a second channel [12], $\Omega^- \pi^+ \pi^+ \pi^-$, where the Ω^- decays into ΛK^-. The invariant mass distribution is shown in Fig. 9. At a mass of (2713.0 ± 5.1) MeV, there are 9 events with an expected background of 3.0 ± 0.6. This observation corresponds to a signal with a significance of 2.0 σ. If I now add both ARGUS observations, and this is shown in Fig. 10, the significance of the Ω_c^0 signal increases to 3.2 σ (23 events with an expected background of 7.4 ± 1.1). The mass of the combined signal is found to be [12]

$$m(\Omega_c^0) = (2716 \pm 4 \pm 2) \text{ MeV},$$

and the decay ratio between the two channels is

$$B(\Sigma^- \pi^+ \pi^+ \pi^-)/B(\Xi^- K^+ \pi^+ \pi^-) = 0.26 \pm 0.15.$$

With this ARGUS observation, all four weakly decaying charmed baryons can now be regarded as well-established.

---

[1]I use the following expression for estimating the significance: $S = (N - B)/\sqrt{\sigma_B^2 + N}$, where N is the number of events in the signal region and $B \pm \sigma_B$ is the expected background and its (statistical plus systematic) error.

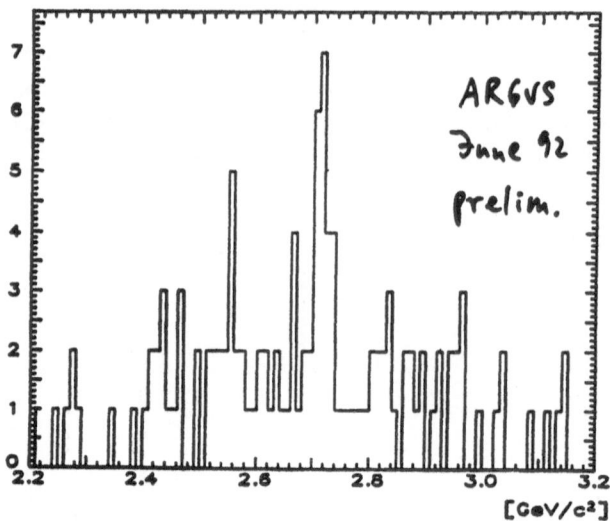

Fig. 10: Summed distribution from Figs. 8 and 9, showing an Ω_c signal at $m = (2716 \pm 4)$ MeV with a significance of 3.2 standard deviations.

2 Beauty Baryons

Using e^+e^- annihilation at $\sqrt{s} = m(Z^0)$ with LEP, ALEPH [13] and OPAL [14] have recently observed $\Lambda\ell^-$ coincidences which they take as evidence for the decay chain

$$Z^0 \to b\overline{b}, \ b \to \Lambda_b X, \ \Lambda_b \to \Lambda_c \ell^- \overline{\nu} X, \ \Lambda_c \to \Lambda X.$$

Background from B meson decays into baryons, $b \to \overline{B} \to \Lambda_c \overline{N}\ell^- \nu$, is drastically reduced by p and p_t cuts, $\Lambda_c \to \Lambda\ell^+ \nu X$ decays give $\Lambda\ell$ coincidences with the wrong charge, and random $\Lambda\ell^-$ combinations are estimated from the observed wrong charge $\Lambda\ell^+$ events. After all subtractions, ALEPH observes 54.5 ± 15 $\Lambda\ell^-$ (and $\overline{\Lambda}\ell^+$) events (e + μ) and OPAL 55 ± 9 (also e + μ). Fig. 11 shows the results of OPAL; the agreement of all observed distributions with those of a Λ_b Monte Carlo decay chain is good. The data can, therefore, be taken as convincing evidence for the production of beauty baryons in e^+e^- annihilation. Λ_b production should dominate, but Fig. 11a does not allow an estimation of the Λ_b mass; the Monte Carlo curves are obtained with $m(\Lambda_b) = 5.62$ GeV. The observed $\Lambda\ell^-$ rates allow a determination of

$$\mathcal{B} = f(b \to \Lambda_b^0 X) \cdot B_1(\Lambda_b^0 \to \Lambda_c^+ \ell^- \nu X) \cdot B_2(\Lambda_c^+ \to \Lambda X)$$

if dominance of Λ_b^0 in all beauty baryon production is assumed. As an average of electron and muons, the groups obtain

$$\mathcal{B} = \begin{cases} (2.9 \pm 0.5 \pm 0.7) \cdot 10^{-3} & \text{(OPAL [14]),} \\ (4.8 \pm 1.1 \pm 1.1) \cdot 10^{-3} & \text{(ALEPH [13]).} \end{cases}$$

The agreement is just good and the average is $\mathcal{B} = (3.3 \pm 0.8) \cdot 10^{-3}$. Taking $\mathcal{B}_1 = (10.5 \pm 0.5)\%$, i. e. the same value as for B meson decays [3], and $\mathcal{B}_2 = 0.40 \pm 0.07$ [15], I obtain from this result

$$f(b \rightarrow \Lambda_b^0 X) = 0.079 \pm 0.024,$$

in good agreement with current fragmentation assumptions.

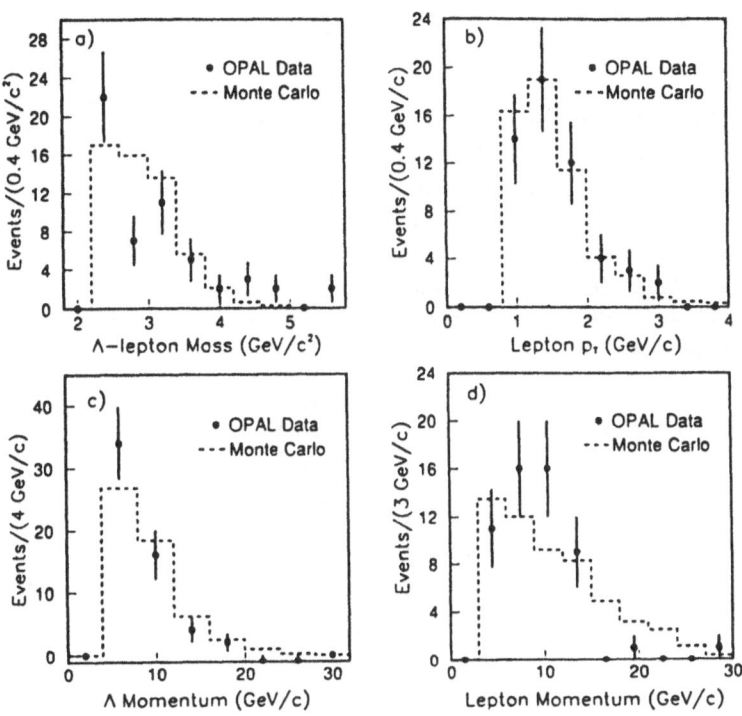

Fig. 11: OPAL results [14] on inclusive $\Lambda\ell^-$ combinations produced in e^+e^- annihilation at LEP, $\sqrt{s} = m(Z)$. a) $\Lambda\ell^-$ invariant mass, b) lepton transverse momentum relative to the jet axis, c) Λ momentum, d) lepton momentum. The dashed histograms show ther expectations from Λ_b decay.

The UA1 group at the CERN $\bar{p}p$ collider reported 16 ± 5 $\Lambda_b^0 \rightarrow J/\Psi \Lambda$ candidates end of 1991 [16]. The mass, $(5640 \pm 50 \pm 30)$ MeV, is in the expected range, the significance is about 3σ (following footnote 1), but the rate is surprisingly high,

$$f(b \rightarrow \Lambda_b^0 X) \cdot \mathcal{B}(\Lambda_b^0 \rightarrow J/\Psi \; \Lambda) = (1.8 \pm 0.6 \pm 0.9) \cdot 10^{-3}.$$

Taking f from the LEP results above and $\mathcal{B} = \mathcal{B}(B \rightarrow J/\Psi K) = (7.3 \pm 1.7) \cdot 10^{-4}$ [3], the UA1 rate is a factor of 31 ± 22 too high. Therefore, confirmation by another experiment would be very desirable.

3 Decays of B Mesons into Baryons

B mesons are the lightest particles which can decay into baryons by weak interactions. Within present errors, B^0 and $B^\pm$ mesons have the same mass, $m(B) = (5.2786 \pm 0.0020)$ GeV [3]. Their lifetime is equal within $\pm 15\%$, the average of both is $(12.9 \pm 0.5) \cdot 10^{-13}$ s [3].

3.1 Λ_c^+ Baryons from B Meson Decays

Both CLEO and ARGUS find evidence for $\Lambda_c^+ \to pK^-\pi^+$ decays in events produced in e^+e^- annihilation at the energy of the $\Upsilon(4S)$ resonance. Fig. 12 shows the recent CLEO result [17]. All $pK\pi$ events with $|\vec{p}(pK\pi)| < 2.5$ GeV/c in the peak around $m(\Lambda_c^+)$ can be attributed to B meson decays since events with the same momentum cut produced at a slightly lower energy in the e^+e^- continuum do not show any peak. Fig. 13 shows the Λ_c^+ momentum distribution in these $\overline{B} \to \Lambda_c^+ X$ events. It is much softer than for two body decays $\overline{B} \to \Lambda_c^+ N$ and still softer than for decays with one or two additional pions. The product decay fraction is found to be

$$\mathcal{B} = \mathcal{B}_1(\overline{B} \to \Lambda_c^+ X) \cdot \mathcal{B}_2(\Lambda_c^+ \to pK^-\pi^+) = \begin{cases} (0.273 \pm 0.064)\% & \text{(CLEO [17])}, \\ (0.31 \pm 0.11)\% & \text{(ARGUS [18])}, \\ (0.282 \pm 0.055)\% & \text{(average)}. \end{cases}$$

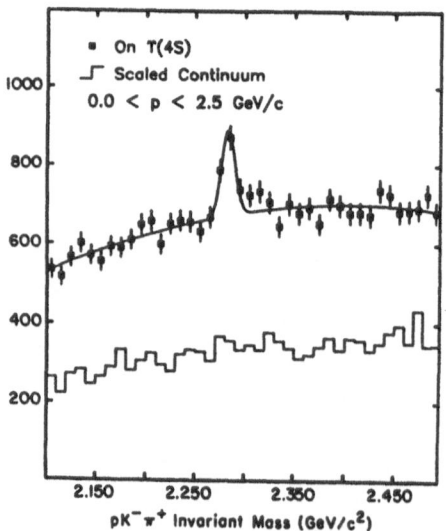

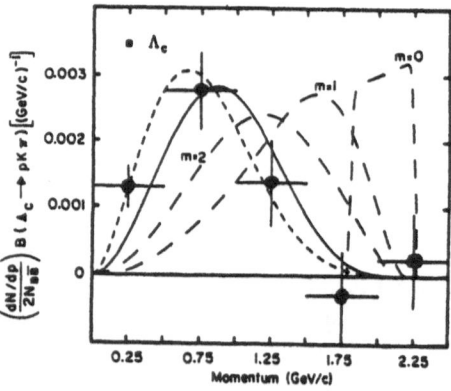

Fig. 12: Invariant mass of $pK^-\pi^+$ from $\Upsilon(4S) \to B\overline{B}$ decays (dots) and from the e^+e^- continuum (histogram), as observed by CLEO [17].

Fig. 13: CLEO results [17] on the momentum distribution of Λ_c candidates. The curves indicate expectations from $\overline{B} \to \Lambda_c \overline{N} X$ decays with various multiplicities of X.

3.2 Additional Measurements for $\mathcal{B}(\overline{B} \to \Lambda_c^+ X)$ and $\mathcal{B}(\Lambda_c^+ \to pK^-\pi^+)$

Inclusive $pK\pi$ observation from B meson decays gives only the product decay fraction $\mathcal{B}_1(\overline{B} \to \Lambda_c^+ X) \cdot \mathcal{B}_2(\Lambda_c^+ \to pK^-\pi^+)$ as shown in the previous subsection. With the help of a large set of additional observations, it is, however, possible to estimate $\mathcal{B}_1$ separately. This has been performed by ARGUS [19, 15, 20] and CLEO [17] using inclusive measurements of

$$\Upsilon(4S) \to B\overline{B} \to \begin{array}{l} pX, \ \Lambda X, \ \Xi^- X, \\ p\overline{p}X, \ (p\overline{\Lambda} + \overline{p}\Lambda)X, \ \Lambda\overline{\Lambda}X, \\ p\ell^+ X, \ p\ell^- X, \ \Lambda\ell^+ X, \ \Lambda\ell^- X. \end{array}$$

Assuming that $b \to c$ transitions dominate $b \to u$ and $b \to s$ also in baryonic decays, that Λ_c^+ is much more frequent than Ξ_c and Ω_c in these decays, and that $\overline{B} \to \Lambda_c\overline{N}X$ dominates $\overline{B} \to DN\overline{N}X$, the two groups find from a fit to their set of measurements

$$\mathcal{B}_1(\overline{B} \to \Lambda_c^+ X) = \begin{cases} (6.8 \pm 0.6)\% & (\text{ARGUS [20]}), \\ (6.4 \pm 1.1)\% & (\text{CLEO [17]}), \\ (6.7 \pm 0.5)\% & (\text{average}). \end{cases}$$

CLEO sets the following limits on the used assumptions: $\mathcal{B}(\overline{B} \to \Xi_c X) < 1.6\%$ and $\mathcal{B}(\overline{B} \to DN\overline{N}X) < 0.6\%$. Since these two contributions could decrease the Λ_c^+ fraction in baryonic B meson decays, the above result is in fact

$$\mathcal{B}(B \to \text{baryons}) = (6.7 \pm 0.5)\%, \ \mathcal{B}(\overline{B} \to \Lambda_c^+ X) = (6.7 \ ^{+0.5}_{-1.2})\%.$$

Combining this with the product in subsection 3.1 gives

$$\mathcal{B}_2(\Lambda_c^+ \to pK^-\pi^+) = (4.2 \ ^{+1.1}_{-0.9})\%,$$

which is larger than the value currently given by the Particle Data Group [3].

4 The Decay $B \to D^*\ell\nu$ and HQET

In this last section I want to present preliminary results of a study with 250 reconstructed $B^0 \to D^{*-}\ell^+\nu$ decays as recently performed by ARGUS [21]. The events $B^0 \to D^{*-}\ell^+\nu$, $\overline{D}^0\pi^-$ are characterized by four variables $q^2, \vartheta, \vartheta^*$, and χ, where $q^\alpha = p^\alpha(B^0) - p^\alpha(D^{*-})$, ϑ is the polar angle of the e^+ or μ^+ in the $\ell^+\nu$ rest frame with respect to the $\ell^+\nu$ boost direction, ϑ^* the polar angle of the $\overline{D}^0$ in the D^{*-} rest frame with respect to the D^{*-} boost direction, and χ is the angle between the two decay planes as sketched in Fig. 14. The $\overline{D}^0$ mesons are reconstructed in two of their decay channels, $K^+\pi^-$ and $K^+\pi^-\pi^+\pi^-$. The complete decay chain $B \to D^* \to D \to K\pi(\pi\pi)$ is reconstructed with the help of the missing mass technique [22]. Fig. 15 shows the B^0 signals in the two D^0 decay channels. Using $\mathcal{B}(D^{*+} \to D^0\pi^+) = (64\pm5)\%$ as indicated by preliminary results of CLEO and ARGUS and D^0 decay fractions of the Particle Data Group [3], the rates in Fig. 15 lead to

$$\mathcal{B}(B^0 \to D^{*-}\ell^+\nu) = (5.2 \pm 0.5 \pm 0.6)\%.$$

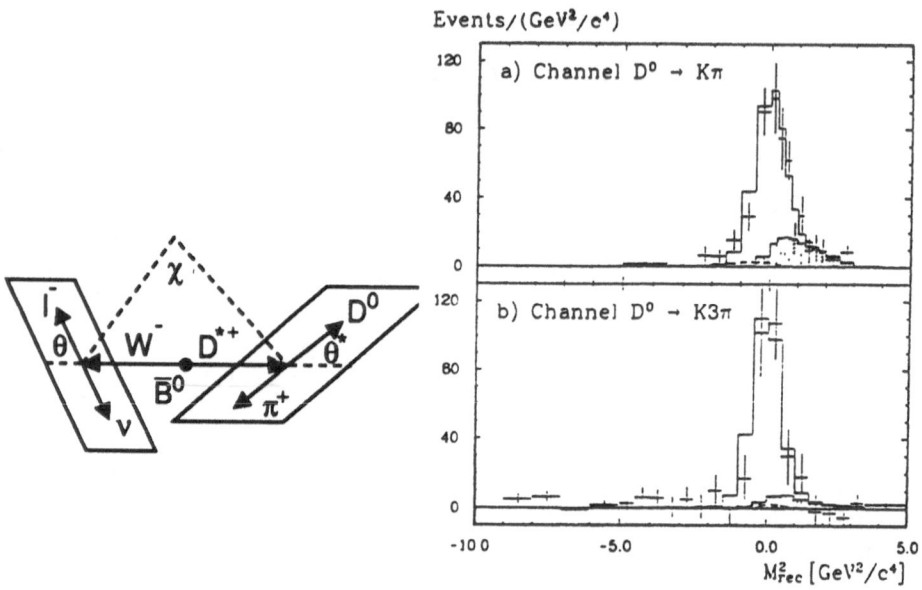

Fig. 14: Definition of the angles as used in the $B \to D^*\ell\nu$ analysis of ARGUS [21].

Fig. 15: Preliminary ARGUS results [21] on the missing mass M_X^2 of $e^+e^- \to \Upsilon(4S) \to B\bar{B}$, $\bar{B} \to D^{*+}\ell^-X$ events in two D^{*+} decay channels. The peaks around zero are from $\bar{B}^0 \to D^{*+}\ell^-\nu$ events, the events in the dashes areas from $\bar{B} \to D^{**}\ell^-\nu$.

An additional result from this analysis, deduced from the background with $MM^2 > 0$ in Fig. 15, is

$$\mathcal{B}(B^0 \to D^{**-}\ell^+\nu) = (3.8 \pm 0.9 \pm 0.6)\%.$$

If we add an older ARGUS result [23],

$$\mathcal{B}(B^0 \to D^-\ell^+\nu) = (1.9 \pm 0.6 \pm 0.5)\%,$$

we find

$$\mathcal{B}(B^0 \to \ell^+\nu D^-, D^{*-}, D^{**-}) = (10.9 \pm 1.2 \pm 1.2)\%,$$

in good agreement with the inclusive result of ARGUS [24], $(10.2\pm0.4\pm0.2)\%$. Three projections of $dN(B^0 \to D^{*-}\ell^+\nu)/dq^2\,d\cos\vartheta\,d\cos\vartheta^*\,d\chi$ are shown in Fig. 16 with $p(\ell^+) > 1$ GeV/c. For $\cos\vartheta^*$ we expect

$$d\mathcal{B}/d\cos\vartheta^* \propto 1 + \alpha \cdot \cos^2\vartheta^*,$$

and a fit to the data leads to

$$\alpha = 1.12 \pm 0.39 \pm 0.19.$$

The other polar angle can show a parity violating effect,

$$d\mathcal{B}/d\cos\vartheta \propto 6 + 3\alpha \cdot \sin^2\vartheta - 4(3+\alpha)A_{FB} \cdot \cos\vartheta,$$

and the fit gives

$$A_{FB} = 0.20 \pm 0.08 \pm 0.06.$$

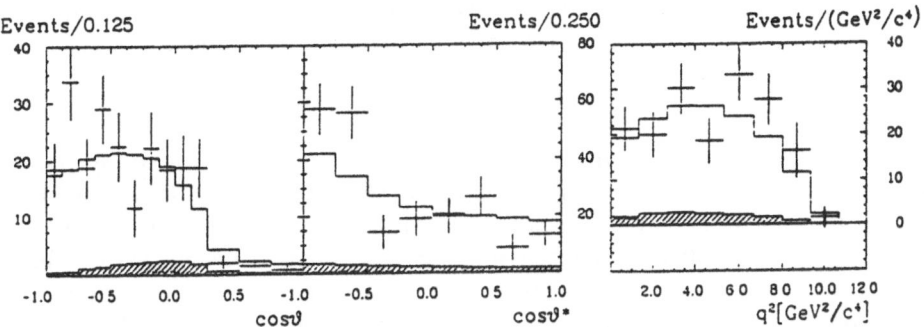

Fig. 16: Preliminary ARGUS results [21] on $dN/d\cos\vartheta$, $dN/d\cos\vartheta^*$, and dN/dq^2 in $\overline{B} \to D^*\ell\nu$ events. The histograms are the best fits including detection efficiencies, the dashed histograms are the background distributions from $\overline{B} \to D^{**}\ell\nu$.

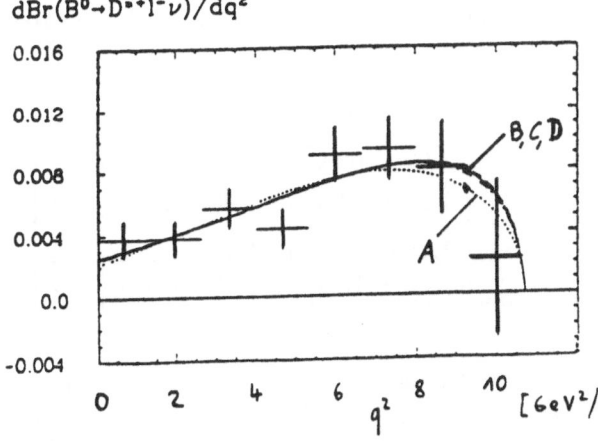

Fig. 17:
Acceptance-corrected q^2 distribution in the preliminary ARGUS analysis [21] of $\overline{B} \to D^*\ell\nu$ events. The four curves are HQET fits; the pertinent expressions and fit results are given in Table 3.

Of major interest is the third projection dN/dq^2. Whereas Fig. 16 shows distributions which are not corrected for acceptance, Fig. 17 shows the corrected $d\mathcal{B}/dq^2$. The curves in this figure are expectations from Heavy Quark Effective Theory (HQET) in four different parametrizations. The Lorentz invariant $y = v_\alpha(B^0)v^\alpha(D^{*-})$ is directly related to q^2 by

$$y = (m_B^2 + m_{D^*}^2 - q^2)/2m_B m_{D^*},$$

and HQET expresses $d\mathcal{B}/dy$ by the Isgur Wise function $\xi(y)$ in the following way [25, 26]:

$$d\mathcal{B}/dy = \frac{G_F^2 |V_{cb}|^2 \tau_B}{48\pi^3} m_{D^*}^3 \cdot (m_B - m_{D^*})^2 \cdot \eta_{QCD} \cdot$$

$$\cdot \sqrt{y^2 - 1} \cdot \xi^2(y) \cdot \left[4y(y+1) \cdot \frac{1 - 2yr + r^2}{(1-r)^2} + (y+1)^2\right],$$

where $\eta_{QCD} = 0.95$ and $r = m_{D^*}/m_B$. All four used parametrizations of $\xi(y)$ give a good fit to the data, the results are given in Table 3. Accepting strict validity of HQET, i. e. $\xi(1) = 1$ without error, the main result is

$$V_{cb} = 0.047 \pm 0.005 \pm 0.004 \pm 0.003,$$

where the first error is from the ARGUS statistics, the second is the systematic experimental error, and the third covers the range of the four HQET parametrizations. In order to accept this as a final result, we have to take $\xi(1) = 1$ with an error of less than 0.05. Since discussions on this precision are not yet conclusive [27], I would only like to comment here on the experimental systematic error of $\pm 8\%$ on V_{cb}. In the presented analysis, this is the quadratic sum of $\pm 2\%$ from the B meson life time, $\pm 4\%$ from $\mathcal{B}(D^* \to D\pi)$, $\pm 3\%$ from $\mathcal{B}(D \to K\pi, K\pi\pi)$, and $\pm 6\%$ from D^{**} background in the missing mass technique.

Table 3: HQET fits to the ARGUS [21] q^2 distribution of $\overline{B}^0 \to D^{*+}\ell^-\nu$ events as shown in Fig. 17. The last column gives the χ^2 value of the fits for 6 degrees of freedom.

Fit	$\xi(y)$	ρ	$10^3 \cdot V_{cb}$	χ^2
A	$1 - \rho^2(y-1)$	$1.09 \pm .08 \pm .07$	$43 \pm 4 \pm 3$	5.1
B	$\frac{2}{y+1} \cdot \exp\left[-(2\rho^2 - 1) \cdot \frac{y-1}{y+1}\right]$	$1.51 \pm .14 \pm .13$	$50 \pm 5 \pm 4$	4.3
C	$[2/(y+1)]^{2\rho^2}$	$1.45 \pm .15 \pm .13$	$49 \pm 6 \pm 4$	4.3
D	$\exp\left[-\rho^2(y-1)\right]$	$1.37 \pm .15 \pm .12$	$48 \pm 6 \pm 4$	4.4

To conclude this section, HQET offers us the most precise way for a determination of the Standard Model parameter $V_{cb} = \sin\vartheta_{23}$. At present, using 250 ARGUS events $B^0 \to D^{*-}\ell^+\nu$, the statistical error with $\pm 10\%$ of V_{cb} is still the dominating one. More data will reduce this part of the error and also the $\pm 6\%$ contribution of the D^{**} background. Since experiments can only determine the product $V_{cb} \cdot \xi(1)$, it will then be essential to control $\xi(1) = 1$ with better than $\pm 5\%$ on the theoretical level.

5 References

[1] E. G. Cazzoli et al. (BNL), Phys. Rev. Lett. 34 (1975) 1125

[2] G. S. Abrams et al. (MARK2), Phys. Rev. Lett. 44 (1980) 10

[3] Review of Particle Properties, Particle Data Group,
Phys. Rev. D 45 (1992) S1

[4] H. Albrecht et al. (ARGUS), Phys. Lett. B 274 (1992) 239

[5] P. Avery et al. (CLEO), Phys. Rev. Lett. 65 (1990) 2842

[6] H. Albrecht et al. (ARGUS), Phys. Lett. B 269 (1991) 234

[7] S. Barlag et al. (ACCMOR), Phys. Lett. B 236 (1990) 495

[8] H. Albrecht et al. (ARGUS), "Evidence for the Decay $\Xi_c^0 \to \Lambda^0 \overline{K}^{0}$",
to be published

[9] S. Henderson et al. (CLEO), Phys. Lett. B 283 (1992) 161

[10] S. F. Biagi et al. (WA62), Z. Physik C 28 (1985) 175

[11] H. Albrecht et al. (ARGUS), DESY preprint 92-052 (1992)

[12] U. Becker (ARGUS), Diplomarbeit, Universität Heidelberg, 1992

[13] D. Decamp et al. (ALEPH), Phys. Lett. B 278 (1992) 209

[14] P. D. Acton et al. (OPAL), Phys. Lett. B 281 (1992) 394

[15] W. Funk (ARGUS), Doktorarbeit, Universität Heidelberg, 1991,
IHEP-HD/91-06

[16] C. Albajar et al. (UA1), Phys. Lett. B 273 (1991) 540

[17] G. Crawford et al. (CLEO), Phys. Rev. D 45 (1992) 752

[18] F. Wartenberg (ARGUS), Diplomarbeit, Universität Hamburg,
DESY-F15 90-02 (1990)

[19] K. R. Schubert, AIP Conference Proceedings 196 (Heavy Quark Physics,
Ithaca 1989, ed. by P. S. Drell and D. L. Rubin) p. 79

[20] H. Albrecht et al. (ARGUS), DESY 92-074 (1992)

[21] K. Reim (ARGUS), private communication

[22] H. Albrecht et al. (ARGUS), Phys. Lett. B 197 (1987) 452

[23] H. Albrecht et al. (ARGUS), Phys. Lett. B 229 (1989) 175,
modified by $\mathcal{B}(\Upsilon 4S \to B^0 \overline{B}^0) = 0.50$ and by $\mathcal{B}(D \to K\pi, K3\pi)$ from ref. 3

[24] H. Albrecht et al. (ARGUS), Phys. Lett. B 249 (1990) 359

[25] N. Isgur and M. B. Wise, Phys. Lett. B 232 (1989) 113

[26] N. Isgur and M. B. Wise, Phys. Lett. B 237 (1990) 527

[27] P. Ball, Phys. Lett. B 281 (1992) 133

[28] S. Barlag et al. (ACCMOR), Phys. Lett. B 283 (1992) 465

Heavy Quark Symmetry and Weak Decays of Heavy Baryons

J.G. Körner

Institut für Physik, Johannes Gutenberg-Universität, Staudingerweg 7, Postfach 3980, D-6500 Mainz, Germany

Abstract. I give an account of the physics ideas that go into the formulation of Heavy Quark Symmetry (HQS) and use HQS ideas to discuss various aspects of the weak semileptonic decays of heavy baryons.

1. Introduction

Much of the motivation to study the weak decay properties of heavy hadrons can be traced back to the need to determine one of the fundamental constants of nature, the Kobayashi-Maskawa (KM) matrix element V_{bc}. It was realized in the last few years that exclusive semileptonic decays of bottom to charm hadrons are much better suited for this purpose than, as had been thought originally, the inclusive semileptonic b → c decays [1]. The reason for this is that the KM matrix element V_{bc} can be regarded as a weak transition charge which can be accurately measured at the zero recoil point, at least in the limit when the bottom and charm quark mass become very heavy. This is so since the associated hadron transition form factor is normalized to one at zero recoil [2] just as in the case of electromagnetic transitions where the charge form factor is normalized to one at $q^2 = 0$. Thus the measurement of the weak transition charge V_{bc} acquires the same status as the measurement of the electric charge, at least in the large mass limit. Much better, when corrections to the large mass limit were studied at a later stage, it was realized that the zero recoil normalization condition remains intact at $O(1/m_Q)$ [3,4], where m_Q is the heavy quark's mass.

Best suited for the determination of the KM element V_{bc} are the mesonic and baryonic ground state to ground state transitions $\bar{B} \to D, D^*$ and $\Lambda_b \to \Lambda_c$, resp., whose flavour diagrams are drawn in Fig.1.

Other decay candidates in the baryon sector are the $1/2^+ \to 1/2^+$ transitions $\Xi_b(b[su]) \to \Xi_c(c[su])$ and $\Omega_b(b\{ss\}) \to \Omega_c(c\{ss\})$ and the $1/2^+ \to 3/2^+$ ttransition $\Omega_b(b\{ss\}) \to \Omega^*_c(c\{ss\})$, where $[q_1q_2]$ and $\{q_1q_2\}$ refer to flavour-antisymmetric spin 0 and flavour-symmetric spin 1 diquark states, respectively. In this report I will mainly concentrate on heavy baryon transitions and among these, on the $\Lambda_b \to \Lambda_c$ transitions. I leave the subject of heavy meson transitions to a companion review [5].

Obviously one needs a bridge to connect the physics at the quark level, where theory is formulated and where V_{bc} is defined, to the particle level, where, after all, the experiments are done. Fortunately, there has been significant progress over the last few years in this program (starting with the papers [6-12]) which I want to report on. The progress is related to the fact that now there exists a systematic expansion of QCD in terms of inverse powers of the heavy quark mass termed the "Heavy Quark Effective Theory (HQET)". The leading term in this expansion gives rise to a new symmetry termed the "Heavy Quark Symmetry (HQS)".

Nature has been very kind to us in that it has divided its six flavoured quarks into a heavy and a light quark sector. The "heavy" c-, b-, t-quarks are much heavier than the QCD scale $\Lambda_{QCD} = 300$ MeV whereas the "light" u-, d-, s-quarks are much lighter than Λ_{QCD}, i.e. one has

$$m_c, m_b, m_t \gg \Lambda_{QCD} \gg m_u, m_d, m_s \tag{1}$$

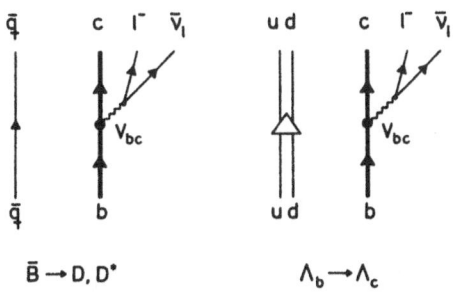

$$\bar{B} \rightarrow D, D^* \qquad \Lambda_b \rightarrow \Lambda_c$$

Fig.1: Flavour diagrams of semileptonic decays $\bar{B} \rightarrow (D,D^*) + \ell^- \bar{\nu}_\ell$ and $\Lambda_b \rightarrow \Lambda_c + \ell^- \bar{\nu}_\ell$

In the heavy quark sector it then makes sense to first consider QCD in the limit where the heavy quark masses become very large and then, in the second stage, to consider power corrections to this limit in terms of a systematic $1/m_Q$ expansion. Likewise one can profitably first study the light quark sector in the zero mass limit, i.e. in the chiral symmetry limit, and then add corrections to the chiral limit at a later stage.

It is quite intriguing that many of the ideas of HQET date back as far as 1937, then of course in the context of QED [13]. In the Block-Nordsieck approach to soft photon radiation it was the electron that was "infinitely" heavy (on the scale of the soft photons) so that the fermionic degrees of freedom could be treated as a classical source of radiation (no e^+e^- pair creation!). In fact, the Block-Nordsieck model was already formulated in terms of an effective theory with the electron degrees of freedom removed from the field theory (see also [14]).

It is quite important to realize that HQS is not a spectrum symmetry but it is a new type of equal veclocity symmetry. That one cannot expect a spectrum symmetry to hold in the heavy quark sector should be quite clear from the fact that there are two orders of magnitude difference between the masses of the c and t quarks! On the other hand, the new type of HQS symmetry at equal velocities takes a little bit of getting used to. But once one has gotten into the habit of thinking in terms of quark and particle velocities the HQS will in fact look quite natural.

We mention that the implications of HQET and HQS have been vigorously studied in the last two-and-a-half years starting with the 1990 papers by Isgur-Wise [6], Bjorken [7], Georgi [10] and our group at Mainz [12]. In the meantime the field is at full blossom with approximately 300 papers published and new papers coming out every week.

To familiarize one-self with the presence of a spin and flavour symmetry at equal velocity it is quite instructive to consider a bottom and charm baryon at rest as shown in Fig.2. The heavy bottom quark and the charm quark at the center are surrounded by a cloud corresponding to the light diquark system. The only communication between the cloud and the center is via gluons. But since gluons are flavour blind the light cloud knows nothing about the flavour at the centre. Also, for infinitely heavy quarks, there is no spin communication between the cloud and the center. Thus one concludes that, in the heavy mass limit, a bottom baryon at rest is identical to a charm baryon at rest regardless of the spin orientation of the heavy quarks, i.e. one has

$$\text{Bottom Baryon at rest} \; \overset{\Uparrow\!\!\Uparrow}{=} \; \text{Charm Baryon at rest} \tag{2}$$

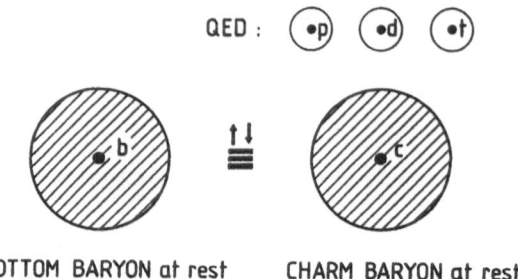

BOTTOM BARYON at rest   CHARM BARYON at rest

Fig.2: Portrayal of bottom and charm baryon wave functions at rest. Upper right corner: wave functions of the hydrogen, deuterium and tritium atoms.

One then just needs to boost the rest configuration by a Lorentz boost from velocity zero to velocity v to conclude

$$\text{Bottom Baryon at velocity v} \; \overset{\Uparrow\!\!\Downarrow}{=} \; \text{Charm Baryon at velocity v} \tag{3}$$

remembering that a Lorentz boost depends only on relative velocities. Eq.(3) exposes the existence of a new spin and flavour symmetry of QCD at equal velocities which holds true in the large mass limit. This is nothing but the advertised Heavy Quark Symmetry HQS.

In fact, everyone should be quite familiar with the existence of such a symmetry in the context of QED. Take a hydrogen, deuterium and tritium atom at rest as also shown in Fig.2. When hyperfine interactions are neglected they possess identical wave function and thus identical atomic properties. The Coulombic interaction between the electron cloud and the nucleus at the centre is sensitive only to the total charge of the nucleus which is the same for all three isotopes.

2. Spin Complexity of Transition Form Factors and Angular Decay Distributions

To start with let us first enumerate the number of form factors that describe the semileptonic $1/2^+ \to 1/2^+$ and $1/2^+ \to 3/2^+$ transitions where J^P denotes the spin (J) and the parity (P) of the heavy baryons. This is easily done in the usual covariant expansion. One has ($q = p_1 - p_2$)

$$1/2^+ \to 1/2^+: \; \left\langle \Lambda_c(p_2) \middle| V_\mu + A_\mu \middle| \Lambda_b(p_1) \right\rangle =$$
$$\bar{u}(p_2)\left\{ \gamma_\mu(F_1^V + F_1^A \gamma_5) + i\sigma_{\mu\nu}q^\nu(F_2^V + F_2^A\gamma_5) \right. \tag{4}$$
$$\left. + q_\mu(F_3^V + F_3^A\gamma_5) \right\}u(p_1)$$

and, equivalently, for the $1/2^+ \to 1/2^+$ transition $\Omega_b \to \Omega_c$. For the $1/2^+ \to 3/2^+$ transition one has

$$1/2^+ \rightarrow 3/2^+: \quad \left\langle \Omega_c^*(p_2) \middle| V_\mu + A_\mu \middle| \Omega_b(p_1) \right\rangle =$$

$$\bar{u}^\alpha(p_2) \Big\{ g_{\alpha\mu}(G_1^V + G_1^A \gamma_5) + p_{1\alpha}\gamma_\mu(G_2^V + G_2^A \gamma_5) \tag{5}$$

$$+ p_{1\alpha}p_{1\mu}(G_3^V + G_3^A \gamma_5) + p_{1\alpha}q_\mu(G_4^V + G_4^A \gamma_5) \Big\} \gamma_5 u(p_1)$$

There are thus (4+2) and (6+2) form factors for the $1/2^+ \rightarrow 1/2^+$ and $1/2^+ \rightarrow 3/2^+$, transitions, respectively. The first number in the brackets counts the number of form factors that can be measured in the zero lepton mass case (typically e and μ) whereas a measurement of the form factors multiplying q_μ ($F_3^{V,A}$ and $G_4^{V,A}$) require non-zero lepton masses (typically the τ). When one wants to define physical observables it is more advantageous to linearly transform the invariant amplitude F_i defined in (7) to helicity amplitudes H_i (see e.g. [15-18]). These again split into the two (4+2) and (6+2) sets mentioned above.

It is quite remarkable that, in the infinite mass limit, HQS tells us that the six form factors in the $\Lambda_b \rightarrow \Lambda_c$ case are all related to one reduced form factor $F_A(\omega)$ which is a function of the ("scaling") velocity transfer variable $\omega = v_1 \cdot v_2$ and which is normalized to one at zero recoil $F_A(\omega=1)=1$. For the transitions involving spin 1 diquarks, the 14 form factors describing the $\Omega_b \rightarrow \Omega_c$ and $\Omega_b \rightarrow \Omega_c^*$ transitions are all related to two reduced form factors $F_L(\omega)$ and $F_T(\omega)$ which satisfy the zero recoil normalization conditions $F_L(1)=F_T(1)=1$. I have intentionally chosen the phrase "reduced form factor" in analogy to the corresponding phrase "reduced matrix element" used in the Wigner-Eckart theorem. We shall later on describe how one actually determines the "Clebsch-Gordan" coefficients that project the general sets of form factors onto the respective reduced form factor $F_A(\omega)$, $F_L(\omega)$ and $F_T(\omega)$. HQS by itself can say nothing about the actual functional form of the reduced form factors except for their normalization at zero recoil $\omega=1$. To obtain their functional form one needs additional dynamical input.

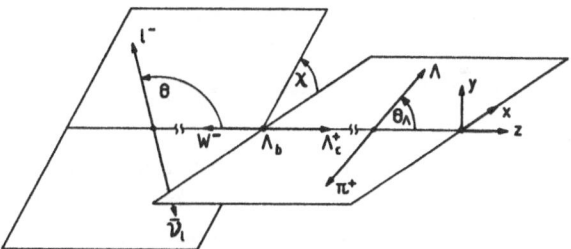

Fig.3: Definition of hadron side polar angle θ_Λ, lepton-side polar angle θ, and azimuthal angle χ in the decay $\Lambda_b \rightarrow \Lambda_c(\rightarrow \Lambda_s\pi) + W^-(\rightarrow \ell \bar{\nu}_\ell)$.

In the following I want to delineate how the form factors can actually be measured in the semileptonic decay processes and how the predictions of HQS can thus be tested. To be specific I shall discuss the $\Lambda_b \rightarrow \Lambda_c$ transition. Fig.3 shows the decay configuation $\Lambda_b \rightarrow \Lambda_c(\rightarrow \Lambda_s\pi) + W^-_{off\text{-}shell}$ $(\rightarrow \ell \bar{\nu}_\ell)$ in the Λ_b rest system. I view the decay process as a two-step process. In the first step the Λ_b decays into the $W^-_{off\text{-}shell}$ on one side and the Λ_c on the other side[1] (back-to-back to the W^-). In the second step these

---

[1] For reasons of conciseness the $W^-_{off\text{-}shell}$ will be referred to as W^- in the following.

further cascade via $W^- \to \ell^- \bar{\nu}_\ell$ (lepton side) and, via $\Lambda_c \to \Lambda_s \pi$ (hadron side). The second-step decays are again analyzed in their respective rest systems in terms of a lepton-side polar angle θ and a hadron-side polar angle θ_Λ. Finally, the relative orientation of the two decay planes defines an azimuthal angle χ as Fig.3 shows. The first step is governed by the weak decay amplitudes H_i or F_i describing the "decay" $\Lambda_b \to \Lambda_c + W^-$. The decay products Λ_c and W^- emerge in highly polarized states. Their polarization density matrices are given in terms of bilinear forms of the decay amplitudes. The second step decays $\Lambda_c \to \Lambda_s \pi$ and $W^- \to \ell^- \bar{\nu}_\ell$, can then in turn be used to analyze the polarization states of the Λ_c and the W^-. In this regard the decay $W^- \to \ell^- \bar{\nu}_\ell$ is an optimal analyzer since it possesses 100% analyzing power.

What has been described in words can be surmised in the form of a joint angular decay distribution for the decay $\Lambda_b \to \Lambda_c (\to \Lambda_s \pi) + W^- (\to \ell^- \bar{\nu}_\ell)$. The joint angular decay distribution will involve the lepton side polar angle θ, the hadron-side polar angle θ_Λ and the relative azimuth χ of the two decay planes. Collecting all kinematical factors one has [17-19]

$$\frac{d\Gamma(\Lambda_b \to \Lambda_c(\Lambda_s\pi) + \ell^- \bar{\nu}_\ell)}{dq^2 d\cos\theta \, d\cos\theta_\Lambda d\chi} = \frac{G^2}{(2\pi)^4} |V_{bc}|^2 \frac{q^2 p}{24 M_1}$$

$$\cdot B(\Lambda_c \to \Lambda_s \pi) \cdot$$

$$\left\{ \frac{3}{8}(1 + \cos\theta)^2 (1 + \alpha_{\Lambda_c} \cos\theta_\Lambda) \left| H_{1/2 \ 1} \right|^2 \right.$$

$$+ \frac{3}{8}(1 - \cos\theta)^2 (1 - \alpha_{\Lambda_c} \cos a\theta_\Lambda) \left| H_{-1/2 \ -1} \right|^2$$

$$+ \frac{3}{4}\sin^2\theta (1 + \alpha_{\Lambda_c} \cos\theta_\Lambda) \left| H_{1/2 \ 0} \right|^2$$

$$+ \frac{3}{4}\sin^2\theta (1 - \alpha_{\Lambda_c} \cos\theta_\Lambda) \left| H_{-1/2 \ 0} \right|^2 \qquad (6)$$

$$- \frac{3}{2\sqrt{2}} \alpha_{\Lambda_c} \cos\chi \sin\theta \sin\theta_\Lambda (1 + \cos\theta) \mathrm{Re}(H_{-1/2 \ 0} H^*_{1/2 \ 1})$$

$$\left. - \frac{3}{2\sqrt{2}} \alpha_{\Lambda_c} \cos\chi \sin\theta \sin\theta_\Lambda (1 - \cos\theta) \mathrm{Re}(H_{1/2 \ 0} H^*_{-1/2 \ 1}) \right\}$$

where $q^2 = (p_1 - p_2)^2$ is the invariant momentum transfer squared and p is the CM momentum of the Λ_c. The $H_{\lambda_f \lambda_w}$ are the aforementioned helicity amplitudes of the decay $\Lambda_b \to \Lambda_c + W^-$ where λ_W is the helicity of the W^- and λ_f is the helicity of the daughter baryon. The decay distribution (6) holds for zero lepton masses. If lepton mass effects are included there are ten more terms in (6) [17]. Furthermore, if one includes also the so-called T-odd contributions that could arise from CP and/or final state interaction effects there are even three more additional terms in (6) when $m_\ell \neq 0$. Thus, when $m_\ell \neq 0$ and T-odd effects are included, there are altogether 19 observables in the decay distribution $\Lambda_b \to \Lambda_c(\to \Lambda_s \pi) + W^- (\to \ell^- \bar{\nu}_\ell)$. Since there are only six independent amplitudes in the decay process a complete or even a partial measurement of the observables would considerably overdetermine the form factor amplitudes.

Let me remind the reader that the analysis of joint angular decay distributions such as the one given in Eq.(6) has by now become a standard fare in the analysis of weak decays of heavy mesons. For example, the well-known amplitude analysis of the decay $D \to K^* + \ell\nu_\ell$ by E691 was based on a full three-fold angular fit to an event sample of ≈ 200 events [20]. A similar analysis was done by E653 for the same decay $D \to K^* (\to K\pi) + \mu\nu_\mu$ (≈ 300 events) where lepton mass effects ($m_\mu \neq 0$) were included in the analysis [21]. For b→c decays, ARGUS (≈ 400 events) [22] and CLEO (≈ 200 events) [23] have done a full amplitude analysis based on the threefold angular decay distribution in the decay B→D*(→ Dπ)+ ℓv_ℓ.

On the theoretical side various aspects of joint angular decay distributions in semileptonic decays have been discussed in the literature. I cite refs. [24-31] for semileptonic meson decays and refs. [18,19,32] for semileptonic baryon decays. Recently there has been a very comprehensive, almost encyclopedic analysis of joint angular decay distributions in the weak semileptonic and nonleptonic decays of heavy mesons and baryons including non-zero lepton mass effects as well as polarization effects [17].

Instead of analyzing the full three-fold angular decay distribution (6) one can also consider single angle distributions. For example, the lepton-side polar angle distribution reads

$$W(\theta) = 1 + 2\alpha' \cos\theta + \alpha'' \cos^2\theta \tag{7}$$

where the asymmetry parameters α' and α'' are given in terms of bilinear forms of the helicity amplitudes. They can be read off from the decay distribution Eq.(6) and read

$$\alpha' = \frac{|H_{1/2\ 1}|^2 - |H_{-1/2\ -1}|^2}{|H_{1/2\ 1}|^2 + |H_{-1/2\ -1}|^2 + 2\left(|H_{-1/2\ 0}|^2 + |H_{1/2\ 0}|^2\right)} \tag{8}$$

$$\alpha'' = \frac{|H_{1/2\ 1}|^2 + |H_{-1/2\ -1}|^2 - 2\left(|H_{-1/2\ 0}|^2 + |H_{1/2\ 0}|^2\right)}{|H_{1/2\ 1}|^2 + |H_{-1/2\ -1}|^2 + 2\left(|H_{-1/2\ 0}|^2 + |H_{1/2\ 0}|^2\right)} \tag{9}$$

On the hadron-side one has the polar angle distribution [18]

$$W(\theta_\Lambda) = 1 + \alpha\alpha_{\Lambda_s} \cos\theta \tag{10}$$

where

$$\alpha = \frac{|H_{1/2\ 1}|^2 - |H_{-1/2\ -1}|^2 + |H_{1/2\ 0}|^2 - |H_{-1/2\ 0}|^2}{|H_{1/2\ 1}|^2 + |H_{-1/2\ -1}|^2 + |H_{1/2\ 0}|^2 + |H_{-1/2\ 0}|^2} \tag{11}$$

and where α_{Λ_s} is the asymmetry parameter in the decay $\Lambda_c \to \Lambda_s + \pi$ which was recently measured by the CLEO [33] and ARGUS [34] collaborations and is given by

$$\alpha_{\Lambda_s} = \begin{cases} -1\ 0^{+.04}_{-.00} & \text{CLEO [33]} \\ -0.96 \pm 0.42 & \text{ARGUS[34]} \end{cases} \tag{12}$$

The two asymmetry parameters α' in EQ.(7) and α in Eq.(10) are sensitive to parity-violating effects, i.e. sensitive to the differences

$$|H_{1/2,\lambda_w}|^2 - |H_{-1/2,-\lambda_w}|^2 \qquad (\lambda_W = 1, 0) \ .$$

They can in fact be utilized to extract information on the chirality of the b→c transition. In the left-chiral case, as predicted by the Standard Model, the c-quark emerges from the weak interaction with dominant negative helicity. This information is handed over to the Λ_c into which it hadronizes.[2] Thus one has

$$|H_{-1/2,-\lambda_w}|^2 > |H_{1/2,\lambda_w}|^2 \qquad (\lambda_W = 1, 0)$$

and consequently the asymmetry parameters α' and α are predicted to be negative irrespective of the details of the underlying quark model dynamics.

---

[2] As mentioned before the Λ_c is made from a c-quark and a spin-zero diquark and thus the helicity of the c-quark is the helicity of the Λ_c. In fact, in the HQS limit, the transfer of the helicity information from the c-quark to the Λ_c is 100% irrespective of whether the fragmentation is direct or indirect [35].

The asymmetry α' and α can be conveniently projected out by defining forward-backward asymmetries. One averages over the events in the respective forward (F) and backward (B) hemispheres of the two decays and then takes the ratios $A_{FB} = (F-B)/(F+B)$. One then has

lepton side:
$$A_{FB} = -\frac{3}{4} \frac{|H_{1/2 1}|^2 - |H_{-1/2 1}|^2}{|H_{1/2 1}|^2 + |H_{-1/2 -1}|^2 + |H_{1/2 0}|^2 + |H_{-1/2 0}|^2} \tag{13a}$$

hadron side:
$$A_{FB} = \frac{1}{2}\alpha\alpha_{\Lambda_c} \tag{13b}$$

where the forward hemispheres are defined w.r.t. the momentum direction of the W^- and Λ_c, i.e. $\pi/2 \le \Theta < \pi$ and $0 \le \Theta_\Lambda < \pi/2$, respectively.

As α and α_{Λ_c} are negative both the lepton-side and hadron-side FB asymmetries (13a) and (13b) are predicted to be positive in the Standard Model. In fact in the diquark model of Ref.[36] one finds

$$A_{FB}(\text{lepton side}) = 0.18 \tag{14a}$$
$$A_{FB}(\text{hadron side}) = -0.35\,\alpha_{\Lambda_c} \tag{14b}$$

The hadron-side FB-asymmetry is predicted to be relatively large on account of the two facts that there are large longitudinal contributions (see Eq.(11)) and that the analyzing power of the decay $\Lambda_c \rightarrow \Lambda_s + \pi$ is large (see Eq.(12)). In addition, the hadron-side FB asymmetry has the advantage of being a true parity-odd spin momentum correlation measure ($\langle \vec{\sigma} \cdot \vec{p}\rangle$-type) and thus does not suffer from the criticism recently raised against using the lepton-side FB asymmetry (parity-even momentum-momentum correlation $\langle \vec{p}_\ell \cdot \vec{p}_{\Lambda_c}\rangle$-type) to conclude for the handedness of the $b \rightarrow c$ current [37].

3. Heavy Quark Symmetry and Heavy Baryon Transition Form Factors

Consider the semileptonic decay of a bottom baryon to a charm baryon as drawn in Fig.4. The bottom quark at four-velocity v_1 makes a transition to a charm quark at four-velocity v_2 by emitting a virtual W^-. The light "spectator" quark system which propagates independently is dragged along to expedite it from the velocity v_1 to v_2 without, however, touching its spin.[3]

The spin neutral velocity kick (or alignment) can be conceived of to result from the exchange of many soft gluons between the c-quark at velocity v_2 and the spectator system which starts off at velocity v_1 and ends up with velocity v_2 in order to align its velocity with the c-quark. Compared to the time scale of the $b \rightarrow c$ transition the alignment process is slow. The exchanged gluons are all of the longitudinal non flip type, i.e. there is no spin information transferred from the heavy side to the light side. This can be made manifest on the heavy side by splitting the gluon's γ_μ coupling into its spin flip and spin non flip components, viz.

$$\gamma_\mu = \underbrace{(\gamma_\mu - v_\mu)}_{\text{flip}} + \underbrace{v_\mu}_{\text{non flip}} \tag{15}$$

---

[3] Remember that in the case of the Λ_Q and the (Ω_Q, Ω_Q^*) the light quark system has spin zero (scalar diquark) and spin 1 (vector diquark).

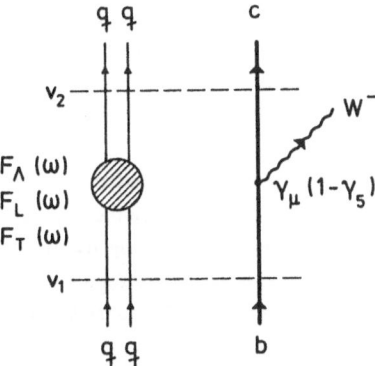

Fig.4: Current-induced transition between heavy baryons. The bottom baryon at velocity v_1 makes a transition to the charm baryon at velocity v_2. There are three independent diquark transition form factors $F_A(\omega)$, $F_L(\omega)$ and $F_T(\omega)$ describing the light scalar and vector diquark transition.

The spin flip coupling $(\gamma_\mu - v_\mu)$ vanishes in the heavy mass limit and one remains with the Bloch-Nordsieck type non flip coupling v_μ. From what was said it is clear that the weak amplitudes $\Lambda_b \to \Lambda_c + W^-$ and $\Omega_c \to (\Omega_c, \Omega_c^*) + W^-$ factorize into a heavy-side and into light-side transition amplitudes. The only information that is exchanged between the heavy- and the light-side is velocity information necessitated by the requirement to reassemble the final charm quark and the light diquark system in the same final charm baryon. The dynamics of the heavy-side transition $b \to c + W^-$ is known. It is specified by the usual SM left-chiral weak coupling with a coupling strength proportional to V_{bc}. The light-side transition invloves the three unknown transition probabilities

$$\left| \text{scalar diquark}; v_1 \right\rangle \to \left| \text{scalar diquark}; v_2 \right\rangle$$
$$\left| \text{vector diquark}; v_1, \lambda_1 \right\rangle \to \left| \text{vector diquark}; v_2, \lambda_2 \right\rangle$$

$(\lambda_1 = \lambda_2 = 0,1)$ where the $\lambda_{1,2}$ are the helicities of the vector diquark.[4] We parametrize the three form factor functions by $F_A(\omega)$, $F_L(\omega)$ and $F_T(\omega)$, where L and T refer to the longitudinal ($\lambda_1 = \lambda_2 = 0$) and transverse $(\lambda_1 = \lambda_2 = \pm 1)$ vector diquark transitions. They can only be a function of $\omega = v_1.v_2$ since the velocity transfer variable ω is the only Lorentz invariant variable that can be constructed in the light-side transitions. At zero recoil, when $v_1 = v_2$ and $\omega = 1$, the diquark goes through unhindered with amplitude 1 and thus we have the normalization condition $F_A(\omega) = F_L(\omega) = F_T(\omega) = 1$. It is clear that one has to identify the $F_i(\omega)$ (i= A,L,T) with the reduced form factor function $F_i(\omega)$ mentioned in Sect.1. One expects the $F_i(\omega)$ to fall when ω moves away from the zero recoil limit as it costs to provide the velocity kick. Pole-type form factors and explicit model calculations confirm this expectation.

Since the zero recoil normalization condition of HQS is of such central importance let us have another look at it from a different point of view. Replace the final-state c-quark in Fig.4 by a b-quark with the same velocity. This is a symmetry operation as shown in Sect.1. At zero recoil, and for the vector current part of the transition, the normalization $F_i(\omega=1)=1$ now is nothing but the well-familiar charge form factor normalization at $q^2=0$ applied to the elastic $\Lambda_b \to \Lambda_c$ and $\Omega_c \to \Omega_b$ transitions. Still

---

[4] Parity relates the helicity transitions $\lambda_1 = \lambda_2 = 1$ and $\lambda_1 = \lambda_2 = -1$.

another way of looking at the zero recoil normalization condition is afforded by considering the b-quark rest configuration in Fig.2. Replacing the b-quark at rest by a c-quark at rest, as happens in the decay at zero recoil, will not affect the wave function of the light diquark system. Thus, the overlap between the wave functions before and after the b→c transition is complete, giving again the zero recoil normalization condition.

Let us now turn to the spin properties of bottom to charm transitions as implied by the spectator diquark picture Fig.4. As has been emphasized before there is complete spin factorization of the heavy-side and light-side transitions. This factorization property was exploited in the helicity matching approach of [15,16,38,39] to derive the HQS heavy baryon form factor structure. The algebraic approach of [40], using spin commutation relations, is quite similar to the helicity matching approach. Finally, the group theoretic approach [41,42] and the Bethe-Salpeter approach [43,44] employ tensor techniques to derive the same heavy baryon form factor structure.

All the above four approaches [15,16,38-44] are of course equivalent. Technically the group theoretic approach of [41,42] is the simplest. The spin wave functions of the Λ-type and Ω-type $J^P = 1/2^+$ ground state baryons are represented by the spinor u and by $u/\sqrt{3}$ $(\gamma_\alpha + v_\alpha)\gamma_5$ and the ground state $J^P = 3/2^+$ Ω^*-type baryon is represented by its Rarita-Schwinger spinor u_α. The HQS form factor structure can then be written down immediately by considering the independent ways of contracting Lorentz indices. One has

$$\langle \Lambda_c | J_\mu | \Lambda_b \rangle = \bar{u} \, F_\Lambda(\omega) \gamma_\mu (1 - \gamma_5) u \qquad (16a)$$

$$\langle \Omega_c, \Omega_c^* | J_\mu | \Omega_b \rangle = (\underbrace{-\frac{1}{\sqrt{3}} \bar{u} \gamma_5 (v_2^\alpha + \gamma^\alpha) + \bar{u}^\alpha}_{\bar{\Omega}_c}) \qquad (16b)$$

$$(F_1(\omega) g_{\alpha\beta} - F_2(\omega) v_{1\alpha} v_{2\beta}) \gamma_\mu (1 - \gamma_5) \underbrace{\frac{1}{\sqrt{3}} (v_1^\beta + \gamma^\beta) \gamma_5 u}_{\Omega_b}$$

Note that one may not use γ-matrices for the contraction as they would bring in spin interactions on the heavy quark legs which are absent in the static approximation. One thus has three universal form factors. The normalization condition for the Λ-type transition is $F_\Lambda(\omega=1)=1$ as before. The normalization condition for the (Ω, Ω^*)-type transitions can be obtained by relating the two form factors $F_1(\omega)$ and $F_2(\omega)$ to the longitudinal and transverse form factors $F_L(\omega)$ in Eq.(16) and $F_T(\omega)$ introduced earlier. One has

$$F_1(\omega) = F_T(\omega) \qquad (17a)$$

$$(\omega^2 - 1) F_2(\omega) = -F_L(\omega) + \omega F_T \qquad (17b)$$

and thus the normalization reads $F_1(\omega)=1$. As Eqs.(16b) or (17b) show the form factor $F_2(\omega)$ does not contribute at zero recoil.

Ref.[43,44] contains a derivation of the form factor structure (16) using Bethe-Salpeter amplitudes for the heavy baryon bound state systems. The form factors are thereby related to wave function overlap integrals which are computable for any given model of the bound state wave functions. Further assumptions on the spin structure of the bound states reduces the number of independent form factors in (16) from three to two and three to one [43].

We mention that the heavy baryon to light baryon form factor structure may be obtained from (16) by allowing for spin interactions of the light active quark [44]. This amounts to the replacement $F_\Lambda \to F_\Lambda' + \gamma F_\Lambda''$, $F_1 \to F_1' + \gamma_1 F_1''$ and $F_2 \to F_2' + \gamma_1 F_2''$. Now there is no normalization condition for the form factors. Also the $\Omega_b \to \Omega_{light}$ and $\Omega_b \to \Omega^*_{light}$ form factors are not related. Phenomenological consequences of the heavy to heavy (including $1/m_c$ corrections) and the heavy to light baryonic form factor structure are presently being worked out [15,18,36,38].

Eq.(16) provides a covariant form of the "Clebsch-Gordan" coefficients that tell us how to project the transition form factors onto the reduced form factors $F_i(\omega)$ (i = Λ,1,2). HQS by itself can say nothing about the actual functional form of the reduced form factors $F_i(\omega)$. To obtain their ω - dependencies one needs additional dynamical input as e.g. provided by the QCD sum rule approach, by lattice calculations or by explicit quark model calculations. In the following I shall briefly discuss an explicit diquark model of heavy Λ-type baryons which, when evaluated in the infinite momentum frame, provides an explicit form of the HQS reduced form factor $F_\Lambda(\omega)$ in the low recoil regime and also $1/m_Q$ corrections to the heavy mass limit [36]. In Sec.5, finally, we consider the large ω or q^2-behaviour of the reduced form factors which can be conveniently studied within the Brodsky-Lepage hard scattering formalism. It is quite remarkable that one retains a modified form of heavy quark symmetry in the large recoil regime [45,46]. The large ω-behaviour of the reduced form factors in the large recoil regime can again be studied within particular models [45,46].

Returning to the low recoil regime the ω-dependence of the mesonic reduced form factor $F(\omega)$ has recently been obtained by Neubert and Rieckert [47] using the heavy meson relativistic oscillator light-cone wave functions of Bauer, Stech and Wirbel (BSW). The $q^2=0$ values of the $(Q_1\bar{q}) \rightarrow (Q_2\bar{q})$ transition form factors were obtained by calculating the wave function overlap integrals for different current components. The overlap integrals were then expanded in a $1/m_Q$ power series with the coefficient functions depending on the mass ratio M_1/M_2 only. Now, since at $q^2=0$ one has $\omega = (M_1/M_2 + M_2/M_1)/2$, the ω-dependence of the coefficient form factor functions can be obtained by varying the mass ratio M_1/M_2. They found their quark model results to be consistent with HQET up to and including the $O(1/m_Q)$ corrections [3], yielding, of course, explicit functional forms for the five $O(1)$ and $O(1/m_Q)$ reduced form factor functions and a value for the dimensionful constant $\overline{\Lambda}$ that appear in the general HQET analysis [3].

Together with B. König, M. Krämer and P. Kroll I have recently extended the Neubert-Rieckert approach to the baryon sector using BSW-type quark-diquark wave functions for the Λ-type heavy baryons [36]. Again the quark model calculation of the $\Lambda_b \rightarrow \Lambda_c$ transitions was found to be consistent with the $1/m_Q$ structure of the HQET [48]. Contrary to the mesonic case, though, one has to restrict oneself to the use of the "good" components of the current transitions only. To illustrate our results I show a plot of the $O(1)$ form factor behaviour of the $\Lambda_b \rightarrow \Lambda_c$ transition form factor $F_\Lambda(\omega) = F(\omega)$ in Fig.5.

The diquark form factor is appropriately normalized to 1 at zero recoil. However, it falls off much faster than the dipole-type form factor as one moves away from the zero recoil point (see Fig.4). The rapid fall-off can be traced back to the rather narrow infinite-momentum-frame wave functions used in [36] that result from adapting conventional three-quark baryon wave functions to the quark-diquark case. The $1/m_Q$ corrections to the $O(1)$ results were found to be quite small, as was the case in mesonic transitions [47]. I refer to [36] for a discussion of phenomenological implications for rates, spectra and asymmetries in $\Lambda_b \rightarrow \Lambda_c$ transitions.

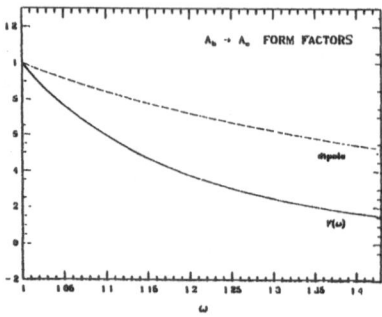

Fig.5: Form factor dependence of the HQET improved $A_b \to A_c$ diquark model form factor at O(1). Also shown is the ω-dependence of a dipole type form factor.

4. Heavy Quark Symmetry at Large Recoil

The heavy quark symmetry predictions are expected to be rather good close to the zero recoil point where not much momentum is transferred to the spectator system. However, as one moves away from the zero recoil point and more momentum gets transferred to the spectator system, hard gluon exchange including spin flip interactions becomes more important and the low recoil heavy quark symmetry discussed in the previous sections can be expected to break down. This is illustrated in Fig.6 where the mismatch between the "kicked" heavy quark momentum and the momentum of the light spectator system becomes progressively larger as one moves away from the zero recoil point.

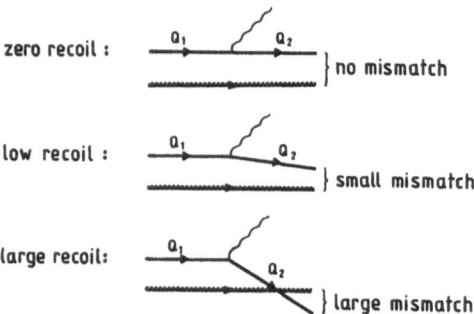

Fig.6: Zero recoil, low recoil and large recoil heavy hadron transitions.

In the large recoil limit the limiting behaviour of the form factors can be conveniently studied in the Brodsky-Lepage formalism [49]. As it turns out the form factors exhibit a new heavy quark symmetry in the large recoil limit which is reminiscent but not identical to the heavy quark symmetry at low recoil. One finds that the transition form factors have the correct large momentum transfer power behaviour as expected from dimensional counting rules.

According to Brodsky and Lepage (BL) [48] the large ω- or q^2-behaviour of the form factors is obtained by convoluting the initial and final state hadron's distribution amplitude with a hard scattering amplitude as shown in Fig.7 for heavy meson transitions $(Q_1\bar{q}) \to (Q_2\bar{q})$.[5] The hard scattering amplitude T_μ is computed in perturbative QCD in the collinear approximation, whereas the distribution amplitudes ϕ_i contain the nonperturbative long distance dynamics.

For the $(Q_1\bar{q}) \to (Q_2\bar{q})$ transitions one obtains

$$\langle(Q_2\bar{q})|V_\mu + A_\mu|Q_1\bar{q})\rangle = \sqrt{m_1 m_2}\; \varepsilon f_1 f_2$$

$$\int dx_1 dy_1\; \phi_2^*(y_1) T_\mu(x_1, y_1, \omega)\phi_1(x_1) \tag{18}$$

where x_1 and y_1 are the longitudinal momentum fractions of the heavy quarks Q_1 and Q_2, $\varepsilon = M_1 - m_1 = M_2 - m_2$ is the flavor-independent heavy meson-heavy quark mass difference and the f_i are the usual wave function at the origin (or meson decay constants) that scale as $f_i \sim 1/\sqrt{M_i}$ [2,50].

To leading order in the heavy mass one obtains

$$\langle(Q_2\bar{q})|V_\mu + A_\mu|Q_1\bar{q})\rangle = \frac{4\pi\alpha_s C_F}{\varepsilon^2(\omega-1)^2}\; f_1 f_2$$

$$\frac{1}{4}\sqrt{M_1 M_2}\; \mathrm{Tr}\left\{(\gamma_5 + \not{z}_2)(\not{y}_2 + 1)\gamma_\mu(1-\gamma_5)(\not{y}_1 + 1)\gamma_5\right\} \tag{19}$$

The second line in Eq.(19) is nothing but the well known HQS "trace" formula for heavy meson transitions (see e.g. [5]) at low recoil.

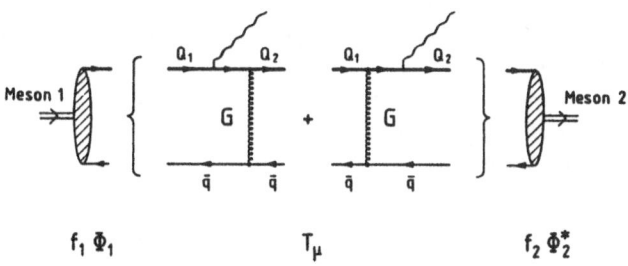

$$f_1\,\Phi_1 \qquad\qquad T_\mu \qquad\qquad f_2\,\Phi_2^*$$

Fig.7: Hard scattering contributions to $(Q_1\bar{q}) \to (Q_2\bar{q})$ mesonic transition form factors.

In order to exhibit the heavy mass structure of the large recoil amplitude Eq.(19) I define Clebsch-Gordan coefficients $\bar{f}_i(\omega)$ that project onto a given transition amplitude $F_i(\omega)$ according to the trace in Eq.(19). In the low recoil regime of HQS one then has generically

$$F_i^{HQS}(\omega) = \bar{f}_i(\omega)\; F^{HQS}(\omega) \tag{20}$$

where $F^{HQS}(\omega)$ is the mesonic HQS reduced form factor function. The large recoil amplitudes (denoted by "BL") have the generic structure

$$\frac{1}{\sqrt{M_1 M_2}} F_i^{BL}(\omega) = \bar{f}_i(\omega)\; F^{BL}(\omega) + 0(\omega/m_Q) \tag{21}$$

where the large ω-behaviour of the reduced BL form factor is given by

$$F^{BL}(\omega) \sim (\omega-1)^{-2} \tag{22}$$

[5] We choose heavy meson transitions to discuss the large recoil behaviour of heavy hadron transitions because mesons are simpler. The large recoil behaviour of heavy baryon transitions is discussed in [46].

In Eq.(21) we have already substituted for the heavy mass scaling behaviour of the wave function at the origin factors f_i. The $1/m_Q$ contributions in (21) contain, among others, spin flip contributions proportional to ω which, when $\omega \gg M_u$, provide for the correct large ω- or q^2-behaviour of the transition form factors.

      The leading terms of the large recoil form factors $F_i^{BL}(\omega)$ in Eq.(21) can be seen to possess the spin and flavor symmetry of HQS if the wave function at origin factors f_i are divided out! We mention that the large recoil heavy baryon transition form factors have a structure identical to Eq.(21) and a dipole-type large ω-behaviour of the reduced BL form factor identical to (22) [46]. In this context it is quite intriguing that the first analysis of the experimental $B \rightarrow D, D^*$ data in the low recoil regime indicates that the reduced form factor has a dipole-type behaviour even at low recoil.

REFERENCES

1. G. Altarelli et al., Nucl.Phys. B208(1982)365
2. M.B. Voloshin and M.A. Shifman, Sov.J.Nucl.Phys. 47(1988)511
3. M.E. Luke, Phys.Lett. B252(1990)447
4. J.G. Körner and G. Thompson, Phys.Lett. B264(1991)185
5. J.G. Körner, Heavy Quark Symmetry and Weak Semileptonics Decays of Heavy Hadrons, Mainz pre print MZ-TH/92-39
6. N. Isgur and M.B. Wise, Phys.Lett. B237(1990)527
7. J.D. Bjorken, invited talk at Les Recontre de Physique de la Vallée d'Aosta, La Thuile, Italy, SLAC-PUB-5278 (1990)
8. A. Falk, H. Georgi, B. Grinstein and M.B. Wise, Nucl.Phys. B343(1990)1
9. B. Grinstein, Nucl.Phys. B339(1990)253
10. H. Georgi, Phys.Lett. B240(1990)447
11. E. Eichten and B. Hill, Phys.Lett. B240(1990)511
12. F. Hussain, J.G. Körner, K. Schilcher, G. Thompson and Y.L. Wu, Phys.Lett. B249(1990)295
13. F. Bloch and A. Nordsieck, Phys.Rev. 52(1937)54
14. N. Boguliubov and D.V. Shirkov, "Introduction to the Theory of Quantized Fields", Interscience Publishes Inc. (New York, 1959)
15. F. Hussain, J.G. Körner and R. Migneron, Phys.Lett. B248(1992)406
16. F. Hussain, J.G. Körner and G. Thompson, Ann.Phys. 206(1991)334
17. P. Bialas, J.G. Körner, M. Krämer and K. Zalewski, Joint Angular Decay Distributions in Exclusive Weak Decays of Heavy Mesons and Baryons, Mainz preprint, MZ-TH/91-06
18. J.G. Körner and M. Krämer, Phys.Lett. B275(1992)495
19. J.G. Körner and H.W. Siebert, Ann.Rev.Nucl.Part.Sci. 41(1991)511
20. E691 Collaboration, J.C. Anjos et al., Phys.Rev.Lett. 65(1990)2630
21. E653 Collaboration, K. Kodama et al., Phys.Lett. B274(1992)246
22. ARGUS Collaboration, reported by Y. Zaitsev, inv. talk at "26th International Conference on High Energy Physics", Dallas 1992
23. CLEO Collaboration, A. Sanghera et al., Cornell preprint CLNS-92/1156(1992)
24. J.G. Körner and G.A. Schuler, Z.Phys. C38(1988)511 (erratum Z.Phys. C41(1988)690)
25. J.G. Körner and G.A. Schuler, Z.Phys. C46(1990)93

26. J.G. Körner and G.A. Schuler, Phys.Lett. B226(1989)185

27. J.G. Körner and G. A. Schuler, Phys.Lett. B231(1989)306

28. J.G. Körner, K. Schilcher and Y.L. Wu, Phys.Lett. B242(1990)119

29. J.G. Körner, K. Schilcher and Y.L. Wu, MZ-TH/92-04, to be publ. in Z. Physik C

30. Hagiwara, A.D. Martin and M.F. Wade, Phys.Lett. B228(1989)144;
    Hagiwara, A.D. Martin and M.F. Wade, Nucl.Phys. B327(1989)569
    F.J. Gilman and R.L. Singleton, Phys.Rev. D41(1990)142

31. G. Köpp, G. Kramer, W.F. Plamer and G.A. Schuler, Z. Phys. C48(1990)327

32. J.D. Bjorken, Phys.Rev. D40(1989)1513

33. CLEO Collaboration, P. Avery et al., Phys.Rev.Lett. 65(1990)2842

34. ARGUS Collaboration, H. Albrecht et al., DESY preprint, DESY 91-091 (1991)

35. F.E. Close, J.G. Körner, R.J.N. Phillips and D.J. Summers, Heavy Quark Symmetry and b-
    Polarization at LEP, MZ-TH/92-14, to be published in Journ.Phys. G

36. B. König, J.G. Körner, M. Krämer and P. Kroll, Diquark Model Calculation of Heavy $\Lambda_b \rightarrow \Lambda_c$
    Transitions including HQET Improvements, Mainz preprint MZ-TH/92-41

37. M. Gronau and S. Wakaizumi, Phys.Lett. B280(1992)79
    M. Gronau and S. Wakaizumi, Phys.Rev. Lett. 68(1992)1814

38. F. Hussain, J.G. Körner and R. Migneron, Phys.Lett. B248(1990)406

39. J.G. Körner, Nucl.Phys. B (Proc.Suppl.) 21(1991)366

40. N. Isgur and M.B. Wise, Nucl.Phys. B348(1991)276

41. H. Georgi, Nucl.Phys. B348(1991)293

42. T. Mannel, W. Roberts and Z. Ryzak, Nucl.Phys. B355(1991)38

43. F. Hussain, J.G. Körner, M. Krämer and G. Thompson, Z. Phys. C51(1991)321

44. F. Hussain, D. Liu, M. Krämer, J.G. Körner and S. Tawfiq, Nucl.Phys. B370(1992)259

45. J.G. Körner and P. Kroll, Heavy Quark Symmetry at Large Recoil, MZ-TH/91-33, to be publ. in
    Phys.Lett. B

46. J.G. Körner and P. Kroll, Heavy Quark Symmetry at Large Recoil: The Case of Baryons,
    MZ-TH/91-34, to be published

47. M. Neubert and V. Rieckert, Heidelberg preprint HD-THEP-91-6 (1991)

48. H. Georgi, B. Grinstein and M.B. Wise, Phys.Lett. B252(1990)456

49. S.J. Brodsky and G.P. Lepage, Phys.Rev. D22(1980)2157

50. H.D. Politzer and M.B. Wise, Phys.Lett. B206(1988)68

COVARIANT BETHE–SALPETER WAVE FUNCTIONS FOR HEAVY HADRONS

F. Hussain

International Centre for Theoretical Physics, Trieste, Italy

1. BETHE–SALPETER AMPLITUDES FOR HEAVY HADRONS

In recent years the dynamics of heavy mesons and baryons has been considerably simplified by the development of the so-called heavy quark effective theory, HQET [1–3] where the heavy quark mass is taken to be infinite. In a series of recent papers [2, 4–8], we have presented a covariant formulation of heavy meson and heavy baryon decays in the leading order of the HQET. This method is based on a Bethe–Salpeter (BS) formulation in the limit of the heavy quark mass going to infinity. In this talk, I would like to briefly review this approach.

The starting point of our investigation was the demonstration that the equal velocity assumption, arising from the heavy quark limit, could be formulated in a covariant manner using the spin–parity projectors developed by Delbourgo, Salam and Strathdee (SDS) [4, 9]. In the zeroth order of HQET, the heavy quark is free and moves with the same four velocity as the hadron, of which it is a constituent:

$$v = \frac{p_Q}{m_Q} = \frac{P}{M} \,.$$ (1)

Since the heavy quark is "free" we can embed the spin of the heavy quark in the usual way in a four-dimensional space of Dirac indices, i.e., the $\left[\left(\frac{1}{2},0\right) \oplus \left(0,\frac{1}{2}\right)\right]$ representation of the Lorentz group.

For example, a meson is represented by a two–index wavefunction $\Phi_{\alpha}^{\beta}(P, p_Q, k)$, where the lower index α, is the heavy quark index and the upper index represents the rest of the light Lorentz structure and $P = p_Q + k$, where P is the momentum of the meson, p_Q the momentum of the heavy quark and k is the momentum of the light quark.

$\Phi_{\alpha}^{\beta}(P, p_Q, k)$ is the Fourier transform of the BS amplitude

$$\Phi_{\alpha}^{\beta}(X, P) = \; < 0 |T\left(\psi_{\alpha}^{h}\left(\frac{x}{2}\right) \; \bar{\psi}_{\ell}^{\beta}\left(-\frac{x}{2}\right) |P, M\right) >$$ (2)

Since the heavy quark is "free" and on–shell we have a Dirac equation on the lower (heavy) index

$$\left(\frac{\not{p}_Q}{m_Q} - 1\right)_{\alpha}^{\alpha'} \Phi_{\alpha'}^{\beta} = 0$$ (3)

but since $\frac{p_Q}{m_Q} = \frac{P}{M} = v$, this equation becomes

$$(\not{v} - 1)_{\alpha}^{\alpha'} \Phi_{\alpha'}^{\beta} = 0$$ (4)

Talk given at the Conference, "Quark Cluster Dynamics" Bad Honnef, 29 June–1 July 1992.

This is the well-known Bargmann–Wigner equation [10]. It is more fashionable nowadays to call it the "velocity super selection rule".

Eq.(4) implies that, in leading order in HQET, all heavy meson wavefunctions are of the form

$$\Phi_\alpha{}^\beta(v,k) = \frac{1}{2}(\not v + 1)_\alpha{}^{\alpha'} \Phi'_{\alpha'}{}^\beta(v,k) . \tag{5}$$

This suggests immediately that $\Phi_\alpha{}^\beta$ should be of the form

$$\Phi_\alpha{}^\beta(v,k) = X_\alpha{}^{\alpha'} A_{\alpha'}{}^\beta(v,k) \tag{6}$$

where $X_\alpha{}^{\alpha'}$ are the spin projection operators developed by SDS. We can prove this by considering an interpolating field in the LSZ framework. Here I will just give a simple diagrammatic argument.

In general, in the quark model, we can decompose the BS amplitude as

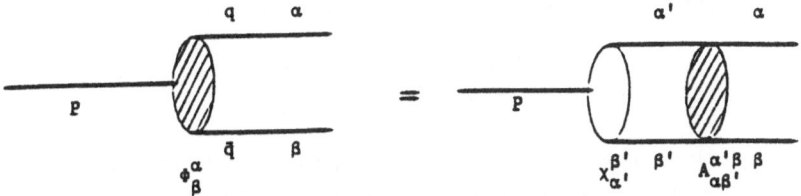

Fig.1. Decomposition of meson Bethe–Salpeter amplitude

that is

$$\Phi_\alpha{}^\beta = X_{\alpha'}{}^\beta A_{\alpha}^{\alpha'}{}_{\beta}^{\beta} \tag{7}$$

where $X_{\alpha'}{}^\beta$ projects out the particular particle (spin) state from the two–body scattering amplitude $A_{\alpha}^{\alpha'}{}_{\beta}^{\beta}$.

Now the picture for the heavy meson is clear. The heavy quark leg is free

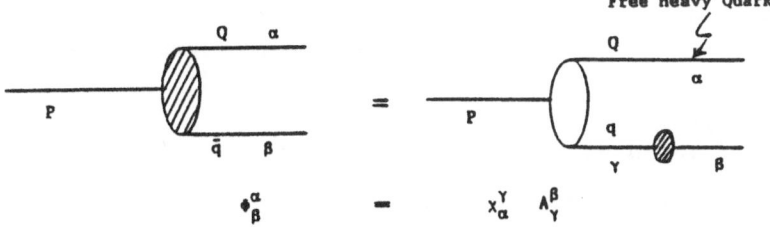

Fig.2. Ansatz for heavy meson B–S amplitude.

that is $A_{\alpha}^{\alpha'}{}_{\beta}^{\beta} \rightarrow \delta_\alpha^{\alpha'} A_{\beta'}^\beta$.

Similarly for heavy hadrons:

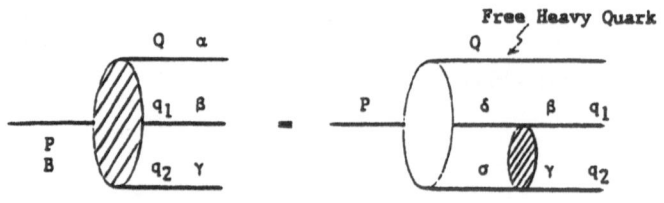

$$B_{\alpha\beta\gamma}(v,q_1,q_2) = \chi_{\alpha\delta\sigma}(v) A^{\delta\sigma}_{\beta\gamma}(q_1,q_2)$$

Fig.3. Ansatz for heavy baryon B–S amplitude.

The spin projection operators χ are nothing but the Bargmann–Wigner wavefunctions. In their classic paper [10], Bargmann and Wigner showed that all types of spinning particles could be described by multispinor fields, that is, products of the fundamental spin $\frac{1}{2}$ representation of the Lorentz group.

The BW wavefunctions are multispinor wavefunctions of some given rank and symmetry type $\chi^{\beta_1\cdots\beta_n}_{\alpha_1\cdots\alpha_n}(v)$ which satisfy the Bargmann–Wigner equations on all the labels

$$(\not{v} - 1)_{\alpha_1}{}^{\alpha'_1} \chi^{\beta_1\cdots\beta_n}_{\alpha'_1\cdots\alpha_n}(v) = 0$$

$$\chi^{\beta_1\cdots\beta_n}_{\alpha_1\cdots\alpha_n}(v) (\not{v} + 1)_{\beta'_1}{}^{\beta} = 0 . \tag{8}$$

There is a simple reason why these χ's describe particles of a given spin. In the rest frame, the Dirac equation is a projection by the operator $(1 + \gamma_0)$. Since γ_0 is the parity operator for spin $\frac{1}{2}$, all that the Dirac equation does is to specify the parity of the fundamental representation. This means that the spin (Dirac) labels take on not four but two independent values. This reduces $SO(4)$ to $SU(2)$. On imposing the Dirac equation on all the labels we get a reducible product representation of $SU(2)$. On fixing to a given symmetry type we get an irreducible representation of $SU(2)$ describing a particle of a given spin.

Example: Consider $\chi(v)$ a totally symmetric spinor of rank 2s. This is the wavefunction of a particle of spin s and parity $(+1)^{2s}$, if they are all lower indices and $(-1)^{2s}$ if all upper indices. For a multispinor, the parity operator is a tensor product of γ_0's.

As a particular example, a rank two symmetric bispinor describes spin one particles. The gamma matrices split into symmetric and antisymmetric matrices as follows:

symmetric   : $\gamma_\mu C, \sigma_{\mu\nu} C,$
antisymmetric : $C, \gamma_5 C, \gamma_\mu \gamma_5 C$.

We can thus write the symmetric bispinor as

$$\chi_{\{\alpha\beta\}} = (\gamma_\mu C)_{\alpha\beta} \phi^\mu + (\sigma_{\mu\nu} C)_{\alpha\beta} \phi^{\mu\nu} . \tag{9}$$

The BW equations reduce to the constraints

$$v_\mu \phi^\mu = 0, \quad \phi^{\mu\nu} = v^\mu \phi^\nu - v^\nu \phi^\mu . \tag{10}$$

Identifying ϕ^μ with the polarization vector ϵ^μ, the spin projector takes the form

$$\chi_{\{\alpha\beta\}} = \left[(1 + \not{v}) C \not{\epsilon} \right]_{\alpha\beta} . \tag{11}$$

Similarly, the antisymmetric bispinor can be written as

$$X_{[\alpha\beta]} = \left[(1 + \not{v})\,\gamma_5\,C\right]_{\alpha\beta} \tag{12}$$

$X_{\{\alpha\beta\}}$ and $X_{[\alpha\beta]}$ have quark number 2 and parity 1.

Similar arguments for quark number zero wavefunctions leads to

$$\Phi_\alpha{}^\beta = \left[(\not{v} + 1)\,\gamma_5\right]_\alpha{}^\beta, \quad \text{spin zero, parity} = -1$$
$$\Phi_\alpha{}^\beta = \left[(\not{v} + 1)\,\not{\xi}\right]_\alpha{}^\beta, \quad \text{spin one, parity} = -1. \tag{13}$$

For baryons consider rank three spinors $X_{\alpha\beta\gamma}(u)$. After the BW equations are imposed each label has two values. The total degrees of freedom are thus eight. We wish to decompose down to its irreducible parts $(\frac{1}{2} \oplus \frac{1}{2} \oplus \frac{3}{2})$.

Put

$$X_{\alpha\beta\gamma} = X_{[\alpha\beta]\gamma} + X_{\{\alpha\beta\}\gamma}. \tag{14}$$

If BW equations are not imposed then $X_{[\alpha\beta]\gamma}$ is reducible. One would have to project out the totally antisymmetric part by

$$X_{[\alpha\beta]\gamma} + X_{[\gamma\alpha]\beta} + X_{[\beta\gamma]\alpha} = 0. \tag{15}$$

However, after BW equations are imposed this condition is identically satisfied. $X_{[\alpha\beta]\gamma}$ represents spin $\frac{1}{2}$ particles. However, $X_{\{\alpha\beta\}\gamma}$ is reducible even after the imposition of the BW equations. We project away the totally symmetric part by

$$X_{\{\alpha\beta\}\gamma} + X_{\{\beta\gamma\}\alpha} + X_{\{\gamma\alpha\}\beta} = 0. \tag{16}$$

This then also represents a spin $\frac{1}{2}$ particle. One can check easily that this has just two degrees of freedom. The complete decomposition is now

$$X_{\alpha\beta\gamma} = X_{[\alpha\beta]\gamma} + X_{\{\alpha\beta\}\gamma} + X_{\{\alpha\beta\gamma\}}. \tag{17}$$

Degrees of freedom = 2+2+4=8 where $X_{\{\alpha\beta\gamma\}}$ is the totally symmetric part.
This decomposition allows a rather nice projection for s–wave heavy baryons.
1. $X_{\{\alpha\beta\gamma\}}$ projects spin $\frac{3}{2}$, Σ_Q^*, Ω_Q^* particles.
2. For Λ_Q and Ξ_Q, the two light quarks are in an antisymmetric $s = 0$ state. Therefore, if we let γ be the heavy quark label, then $X_{[\alpha\beta]\gamma}$ is the projector we need. It has the form

$$X_{[\alpha\beta]\gamma} = \left[(1 + \not{v})\,\gamma_5\,C\right]_{\alpha\beta} u_\gamma, \quad (\not{v} - 1)u = 0. \tag{18}$$

3. For the heavy Σ_Q and Ω_Q baryons, the light quarks are in a symmetric $s = 1$ state. With γ the heavy label, the other spin $\frac{1}{2}$ X, $X_{\{\alpha\beta\}\gamma}$ is the correct projector.

There are two ways to get an explicit form for $X_{\{\alpha\beta\}\gamma}$.
(i) Expand $X_{\{\alpha\beta\}\gamma}$ in terms of symmetric gamma matrices, in α and β, and impose the BW equations. This gives

$$X_{\{\alpha\beta\}\gamma} = \left[(1 + \not{v})\phi_\mu\,u\right]_\gamma \left[(1 + \not{v})\gamma^\mu\,C\right]_{\alpha\beta}. \tag{19}$$

On now imposing tracelessness conditions we find

$$\begin{aligned}X_{\{\alpha\beta\}\gamma} &= \left[(1 + \not{v})\gamma_\mu\gamma_5 u\right]_\gamma \left[(1 + \not{v})\gamma^\mu\,C\right]_{\alpha\beta} \\ &= \left[(\gamma_\mu + v_\mu)\gamma_5 u\right]_\gamma \left[(1 + \not{v})\gamma^\mu\,C\right]_{\alpha\beta}.\end{aligned} \tag{20}$$

The first square bracket on the right–hand side of this equation is the wavefunction proposed by Georgi [11].

(ii) Alternatively start with the observation that the trace condition is easily solved in terms of antisymmetric objects

$$\chi_{(\alpha\beta)\eta} = \phi_{[\alpha\eta]}\,\phi_\beta + \phi_{[\beta\eta]}\,\phi_\alpha \tag{21}$$

which on imposing BW equations yields

$$\chi_{(\alpha\beta)\eta} = \left[(1+\not v)\gamma_5\,C\right]_{\alpha\eta}\,u_\beta + \left[(1+\not v)\gamma_5\,C\right]_{\beta\eta}\,u_\alpha\,. \tag{22}$$

This is the form proposed by us [7, 8].

Of course, the two forms Eq.(20) and Eq.(22) must be identical as they represent the same Lorentz structure. It is a matter of some gamma algebra to explicitly show this [8].

Upto now I was only considering S–wave states. We can apply the same techniques to P–wave and higher orbital states. Only the algebra is a bit more complicated. For mesons we now list the complete S–wave and P–wave BS amplitudes incorporating these projectors [12]:

S–wave:

$$\Phi_\alpha{}^\beta(v,k) =$$

$$^1S_0(0^{-+}) \;:\; \frac{1}{2}\sqrt{M}\,(\not v+1)\,\gamma_5\,\phi_s(v,k)$$

$$^3S_1(1^{-+}) \;:\; \frac{1}{2}\sqrt{M}\,(\not v+1)\,\not\epsilon\,\phi_s(v,k) \tag{23}$$

P–wave:

$$\Phi_\alpha{}^\beta(v,k) =$$

$$^3P_0(0^{++}) \;:\; -\frac{1}{2\sqrt{3}}\sqrt{M}\,(\not v+1)(\not k - k\cdot v)\phi_p(v,k)$$

$$^3P_1(1^{++}) \;:\; -\frac{i}{2\sqrt{2}}\sqrt{M}\,(\not v+1)\,\epsilon^{\mu\nu\rho\sigma}\,v_\mu\,\epsilon_\nu\,k_\rho\,\gamma_\sigma\,\phi_p(v,k)$$

$$^3P_2(2^{++}) \;:\; \frac{1}{2}\sqrt{M}\,(\not v+1)\,\gamma^\mu\,\epsilon_{\mu\nu}\,k^\nu\,\phi_p(v,k)$$

$$^1P_1(1^{+-}) \;:\; \frac{1}{2}\sqrt{M}\,(\not v+1)\,\gamma_5\,k\cdot\epsilon\,\phi_p(v,k)\,. \tag{24}$$

In Eqs.(23) and (24) we have factored out the heavy mass scale $\sqrt{M}$. ϕ_s and ϕ_p are the S–wave and P–wave orbital functions. These, in general, will be bispinors.

These states were constructed using the analogue of the LS–coupling scheme, in which charge conjugation parity is obvious. This is O.K. for heavy $Q\bar{Q}$ quarkonium states but no longer useful for $Q\bar{q}$ states. Here the appropriate degenerate heavy light states are determined by coupling orbital angular momentum L with the spin of the light quark just as in the hydrogen atom. For P–wave states, with $\ell = 1$, one has two degenerate multiplets

$$J^P = \left(0^+,1^+_{1/2}\right),\qquad J^P = \left(1^+_{3/2},2^+\right)\,.$$

Subscript on 1^+ states labels the total angular momentum of the light quark system. The $J^P = 1^+$ mesonic P–wave states $|1^+_{1/2}>$ and $|1^+_{3/2}>$ are linear combinations of the $J = 1$, P–wave spin states:

$$|1^+_{1/2}> = \sqrt{\frac{1}{3}}\,|1^{+-}> + \sqrt{\frac{2}{3}}\,|1^{++}>$$

$$|1^+_{3/2}> = \sqrt{\frac{2}{3}}\,|1^{+-}> - \sqrt{\frac{1}{3}}\,|1^{++}>\,. \tag{25}$$

where $|1^+_{1/2}>$ and $|1^+_{3/2}>$ are not C–eigenstates.

2. WEAK TRANSITIONS

Given the wavefunctions we can now compute current induced transitions between heavy mesons. The general form of the matrix element is [12]:

$$M_\mu = < M_2(v_2)|J_\mu^{V-A}|M_1(v_1) >$$
$$= \int d^4 \, k_1 \, d^4 \, k_2 \, Tr \left\{ \bar{\Phi}_2(v_2, k_2) \, T_\mu(k_1, k_2; v_1, v_2) \, \Phi_1(v_1, k_1) \right\} \qquad (26)$$

with $\bar{\Phi} = \gamma_0 \, \Phi^+ \, \gamma_0$.

To lowest order in perturbation theory we get the Mandelstam–Nishijima [13] formula for transitions between bound states with

$$T_\mu(k_1, k_2; v_1, v_2) = (\not{k} - m) \otimes \gamma_\mu(1 - \gamma_5) \delta(k_1 - k_2) \qquad (27)$$

In the heavy quark limit, one can go beyond perturbation theory. One finds quite generally [12, 14] that in this limit

$$T_\mu(k_1, k_2; v_1, v_2) = T(k_1, k_2; v_1, v_2) \otimes \gamma_\mu(1 - \gamma_5) . \qquad (28)$$

T connect light quark legs whereas the weak current connects the heavy quarks, Fig.4

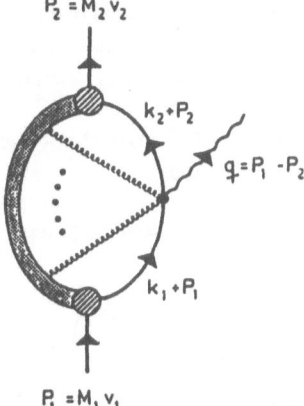

Fig.4. Current induced transitions between heavy mesons in the HQET. The curly lines represent the Wilson line glue contributions that arise on transforming to free heavy quarks.

Because of this factorization, the matrix element can always be written as

$$M_\mu = Tr \left\{ \bar{X}_2 \, \gamma_\mu(1 - \gamma_5) X_1 \int d^4 \, k_2 \, d^4 \, k_2 \, A_1(k_1) T(k_1, k_2; v_1, v_2) \bar{A}_2(k_2) . \qquad (29)$$

Eq.(29) is obtained by substituting Eq.(28) in Eq.(26) and recalling that $\Phi = XA$.

S–wave $\rightarrow$ S–wave transitions

$\Phi = \chi A_S$ with $\chi = (1 + \not{v})\gamma_5$ or $(1 + \not{v})\,\not{\epsilon}$, but with A_S the same for both 0^- and 1^- states.

Evidently, for all the transitions $0^- \rightarrow 0^-, 0^- \rightarrow 1^-$ and $1^- \rightarrow 1^-$ we have the same integral

$$\int d^4 k_1 \, d^4 k_2 \, A_1(k_1) \, T \bar{A}_2(k_2) \cdot .$$

This is a scalar Dirac matrix. It can only be proportional to $1, \not{v}_1, \not{v}_2$ or $\not{v}_1 \not{v}_2$. However, because of the projectors χ all of these collapse to $\mathbb{1}$. Hence

$$M = Tr \left\{ \bar{\chi}_2 \, \gamma_\mu (1 - \gamma_5) \chi_1 \right\} \, F(q^2) . \tag{30}$$

There is a single universal form factor $F(q^2)$ describing all S–wave to S–wave transtions. This was first suggested in 1962 by Durand, de Celles and Marr [15]. Nowadays, this is written as $\xi(w)$, with $w = v_1 \cdot v_2$.

S–wave $\rightarrow$ P–wave transitions

Writing $\Phi = \chi_\nu \, k^\nu \, A_P$ for P–waves one finds for $S \rightarrow P$–wave transitions

$$M_\mu = Tr \left\{ \bar{\chi}_2^\nu \, \gamma_\mu (1 - \gamma_5) \chi_1 (F_j(w) \, v_{1\nu} + F_j'(w) \, \gamma_\nu) \right\} \tag{31}$$

where the index $j = \frac{1}{2}$ and $\frac{3}{2}$ for the degenerate multiplets $(0^+, 1_{1/2}^+)$ and $(1_{3/2}^+, 2^+)$ respectively. In fact, it turns out that there are only two linear combinations of these F_j's which describe the S to P–wave transitions. We denote the two reduced form factor functions that describe the $(0^-, 1^-) \rightarrow (0^+, 1_{1/2}^+)$ and $(0^-, 1^-) \rightarrow (1_{3/2}^+, 2^+)$ transitions by $\xi_{1/2}^*(w)$ and $\xi_{3/2}^*(w)$, respectively.

$$\xi_{1/2}^*(w) = (w + 1) \, F_{1/2}(w) - 3 F_{1/2}'(w)$$
$$\xi_{3/2}^*(w) = F_{3/2}(w) . \tag{32}$$

We now collect together the results for all the S–wave to S–wave and S–wave to P–wave transitions, for example, in $b \rightarrow c$ transitions.

(i)

$$\begin{pmatrix} 0^- \\ 1^- \end{pmatrix} \rightarrow \begin{pmatrix} 0^- \\ 1^- \end{pmatrix}$$

$$M_\mu = \frac{1}{4} \sqrt{M_1 M_2} \, Tr \left(\begin{pmatrix} \gamma_5 \\ \not{\epsilon}_2^* \end{pmatrix} (\not{v}_2 + 1) \, \gamma_\mu (1 - \gamma_5)(\not{v}_1 + 1) \begin{pmatrix} \gamma_5 \\ \not{\epsilon}_1 \end{pmatrix} \right) \xi(w) \tag{33}$$

(ii)

$$\begin{pmatrix} 0^- \\ 1^- \end{pmatrix} \rightarrow \begin{pmatrix} 0^+ \\ 1_{1/2}^+ \end{pmatrix}$$

$$M_\mu = \frac{1}{\sqrt{3}} \sqrt{M_1 M_2} \, \frac{1}{4} \, Tr \left(\begin{pmatrix} 1 \\ \not{\epsilon}_2^* \gamma_5 \end{pmatrix} (\not{v}_2 + 1) \, \gamma_\mu (1 - \gamma_5)(\not{v}_1 + 1) \begin{pmatrix} \gamma_5 \\ \not{\epsilon}_1 \end{pmatrix} \right) \xi_{1/2}^*(w) \tag{34}$$

(iii)

$$\begin{pmatrix} 0^- \\ 1^- \end{pmatrix} \rightarrow \begin{pmatrix} 1_{3/2}^+ \\ 2^+ \end{pmatrix}$$

$$M_\mu = \frac{1}{4} \sqrt{M_1 M_2} \; Tr \left(\left(\begin{matrix} -\sqrt{\frac{1}{6}} \left((\omega+1) \not\in_2^* + 3 v_1 \cdot \in_2^* \right) \gamma_5 \\ v_{1\nu} \in_2^{*\nu\nu'} \gamma_\nu \end{matrix} \right) \times \right.$$

$$\left. \times (\not\psi_2 + 1) \; \gamma_\mu (1 - \gamma_5)(\not\psi_1 + 1) \begin{pmatrix} \gamma_5 \\ \not\in_1 \end{pmatrix} \right) \xi_{3/2}^*(\omega) \; .$$

3. CONCLUSIONS

In leading order in the HQET, i.e., in the limit $m_b \to \infty, m_c \to \infty$ one finds a dramatic reduction in the number of reduced form factors or independent amplitudes describing the transitions.

For $(0^-, 1^-) \to (0^-, 1^-)$ transitions, one has a reduction from 20 to 1 independent amplitude. Similarly, for $(0^-, 1^-) \to (0^+, 1^+_{1/2})$ there is a reduction from 20 to 1 and for $(0^-, 1^-) \to (1^+_{3/2}, 2^+)$ from 30 to 1.

References

1. E. Eichten and F. Feinberg, Phys. Rev. D23 (1981) 2724;
    G.P. Lepage and B.A. Thacker, Nucl. Phys. B (Proc. Suppl.) 4 (1988) 199;
    E. Eichten, ibid 170;
    E. Eichten and F. Feinberg, Phys. Rev. Lett. 43 (1979) 1205;
    W.E. Caswell and G.P. Lepage, Phys. Lett. 167B (1986) 437;
    M.B. Voloshin and M.A. Shifman, Sov. J. Nucl. Phys. 47 (3) (1988) 511
2. F. Hussain, J.G. Körner, K. Schilcher, G. Thompson and Y.C. Wu, Phys. Lett. 249B (1990) 295
3. J.G. Körner and G. Thompson, Phys. Lett. 264B (1991) 185
4. F. Hussain, J.G. Körner and G. Thompson, Ann. Phys. (NY) 206 (1991) 534
5. F. Husain, J.G. Körner and R. Migneron, Phys. Lett. B248 (1990) 406;
    erratum Phys. Lett. B252 (1990) 723
6. F. Hussain and J.G. Körner, Z. Phys. C51 (1991) 607
7. F. Hussain, J.G. Körner, M. Krämer and G. Thompson, Z. Phys. C51 (1991) 321
8. F. Hussain, D.S. Liu, J.G. Körner, M. Krämer and S. Tawfiq, Nucl. Phys. B370 (1992) 259
9. A. Salam, R. Delbourgo and J. Strathdee, Proc. Roy. Soc. A284 (1965) 146
10. V. Bargmann and E.P. Wigner, Proc. Nat. Acad. Sci. 34 (1948) 211
11. H. Georgi, Nucl. Phys. B348 (1991) 293
12. S. Balk, F. Hussain, J.G. Körner and G. Thompson, ICTP, Trieste, preprint No.IC/91/397, MZ-TH 92/22, to appear in Z. Phys. C.
13. K. Nishijima, Prog. Theor. Phys. 10 (1953) 549; 12 (1954) 279; 13 (1956) 305;
    S. Mandelstam, Proc. Roy. Soc. A253 (1955) 248
14. S. Balk, J.G. Körner and G. Thompson, in preparation
15. L. Durand III, P.C. de Celles and R.B. Marr, Phys. Rev. 126 (1962) 1892

Overview on Exotic Meson States

H. Koch

Institute of Experimental Physics, Ruhr-Universität Bochum, D-4630 Bochum, Germany

Abstract: Candidates for exotic meson-like states, like Glue-Balls, Hybrids or Multiquark-States, are discussed. Special emphasis is given to the first results from the Crystal Barrel experiment at LEAR.

Key words: Exotic Mesons, Glue Balls, Hybrids, Crystal Barrel Experiment.

1. Introduction

This report is not meant to cover the field of exotic states completely. Rather, it concentrates on the area, which is easily accessible to the Crystal Barrel experiment at LEAR, i.e. states with masses $\leq$ 2.3 GeV. More complete reports can be found elsewhere [1-5].

2. Definition of Exotic Meson States

Fig. 1 gives an overview on the lightest $q\bar{q}$-states (nonetts) as predicted by theory, with experimentally found particles [6] attached to the scheme. Additionally, some candidates for exotic states not fitting into the $q\bar{q}$-nonett-picture are given. According to their quark/gluon content, three kinds of exotics will be discussed in the following:

- *Glue Balls*: They have no quark content and consist only of two or three constituent gluons.
- *Hybrids*: These are $\bar{q}q$-states with an additional constituent gluon, which can exhibit dynamical excitations.
- *Multiquark-states*: These are states consisting of two (three) quarks and two (three) antiquarks, or even more.

Another classification uses the quantum numbers of the exotics:

- *Exotics of the first kind*: These are states with $Q \geq 2$, $S \geq 2$ or $I \geq 2$.

They are obviously inconsistent with any $q\bar{q}$-picture. No firm candidate for such date exists yet.

- *Exotics of the second kind*: These are states with quantum-number combinations (J^{PC}) not allowed in a $q\bar{q}$-meson-model (*Exotic quantum numbers*). One example would be 0^--states, which could decay e.g. into $\pi^\circ\omega$, $\pi^\circ\phi$, $\eta\phi$, or $\eta\omega$. Other examples are 1^{-+}-($\pi^\circ\eta$, $\pi^\circ\eta'$, $\phi\phi$, $\eta\eta'$), 2^{+-}-($\pi^\circ\omega$, $\eta\phi$,..), and 3^{-+}-($\pi^\circ\eta$, $\phi\phi$, $\eta\eta'$,...) states. Two candidates for such states with $J^{PC} = 1^{-+}$ have been found by one experiment [7,8], but could not yet be confirmed by any other experiment.

- *Exotics of the third kind*: These states are sometimes also called *crypto-exotics*. They have the same J^{PC} combinations as $q\bar{q}$-states, but are either superfluous (the corresponding $q\bar{q}$-nonett is already filled) or have properties, which cannot be easily explained in terms of a q̄q-model. Several of them are known [6]. Some examples are:

a_0 (980), f_0 (975): These states have masses and widths, which are in clear disagreement with all q̄q-model predictions

f_0 (1590): This state shows unusual decay modes, e.g. a very high $\eta\eta'$ decay probability.

f_2 (1515): This particle has an unusual production mode. It has been dominantly seen only in antiproton-proton annihilation reactions and nowhere else.

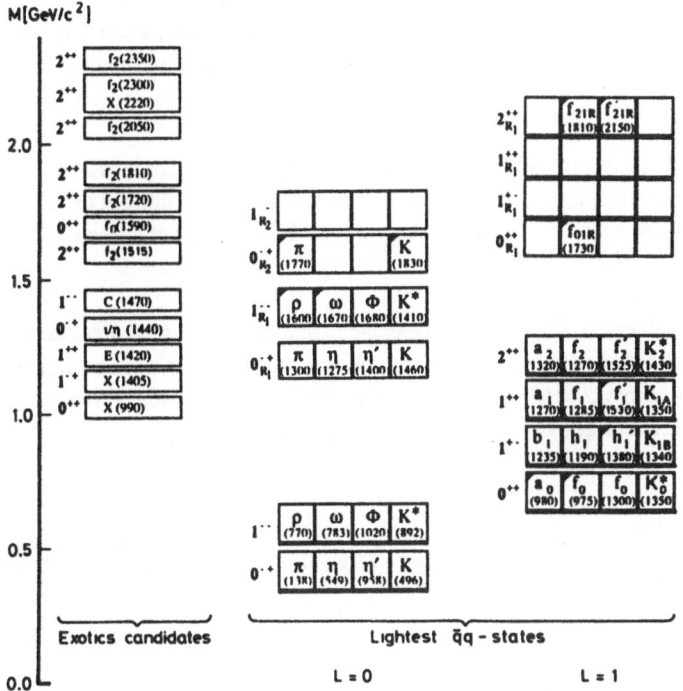

Fig. 1: Right side: Lightest q̄q-states with angular momenta L=0,1.
   (a_0 (980) and f_0 (975) appear here as members of the 0^{++} -
   q̄q-nonett, but rather are exotic states with high probability).
   Left side: Candidates for exotic states

3. How to look for exotics

The disentangling between exotic states and ordinary q̄q-states is easy only in the case of exotics of the first and the second kind. The disentangling between q̄q-states and crypto exotics is generally not unique and is based on expectations on the decay- and production- properties of such states.

3.1 Distinctive decays

3.1.1 Glue Balls

- The decays of Glue Balls should be flavour blind. That means, that (after phase space- and isospin-corrections) decays into e.g. $\pi\pi$, $K\overline{K}$, and $\eta\eta$ should be equally frequent.
- Glue Ball states might have widths as low as 10 MeV due to the OZI-rule, which is relevant for all Glue-Ball decay modes.
- The coupling strength of Glue Balls to photons is zero in first order.
- The decays into η,η' should be enlarged in comparison to $q\bar{q}$-states. This assumption is based on the radiative Ψ-decay experiments, where η and η' particles are abundantly observed.
- The $4\pi^0$-decay channel should be significantly stronger than in $\bar{q}q$-mesons, where it is suppressed because of the $\rho\rho$-dominance, which only shows up in 4π-channels with *charged* pions.

3.1.2 Hybrids

Various model calculations agree, that hybrid decays into $(q\bar{q})_{L=1} + (\bar{q}q)_{L=0}$ should be favoured over decays into $(q\bar{q})_{L=0} + (q\bar{q})_{L=0}$. Orbitally excited mesons with L=1 are difficult to measure, and it is claimed, that this is one of the reasons, why no hybrid could be unambiguously identified yet.

3.1.3 Multi-Quark-States

Here, e.g., decays into $\pi^0\phi$ or $\rho^0\phi$ would be interesting to measure. These final states have I=1 and contain a $s\bar{s}$-quark pair. In $\bar{q}q$-systems with I=1 such decays would be suppressed because of the OZI-rule, in the I=0 case they would be suppressed because of the isospin violation. However, they would be easily allowed for a four-quark-state, e.g. in a $(\bar{u}u\,(s\bar{s}))$-system.

3.2 Distinct production

In the following, several production mechanisms for exotic particles are discussed (Fig. 2) and their relative merits are compared.

3.2.1 Hadronic peripheral production with CEX (low t)

An example here are processes like $\pi^-(K^-) + p \to X + n\,(\Lambda)$. The X (H or G) may decay, e.g., into $\pi^0\eta$ (Fig. 2a) or even with OZI suppression into $\phi\phi$ (Fig. 2b). Such hadronic processes were investigated in many experiments. They have contributed a lot to our insight into $q\bar{q}$- and exotic-states, but suffer from low production rates. Usually, the use of specific triggers does not allow the determination of branching ratios within the same measurement. Additionally, the detection efficiency is often not uniform, and models for the J^P- analysis are needed.

3.2.2 Hadronic central production (higher t)

Typical examples are back-scattering processes like $\pi^- + p \to p + H$ (Fig. 2c), glue sea interactions (Fig. 2d) or double pomeron exchanges (Fig. 2e) at higher energies.

3.2.3 Radiative or mesonic ᴪ-decays

The investigation of these processes which are dominated by two- or three- gluon exchanges (Fig. 2f,g) has had a vital impact on meson- and exotics-spectroscopy in the last decade. One advantage is, that only *one* partial wave in the initial state dominates. Runs with minimum bias trigger are possible, so that branching ratios can be easily extracted. The experiments still suffer from statistics, which is not yet high enough to allow very sensitive statements.

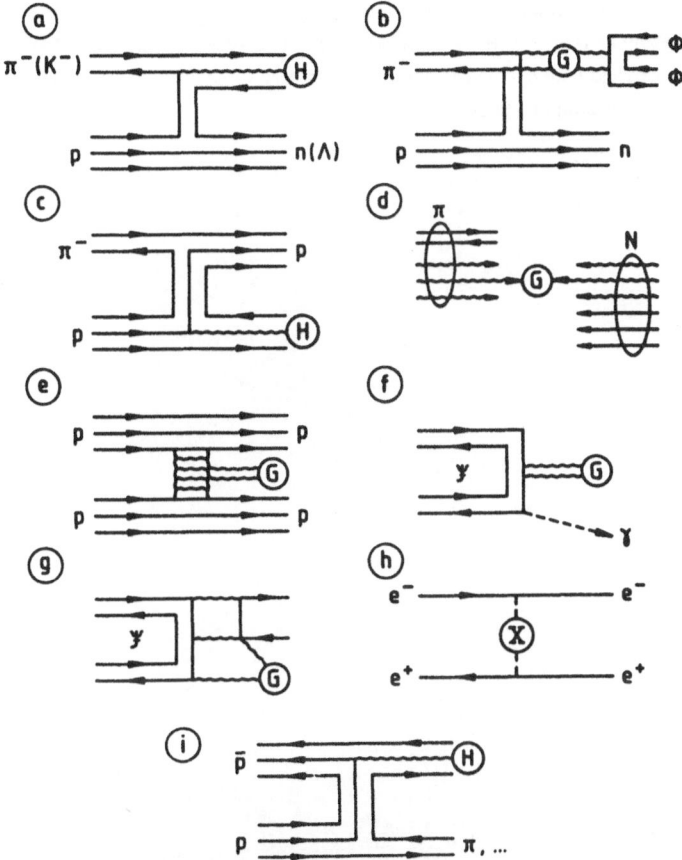

Fig. 2: Overview on possible production mechanisms for Exotics
(X=any Exotics, G=Glue-Ball, H=Hybrid; Quark (solid line),
Glue (wavy line), Gamma (dashed line))
(a) Hadronic peripheral production (low t) with charge exchange
(b) Some as (a), but with OZI suppressed decay-mode
(c) Hadronic central production (high t), Backscattering
(d) Hadronic central production (high t), Glue seas interaction
(e) Hadronic central production (high t), Double Pomeron Exchange
(f) Radiative ᴪ-decay
(g) Mesonic ᴪ-decay
(h) Two photon production
(i) p̄p (d) annihilation at rest or in flight

3.2.4 Two Photon Production

These measurements (Fig. 2h) are very helpful in determining the $q\bar{q}$-contents of meson resonances. The appearence of a state (X) in these reactions signals the absence of glue, as it does not couple to photons in first order. The comparison between the production of a state in 2γ- and radiative ψ-reactions yields a powerful criterium (stickiness) for the quark- or glue-content of this state.

3.2.5 $\bar{p}p(d)$-annihilation at rest and in flight

A typical reaction would be $\bar{p}+p \to X(H) + \pi,\eta,\eta'$ (Fig. 2i). This type of experiments has been started more than three decades ago with bubble chambers and was very successfull in detecting new mesons and in determining their quantum numbers.

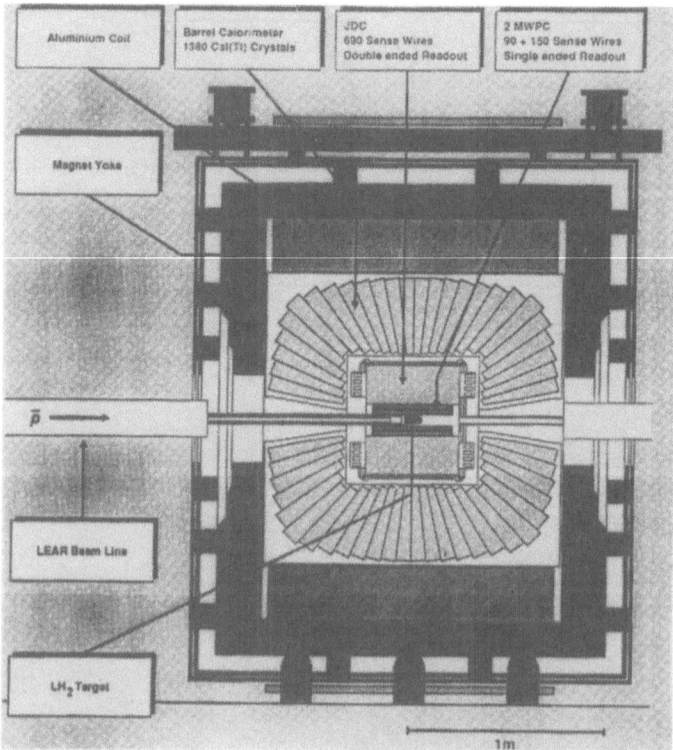

Fig. 3: The Crystal Barrel Detector at LEAR
Members of the Crystal Barrel Collaboration: Univ. of Bochum; Academy of Science, Budapest; CERN, Geneva; Univ. of Hamburg; Univ. of Karlsruhe; Queen Mary and Westfield College, London; Univ. of Mainz; Univ. of Munich; Rutherford and Appleton Laboratory, Chilton; Centre de Recherches Nucléaires, Strasbourg; UC Berkeley; US Los Angeles; Univ. of Zürich

At LEAR, these experiments came into in a new era, where very high statistics and very clean data samples can be collected. As a typical example for an experimental set-up at LEAR, Fig. 3 shows the Crystal Barrel detector [9], with which within the last two years more than 6×10^7 annihilation events were measured, mostly at rest, but also in flight up to an antiproton momentum of 1.9 GeV/c. The detector is nearly hermetic for the detection of photons, and has a solid angle of 70% $\bullet$ 4π for the charged annihilation products. Thus, it allows a complete measurement of all kinds of annihilation reactions with excellent mass resolution and nearly constant efficiency over the full phase space.

4. Candidates for Exotics

With two exceptions (M(1405), C(1480)), only candidates for exotics are discussed the evidence and the quantum numbers of which are beyond any doubt. They are ordered according to their J^{PC}-quantum numbers and subsequently discussed in the following.

4.1. 0^{++}

4.1.1 a_0 (980), f_0 (975)

These states were originally believed to be members of the 0^{++} $q\bar{q}$-nonett, but their masses and widths (too low) do not at all coincide with theoretical expectations. On the other hand, two 0^{++}-states at 1300 and 1500 MeV with broad widths have been found recently [7,10], which could replace a_0 and f_0 in the nonett. One of them (f_0(1400))was seen very clearly in the Crystal Barrel experiment in the $\eta\eta$-decay channel [11]. Therefore we suppose, that a_0 and f_0 are not $\bar{q}q$-states but something else. Their properties can be well explained, if they were $K\bar{K}$- molecules or $2q2\bar{q}$-states.

4.1.2 f_0 (1560)

This state was seen in peripheral and central production hadronic experiments [12,13]. Its decay pattern shows very unusual features, e.g. a very high $\eta\eta'$ branching ratio. It was recently confirmed by the Crystal Barrel experiment (Fig. 4) in the analysis of the reaction $(\bar{p}p)_{Rest} \rightarrow f_0 (1560)\pi^0$, $f_0 (1560) \rightarrow \eta\eta$ [11]. The mass and width of the state were determined to be (1560 ± 25) MeV and (245 ± 50) MeV, respectively, with $J^{PC} = 0^{++}$. The state is the one of the strongest candidates for a Glue Ball because of its peculiar decay properties. The search for more decay modes of this state in the antiproton-proton annihilation goes on.

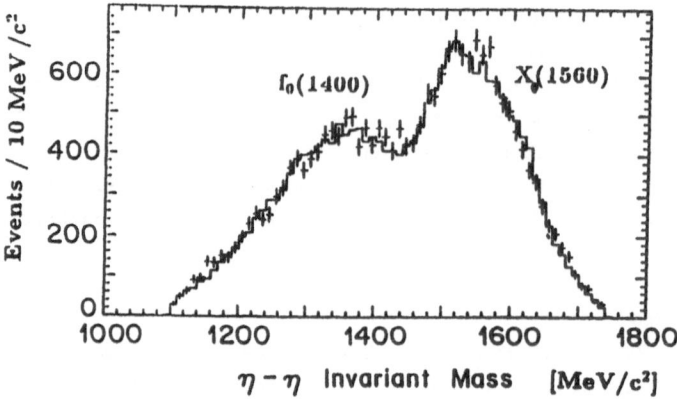

Fig. 4: ($\eta\eta$)-invariant masses of the reaction $(\bar{p}p)_{Rest} \to \pi^0 \eta\eta$, as measured by the Crystal Barrel collaboration [11]. The full line corresponds to the best fit to the data.

4.2 0^{-+}

4.2.1 X(1420), X(1490)

Several experiments in the past have observed interesting structure in the $K\overline{K}\pi$-channel in the mass region between 1400 and 1500 MeV. The most recent analysis of the K^*K-channel (Mark III-collaboration [14]) claims, that three narrow states exist in this energy region: Two with the quantum numbers 0^{-+} (1420,1490 MeV), one with the quantum number 1^{++} (1440 MeV). One of the 0^--states is probably exotic and can be explained as a hybrid or a glue ball. The appearance of a $K\overline{K}\pi$-resonance in this mass region has been firstly detected by a bubble chamber experiment on antiproton-proton annihilations as early as 1963. Crystal Barrel will try to gain enough statistics to contribute significantly to the disentangling of the states in this mass region.

4.3 1^-

C(1480)

The Lepton-F-collaboration has seen a 1^{--} state in the reaction $\pi^- + p \to C(1480)+n$, $C(1480) \to \pi^0\phi$. There are many reasons to exclude that state to be a radial excitation of the ρ, and an interpretation in terms of a $2q2\bar{q}$-state, e.g. $(u\bar{u} - d\bar{d})s\bar{s}$, is quite reasonable. The Crystal Barrel collaboration has searched for the mass degenerate isoscalar partner of this state decaying into $\eta\omega$, but has no evidence yet.

4.3. 1^{-+}

M(1406)

The GAMS collaboration claims to have detected a state M(1406) with the quantum numbers $J^{PC}=1^{-+}$ (exotic of the second kind) in the reaction $\pi^- + p \to M(1406)+n$, $M \to \pi^0\eta$ [7]. The determination of J^P in the partial wave analysis seems not to be unambiguous [16], so that the exact nature of this state is not yet certain. The Crystal Barrel collaboration has searched for this state in the reaction $(\bar{p}p)_{Rest} \to M\pi^0$, $M \to \pi^0 \eta$. Nearly 30000 ($\pi^0 \pi^0 \eta$)- events were collected, but the M-state was not prominently seen. The analysis of the data goes on, also looking for a higher mass state (1910 MeV) with possibly the same exotic quantum numbers [8]

4.4. 1^{++}

$f_1(1420)$

As already a discussed before, a 1^{++} state has been seen in the $K^*\bar{K}$-decay channel at 1440 MeV in radiative ψ-decays. It was also observed in various hadronic reactions at 1420 MeV. It is probably no $\bar{q}q$-state, because there is another state (f_1 (1510)) which fits better the quark model predictions. It is also no good candidate for a glue ball ($\gamma\gamma$-width too large in comparison with model calculations) and no good candidate for a hybrid because of its low mass. It could rather be a $K^*\bar{K}$-molecule as predicted by Weinstein/Isgur or Longacre [17,18].

4.5. 2^{++}

$X_2(1515)$

There is overwhelming evidence for a resonance with $J^{PC}=2^{++}$ above 1500 MeV. Probably it has been seen for the first time in a $(\bar{p}d)_{Rest}$ -experiment at Brookhaven in the ($\rho\rho$)-decay-channel, then very clearly in the ($\pi^+\pi^-$)-channel by the ASTERIX collaboration [19] and recently even more clear in the ($\pi^0\pi^0$)-channel in the Crystal Barrel experiment [20].

Fig. 5 shows the ($\pi^0\pi^0$)-mass projection of the Dalitz Plot of the annihilation channel $(\bar{p}p)_{Rest} \to \pi^0\pi^0\pi^0$. The state X_2 with the mass of (1515 ± 10) MeV and the width of (120 ± 10) MeV is a very good candidate for an exotic particle. It cannot be identical to the $f_2'(1525)$-$\bar{q}q$-state because of it negligible decay probability to $K\bar{K}$. It also cannot be a radial excitation of the f_2 (1270), which is predicted at higher masses. It is improbable, that it is a Glue Ball, because it was not observed in radiative ψ-decays, but it fits well into the picture of a multi-quark state [21]. Other decay channels are predicted by this model and are presently checked by the Crystal Barrel collaboration.

Until recently, $f_2(1720)$ was another strong candidate for an exotic particle. It has been seen in radiative ψ-decays and in double Pomeron exchange reactions. A recent analysis, however, casts doubt on the right assignment of its spin, which might well be $J^P=0^+$ instead of 2^+. At present, the nature of the particle is unclear. The Crystal Barrel collaboration will check for this particle in $\bar{p}p(d)$-annihilations.

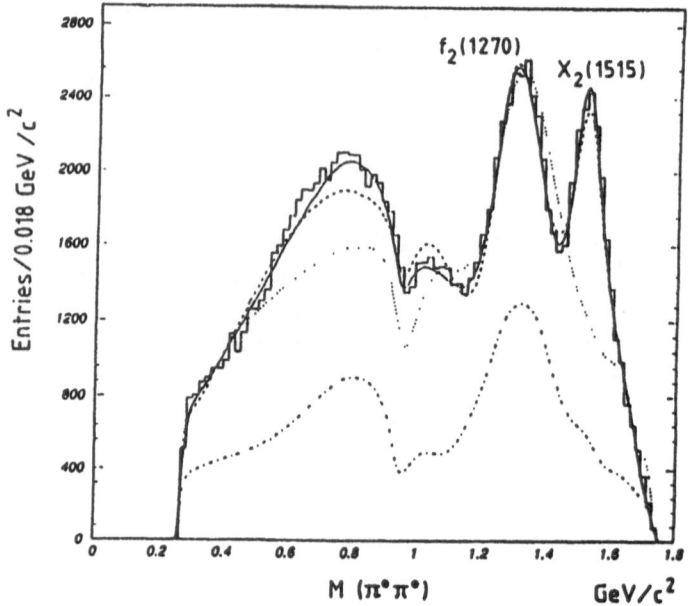

Fig. 5: ($\pi^0\pi^0$)-invariant masses of the reaction $(\bar{p}p)_{Rest} \to \pi^0 \pi^0 \pi^0$, as measured by the Crystal Barrel collaboration [20]. The full line corresponds to the best fit to the data.

5. Outlook

The advent of modern 4π-detectors, e.g. the Crystal Barrel, has shown, that considerable improvement in the understanding of the mesonic spectrum is possible, provided, high statistics and very clean data samples are available. Very complicated many body final states can be properly disentangled, and the combinatorical problems are handable due to the big computer power available today (kinematical fitting). Great care has to be taken with the analysis of the data because of the overlapping of resonances, their interferences and the handling of thresholds. The Crystal Barrel program is a good example, how progress in the complicated field of low energy meson spectroscopy can be achieved. Very recent data at 1.9 GeV/c $\bar{p}$-momenta show copious production of $(\bar{q}q)_{L=1}$ mesons, may be a smoking gun towards the detection of Hybrids. Further progress is expected from dedicated machines like Super LEAR, a τ/c-factory or Daphne.

References

[1] F. Close, Rep. Prog. Phys. **51**(1988)833

[2] T. Burnett and S.R. Sharpe, Ann. Rev. Nucl. Part. Sci. **40**(1990)327

[3] S. Godfrey, Proc. of the Super LEAR Workshop, held at the Univ. of Zürich, 9.-12. Oct. 1991, ed. C. Amsler and D. Urner, Inst. of Phys. Conf. Series #124

[4] L.G. Landsberg, Soviet Phys. Uspekhi, Vol. 33, #3, March 1990

[5] K. Koenigsmann, Proc. of the XI. Int. Conf. on Physics in Collision, held in Colmar, 20-22 June 1991; CERN-Preprint CERN - PPE/91-160

[6] Particle Data Group, P.L. **B239**(1990)1

[7] D. Alde et al., P.L **B205**(1988)397

[8] D. Alde et al., P.L. **B216**(1989)447

[9] E. Aker et al., The Crystal Barrel Spectrometer at LEAR, to be published in NIM(1992)

[10] M. Feindt, J. Harjes, in Proc. of the Rheinfels Workshop, Nucl. Phys. **B21**(Proc. Suppl.), 61(1990)

S.J. Lindenbaum, R.S. Longacre, Preprint BNL - 45878(1991)

T.A. Armstrong et al., CERN-Preprint CERN-PPE/91-40, subm. to Z. Phys. C

D. Morgan, M.R. Pennington, Z. Phys. **C48**(1990)623

[11] C. Amsler et al., Proton-Antiproton-Annihilation into $\eta\eta\pi^0$, to be published in P.L.(1992)

[12] D. Alde et al., Nucl. Phys. **B269**(1986)485

[13] D. Alde et al., P.L. **B201**(1988)160

[14] Z. Bai et al., P.R.L. **65**(1990)2507

[15] S.I. Bityukov et al., P.L. **B188**(1987)383

[16] S.A. Sadovsky, IHEP-Reprint 91-75

[17] J. Weinstein, N. Isgur, P.R. **D41**(1990)2236

[18] R.S. Longacre, P.R. **D42**(1990)874

[19] B. May et al., Z. Phys. **C46**(1990)191 and 203

[20] E. Aker et al. P.L. **B260**(1991)249

[21] C.B. Dover, T. Gutsche and A. Faessler, P.R. **C43**(1991)379

Scalar Mesons, Diquarks and Gribov Confinement

F E Close

Rutherford Appleton Laboratory, Chilton, Didcot, Oxon, OX11 OQX, England

Abstract   The scalar mesons $f_0(975)$ and $a_0(980)$ can be produced in ϕ radiative decays and in $\gamma\gamma$ collisions. The ratio of the f_0 and a_0 production rates may determine the substructure of these mesons.

1   Introduction

20 years after the discovery of QCD and 30 years after the invention of the quark model we still do not know whether glueballs or hybrid states exist. This is an unsatisfactory gap in our knowledge. There are as many predictions for the spectrum of glueballs as there are theorists who have worked on the problem; however they agree on one thing — the lightest glueball is predicted to be a scalar, $J^{PC} = 0^{++}$.

In order to identify this, and other gluonic states, it is a good strategy to understand better the spectroscopy of the conventional $q\bar{q}$ hadrons, in particular those with $c = +$. To do this, $\gamma\gamma$ physics is particularly useful and recently we have developed tests that will help to identify $q\bar{q}$ and $q\bar{q}g$ states[1,2,3]. This has already had positive results in that it has enabled two groups[4] to isolate the candidate $f_0(1300)$ as a $q\bar{q}$ state coupling to $\gamma\gamma$ with canonical strength.

There are also two, nearly degenerate, scalars $f_0(975)$ and $a_0(980)$ with $I = 0, 1$ respectively whose internal structure has long been controversial: they may be $q\bar{q}$, KK molecules[5], $qq\bar{q}\bar{q}$[6,7] or excitations of negative energy quarks[8]. I will show how radiative decays of the ϕ meson (at DAΦNE or VEPP) and gamma-gamma physics can discriminate among the various hypotheses. In particular the ratio of branching ratios

$$\frac{\Gamma(\phi \to \gamma a_0)}{\Gamma(\phi \to \gamma f_0)} \quad ; \quad \frac{\Gamma(\gamma\gamma \to a_0)}{\Gamma(\gamma\gamma \to f_0)}$$

can provide rather sharp tests of their substructure.

2 Present Phenomenology of the a_0, f_0 States

Much of the interest in this conference is in the possible existence of diquarks and bound states of multiple quarks and antiquarks. One of the main contenders for the dynamical origin of the two best known scalar mesons, the $f_0(975)$ and $a_0(980)$, is that they are $qq\bar{q}\bar{q}$ states. The original proposal[6] for this interpretation, based on the bag model, also predicted many other states which have not been seen (although this shortcoming is now understood to some degree[7]). The $qq\bar{q}\bar{q}$ interpretation of these two states was later revived in a different guise within the quark potential model as the "$K\bar{K}$ molecule" interpretation[5]. Since providing a test of this particular interpretation is one of the main results to be presented here, I shall first briefly elaborate on these two models of multiquark states, (see also Ref. 9 from which this is drawn).

In the naive bag model the $qq\bar{q}\bar{q}$ states consist of four quarks confined in a single spherical bag interacting via one gluon exchange. It is obvious that such a construction will lead to a rich spectroscopy of states. Although it is not clear how to treat or interpret the problem of the stability of this spectrum under fission into two bags[7], it is very interesting that the dynamics of this model predicts that the lowest-lying such states will (in the SU(3) limit) form an apparently ordinary ("cryptoexotic") nonet of scalar mesons. It is, moreover, not impossible that a better understanding of bag stability could solve both the problem of unwanted extra predicted states and also a problem with the a_0 itself: in the naive model it can "fall apart" into $\pi\eta$ so that it is difficult to understand its narrow width given the presently accepted pseudoscalar meson mixing angle (see footnote 22 in the first of Refs. 6). In the absence of an understanding of how to overcome these difficulties, I shall not consider the bag picture further in this paper, though ref. 10 has suggested a possible way out of the $a_0 \rightarrow \pi\eta$ problem.

In the potential model treatment[5] it is found that the low-lying $qq\bar{q}\bar{q}$ sector is most conveniently viewed as consisting of weakly interacting ordinary mesons: the resulting spectrum is normally a (distorted) two particle continuum. Within the ground state u,d,s meson-meson systems, the one plausible exception to this rule is found in the $K\bar{K}$ sector (i.e. the $K\bar{K}$ channel and those other channels strongly coupled to it): the $L = 0$ (i.e. $J^{PC_n} = 0^{++}$) spectrum seems to have sufficient attraction to produce weakly bound states in both I=0 and I=1. There are a number of phenomenological advantages to the identification of these two states with the $f_0(975)$ and $a_0(980)$. Among them are:

1) It is immediately obvious why the $f_0(975)$ and $a_0(980)$ are found just below $K\bar{K}$

threshold: they bear much the same relationship to it that the deuteron bears to np threshold.

2) The problem of the f_0 and a_0 widths is solved. If these states were $^3P_0$ quarkonia with flavours corresponding to ω and ρ (as suggested by their degeneracy), then $\Gamma(f_0 \to \pi\pi)/(\Gamma(a_0 \to \pi\eta)$ would be about 4 in contrast to the observed value of about $\frac{1}{2}$. At least as serious is the problem in the quarkonium picture with the absolute widths of these states: models[11-13] predict, for example,

$$\Gamma(f_0 \to \pi\pi) \simeq (3-6)\Gamma(b_1 \to (\omega\pi)_S) \qquad (2.1)$$
$$\simeq 500 - 1000 \ MeV \qquad (2.2)$$

versus the observed partial width of 25 MeV. We have already noted the problem in the bag model $qq\bar{q}\bar{q}$ interpretation with $a_0 \to \pi\eta$. In the $K\bar{K}$ molecule picture the narrow observed widths are a natural consequence of weak binding: $(K\bar{K})_{I=0} \to \pi\pi$ and $(K\bar{K})_{I=1} \to \pi\eta$ occur slowly because the $K\bar{K}$ wavefunction is diffuse.

3) Both the f_0 and a_0 seem to bear a special relationship to $s\bar{s}$ pairs: their $K\bar{K}$ "couplings" are very large and they are observed in channels which would violate the Okubo-Zweig-Iizuka rule[14] for an $\omega, \rho-$like pair of states[15].

4) The $\gamma\gamma$ couplings of the f_0 and a_0 are about an order of magnitude smaller than expected for $^3P_0$ quarkonia[1], but consistent with the expectations for $K\bar{K}$ molecules[16].

Although these observations argue against the viability of the $^3P_0$ quarkonium interpretation of $f_0(975)$ (and probably also $a_0(980)$), they are insufficient to rule it out completely. (Moreover, a unitarized variant of the quark model[17], in which the scalar mesons are strongly mixed with the meson-meson continuum, avoids several of these problems).

In a recent paper[9] we pointed out a simple (and to us unexpected) experimental test which sharply distinguishes among these alternative explanations. We showed that the rates for $\phi \to f_0 (975) \ \gamma \to \pi\pi\gamma$ and $\phi \to a_0 (980) \ \gamma \to \pi\eta\gamma$ in the quarkonium, glueball, and $K\bar{K}$ molecule interpretations differ significantly; furthermore, the ratio of branching ratios

$$\frac{\phi \to a_0(980)\gamma}{\phi \to f_0(975)\gamma}$$

also may prove to be an important datum in that it can have a model-dependent value anywhere from zero to infinity (see Table 2)!

In the quarkonium interpretation, $\phi \to f_0(975) \ \gamma$ and $\phi \to a_0 (980) \ \gamma$ are simple electric dipole transitions quite similar in character to several other measured electric multipole transitions, including not only the light quark transitions a_2 (1320) $\to \pi\gamma, K^*$ (1420) $\to K\gamma, a_1$ (1275) $\to \pi\gamma$, and b_1 (1235) $\to \pi\gamma$, but also such decays as $\chi_{co} \to \psi\gamma$ and $\chi_{bo} \to \Upsilon\gamma$. From the comparison between theory and experiment given in Ref. 11, we expect that the quark model predictions for these processes given in Table 1 are reliable to within a factor of two. Thus if the f_0 is an $s\bar{s}$ quarkonium, the branching ratio for $\phi \to S\gamma$ would typically be of the order of 10^{-5}.

If the f_0 (975) is a glueball (in Ref. 18 there is a glueball component of the "S^* effect", dubbed the S_1(991), which couples to $\pi\pi$ and is responsible for the resonant behaviour seen in $\pi\pi$ phase shift analyses; the other component, dubbed the S_2(998) is practically uncoupled to $\pi\pi$) then one would naturally expect $\phi \to f_0$ (975) $\gamma \to \pi\pi\gamma$ to be even smaller than in the quarkonium interpretation since the decay would be OZI-violating. The remarks made above on the strong decay widths of the quarkonium states would suggest that quarkonium — glueball mixing, through which we presume the OZI-violation would proceed, must be small in order that the f_0 (975) remain narrow. Thus we can crudely estimate the glueball — quarkonium mixing angle to be less than $[\Gamma(f_0 \to \pi\pi)/\Gamma(^3P_0 \to \pi\pi)]_{\frac{1}{2}}$ so that if f_0 (975) is a glueball

$$\Gamma(\phi \to f_0 \text{ (glueball) } \gamma) \leq \frac{\Gamma(f_0 \to \pi\pi)}{\Gamma(^3P_0 \to \pi\pi)}\Gamma(\phi \to f_0 \text{ (quarkonium) } \gamma) \quad (2.3)$$

$$\leq \frac{1}{20}\Gamma(\phi \to f_0 \text{ (quarkonium) } \gamma) \quad (2.4)$$

Thus if f_0 (975) is a glueball, this branching ratio should be more than an order of magnitude smaller than it would be to a ϕ-like quarkonium.

If the f_0 is a quarkonium consisting only of nonstrange flavours, with a_0 its isovector quarkonium partner, these states will be OZI decoupled in the ϕ radiative decay. The OZI-violating production rate via a $K\bar{K}$ loop, viz. $\phi \to \gamma K\bar{K} \to \gamma a_0$, may be calculated. This calculation reveals some interesting points of principle which shed light on the role of finite hadron size in such loop calculations (see Ref. 9).

Interesting questions arise in the case of $qq\bar{q}\bar{q}$ or $K\bar{K}$ bound states ("molecules"). The quark contents of these two systems are identical but their dynamical structures differ radically. The situation here has its analog in the case of the deuteron which contains size quarks but is not a "true" six-quark bound state. The essential feature is whether the multiquark system is confined within a hadronic system with a radius of order $(\Lambda_{QCD})^{-1}$ or is two identifiable colour singlets spread over a region significantly greater than this (with radius of order $(\mu E)^{1/2}$ associated with the interhadron binding energy E for a system of reduced mass μ). In the former case the branching ratio may be as large as 10^{-4} (see Ref. 19 and section 4 of Ref. 9); the braching ratio for a diffuse $K\bar{K}$ molecular system can be much smaller, as discussed in ref. 9 and summarised in table 2.

3 Ratios of Branching Ratios as Tests of Substructure

The ratio of branching ratios is also interesting. The ratio of $\Gamma(\phi \to \gamma a_0)/\Gamma(\phi \to \gamma f_0)$ is approximately zero if they are quarkonia (the f_0 being $s\bar{s}$ and the a_0 being OZI decoupled), it is approximately unity if they are $K\bar{K}$ systems, while for $q^2\bar{q}^2$ the ratio is sensitively dependent on the internal structure of the states. This sensitivity in $qq\bar{q}\bar{q}$ arises because $\phi \to S\gamma$ is an $E1$ transition whose matrix element, being proportional to $\Sigma e_i \vec{r}_i$, probes the electric charges of the constituents weighted by their vector distance from the overall centre of mass of the system. Thus, although the absolute transition rate for $S = qq\bar{q}\bar{q}$ depends on unknown dynamics, the ratio of a_0 to f_0 production will be sensitive to the internal spatial structure of the scalar mesons through the relative phases in $I = 0$ and 1 wavefunctions and the relative spatial distributions of quarks and antiquarks.

For example, suppose that the state's constituents are distributed about the centre of mass with the structure $(q\bar{s})(\bar{q}s)$, where q denotes u or d and (ab) represents a spherically symmetric cluster. Then

$$\left\{ \begin{matrix} f_0 \\ a_0 \end{matrix} \right\} = \frac{1}{\sqrt{2}}[(u\bar{s})(\bar{u}s) \pm (d\bar{s})(\bar{d}s)] \tag{3.1}$$

and the $E1$ matrix element will be

$$M \sim [(e_u + e_{\bar{s}}) \pm (e_d + e_{\bar{s}})] = e_{K^+} \pm e_{K^0}$$

and hence the ratio $\Gamma(\phi \to \gamma f_0)/\Gamma(\phi \to \gamma a_0)$ will be unity. The quarks are distributed *as if* in a $K\bar{K}$ molecular system (which is a specific example of this configuration) and only the absolute branching ratio will distinguish $q^2\bar{q}^2$ from $K\bar{K}$.

If the distribution is $(q\bar{q})(s\bar{s})$ then the matrix element

$$M \sim [(e_q + e_{\bar{q}}) - (e_s + e_{\bar{s}})] = 0.$$

Here the quark distributions mimic $\pi^0\eta$ (in the a_0) or $\eta\eta$ (in the f_0). In this case the absolute branching ratios will be suppressed. Most interesting is the case where $S = D\bar{D}$, where D denotes a diquark, i.e. where

$$\left\{ \begin{matrix} f_0 \\ a_0 \end{matrix} \right\} = \frac{1}{\sqrt{2}}[(us)(\bar{u}\bar{s}) \pm (ds)(\bar{d}\bar{s})] \tag{3.2}$$

in which case

$$M \sim [(e_u + e_s) \pm (e_d + e_s)]$$

so that

$$\frac{\Gamma(\phi \to \gamma a_0)}{\Gamma(\phi \to \gamma f_0)} = (\frac{1+2}{1-2})^2 = 9.$$

The absolute rate in this case depends on an unknown overlap between $K\bar{K}$ and the diquark structure; nonetheless the dominance of a_0 over f_0 would be rather distinctive. For convenience these possibilities are summarised in Table 2.

It is even possible that finer details of a diquark substructure may be resolved. As the strange quark is more massive than the non-strange, one may anticipate that

$$< r >_s \, < \, < r >_{u,d}$$

Thus if

$$\frac{< r >_s}{< r_q >} = x$$

then the ratio

$$\frac{\Gamma(\phi \to \gamma a_0)}{\Gamma(\phi \to \gamma f_0)} = (\frac{3}{1+2x})^2$$

for a KK-like "bag" and for a $D\bar{D}$-like "bag" one has

$$\frac{\Gamma(\phi \to \gamma a_0)}{\Gamma(\phi \to \gamma f_0)} = (\frac{3}{1-2x})^2.$$

Starting from the DD configuration there is a continuous transition as x journeys from 1 (the "symmetry limit") to -1 (which corresponds to a KK-like structure). It is when $x = \frac{1}{2}$ that the "distorted diquark" phase kills the f_0 production. If $x \to 0$ one has the "transition" configuration where $s\bar{s}$ are at the origin and thereby "frozen out" of the $E1$ transition. This gives a ratio of 9 (which will of course also be the ratio for a $u\bar{u} \pm d\bar{d}$ picture, as in Gribov's theory, Ref 8). I summarise these results in diagrammatic form in table 3.

quarkonium	formula	ϕ branching ratio
$f_0 = \sqrt{\frac{1}{2}}(u\bar{u} + d\bar{d}) \, {}^3P_0$	$0^{a)}$	$\lesssim 10^{-6}$
$f_0 = s\bar{s} \, {}^3P_0$	$\frac{4\alpha \lvert d_{f_0} \rvert^2 \omega^3}{243}$	$\simeq 1 \times 10^{-5}$
$a_0 = \sqrt{\frac{1}{2}}(u\bar{u} - d\bar{d}) \, {}^3P_0$	$0^{b)}$	$\lesssim 10^{-6}$

Table 1: ϕ photodecays to quarkonia

a) proceeds through $\omega - \phi$ and $f_0 - f_0'$ mixing
b) proceeds through $\omega - \phi$ mixing only

scalar meson constitution	$\frac{\Gamma(\phi\to a_0\gamma)}{\Gamma(\phi\to f_0\gamma)}$	absolute	comments branching ratios
$K\bar{K}$ molecule	$1^{(a)}$	$a_0 \simeq f_0 \simeq 4\times 10^{-5}$	$K\bar{K}$ dominates loop diagrams see ref. 9
$q^2\bar{q}^2$: $K\bar{K}$-like "bag"	$1^{(b)}$		
$D\bar{D}$-like "bag"	$9^{(c)}$	rates probably	see section 5
$(n\bar{n})(s\bar{s})$-like "bag"	-	$< 10^{-6}$, see section 5	of ref. 9
$(q\bar{q})^3 P_0$: $f_0 = n\bar{n}$	-		
$f_0 = s\bar{s}$	$\simeq 0$	see Table 1	see Table 1
f_0 glueball, a_0 quarkonium	-	$\leq 10^{-6}$	see text

Table 2: Some qualitative implications of $\frac{\Gamma(\phi\to a_0\gamma)}{\Gamma(\phi\to f_0\gamma)}$

a) neglecting $I = 0, 1$ mixing effects.
b) if $\frac{\langle r_s\rangle}{\langle r_q\rangle} \equiv x$, then this ratio is $(\frac{3}{1+2x})^2$.
c) if $\frac{\langle r_s\rangle}{\langle r_q\rangle} \equiv x$, then this ratio is $(\frac{3}{1-2x})^2$.

Spatial Configuration	Description	$\Gamma(\phi\to\gamma a_0)/\Gamma(\phi\to\gamma f_0)$
	$D\bar{D}$	9
	distorted $D\bar{D}$	∞
	frozen $s\bar{s}$	9
	distorted KK	2
	KK	1

Table 3: Strange quarks □ and antiquarks □, and nonstrange quarks • and antiquarks o form a $qs\bar{q}\bar{s}$ state. The spatial configuration relative to the c.m. is indicated in column 1 and a qualitative description of the state is in column 2 (D refers to "diquark"). The ratio of ratios is in column 3.

4   Scalar Mesons in $\gamma\gamma$ Physics

The ϕ radiative is sensitive to the constituent charges (via the γ) and the $s\bar{s}$ content (via the ϕ); in $\gamma\gamma$ coupling one has a complementary probe of the internal charge structure of the produced mesons. The combination of $\phi \to \gamma S$ and $\gamma\gamma \to S$ may definitively establish the nature of the a_0 and f_0 states.

Unlike the ϕ radiative decays, there are data on $\gamma\gamma$ couplings to the scalars. The data were [18]

$$\Gamma(\gamma\gamma \to a_0) = 0.19^{+0.17}_{-0.14} \, kev$$
$$\Gamma(\gamma\gamma \to f_0) = 0.30 \pm 0.10 \, kev \tag{4.1}$$

and tantalisingly similar to Barnes' [16] calculation of the width for KK molecules (0.2 kev), and an order of magnitude smaller than expected for a $q\bar{q}$ $^3P_0$ [1-3] (though there is a possible loophole in the latter case since in the $m_q \to 0$ limit the $^3P_0\gamma\gamma$ width is predicted to vanish). If the KK loop dominates the interaction ($\gamma\gamma \to KK \to S$) then the above widths are very natural [19] but possible contributions from $\pi\pi$ in the case of the f_0 confuses this simple result. The modelling of the $\gamma\gamma$ width needs more detailed study. However a recent analysis by Morgan and Pennington[20] has modified these numbers. The current world average[21] is

$$\Gamma(\gamma\gamma \to a_0)B(a_0 \to \eta\pi) = 0.24 \pm 0.08 \, kev$$
$$\Gamma(\gamma\gamma \to f_0) = 0.56 \pm 0.11 \, kev \tag{4.2}$$

and the absolute magnitude of a_0, and the ratio of the two, depends on the poorly determined $B(a_0 \to \eta\pi)$. This has often been approximated by unity (as in ref. 16 and the numbers cited above in eq. 4.1) whereas the value could easily be below 50% (ref. 21). Thus one cannot eliminate equality for the two $\gamma\gamma$ widths. Given that $B(a_0 \to \eta\pi) \leq 1$ we can at least limit the ratio

$$\frac{\Gamma(\gamma\gamma \to f_0)}{\Gamma(\gamma\gamma \to a_0)} \leq \frac{21 \pm 4}{9 \pm 3} \tag{4.3}$$

The ratio of $\gamma\gamma$ ratios is a direct probe of the substructure.

In the case of a KK molecule, dominated by the KK loop, the ratio will be unity. Contrast this with a simple $q\bar{q}$ picture. If

$$a_0 = (u\bar{u} - d\bar{d})1/\sqrt{2}$$
$$f_0 = (u\bar{u} + d\bar{d})/\sqrt{2}$$

then

$$\frac{\Gamma(\gamma\gamma \to f_0)}{\Gamma(\gamma\gamma \to a_0)} = \frac{25}{9}$$

due to the amplitude being proportional to the squared charges of the constituents. This would apply to Gribov's picture where the f_0, a_0 are excitations of negative energy light flavours. Within the large errors this ratio too is still allowed though hardly favoured. Hence improvement in the $\gamma\gamma$ widths (and in particular the $\eta\pi$ branching ratio of the a_0) could be rather crucial.

If the a_0, f_0 are compact four quark states, the ratio of widths is less well defined and depends on the nature of the virtual intermediate state between the two photons. A coherent production involving only the ground state (Born) meson is one extreme; a sum over a complete set of intermediate states, which effectively may be modelled by incoherent Compton scattering is another extreme.

For the coherent case one has for $\Gamma(\gamma\gamma \to f_0)/\Gamma(\gamma\gamma \to a_0)$

$$
\begin{array}{cc}
K\bar{K} & 1 \\
D\bar{D} & \dfrac{25}{9}
\end{array}
$$

For the incoherent case one would have $\frac{25}{9}$ for all cases in the limit where the $s\bar{s}$ production is frozen out; if $s\bar{s}$ is produced by gluons with the same strength as the non-strange, the ratio rises to $\frac{49}{9}$. This latter value is probably ruled out empirically.

Finally one has the possibility that each photon produces a single $q\bar{q}$. In this case the matrix element is proportional to $(e_u \pm e_d)e_s$ and hence the f_0/a_0 ratio will be $\frac{1}{9}$. This is ruled out unless the $B(a_0 \to \eta\pi)$ is very small and the $\Gamma(\gamma\gamma \to a_0)$ an unlikely 5 keV. Thus in these simple examples alone the ratio can vary from $\frac{1}{9}$ to $\frac{49}{9}$. More realistic modelling seems to be needed.

In any event, the combination of $\gamma\gamma \to S$ and $\phi \to \gamma S$ data for both a_0 and f_0 can make important contributions to clarifying the structure of these states.

References

[1] T. Barnes, F.E. Close and Z.P. Li, Phys. Rev. **D43**, 2161 (1991).

[2] F.E. Close and Z.P. Li, Z. Phys. **C54**, 147 (1992).

[3] E. Ackleh, T. Barnes and F.E. Close, Phys. Rev. **D**, (in press).

[4] Crystal Ball, J. Bienlein et al, DESY 91-145, Proc 9th Photon Photon Workshop, La Jolla 1992; CELLO, J. Harjes, Proc Hadron 91.

[5] J. Weinstein and N. Isgur, Phys. Rev. Lett. **48**, 659 (1982), Phys. Rev. **D27**, 588 (1983), Phys. Rev. **D41**, 2236 (1990).

[6] R. Jaffe, Phys. Rev. **D15**, 267, 281 (1977); Phys. Rev. **D17**, 1444 (1978).

[7] R. Jaffe and F. Low, Phys. Rev. **D19**, 2105 (1979).

[8] V. Gribov, private communication.

[9] F.E. Close, N. Isgur and S. Kumano, Nucl. Phys. B (in press), RAL-92-026, CEBAF 92-13, IUNTC 92-16.

[10] N.N. Achasov, S.A. Devyanin, and G.N. Shestakov, Phys. Lett. **96B**, 168 (1980); Sov. J. Nucl. Phys. **32**, 566 (1980); Sov. Phys. Usp. **27**, 161 (1984).

[11] S. Godfrey and N. Isgur, Phys. Rev. **D32**, 189 (19785.

[12] R. Kokoski and N. Isgur, Phys. Rev. **D35**, 907 (1987).

[13] A. LeYaouanc, L. Oliver, O. Pene, and J.C. Raynal, Phys. Rev. **D8**, 2223 (1973); **D9**. 1415 (1974); **D11**, 1272 (1975); M. Chaichan and R. Kogerler, Ann. Phys. (N.Y.) **124**, 61 (1980); S. Kumano and V.R. Pandharipande, Phys. Rev. **D38**, 146 (1988).

[14] S. Okubo, Phys. Lett. **5**, 1975 (1963); Phys. Rev. **D16**,2336 (1977); G. Zweig, CERN Report No. 8419 TH 412, 1964 (unpublished); reprinted in Developments in the Quark Theory of Hadrons, ed. D.B. Lichtenberg and S.P. Rosen (Hadronic Press, Nonantum, MA, (1980); J. Iizuka, K. Okada and O. Shito, Prog. Theor. Phys. Suppl. **37**, 38 (196).

[15] L. Kopke (representing the Mark III Collaboration) in Proceedings of the XXIII Int. Conf. on High Energy Physics, Berkeley, ed. S. Loken (World Scientific, 1987), p. 692.

[16] T. Barnes, Phys. Lett. **B165**, 434 (1985).

[17] N.A. Tornquist, Phys. Rev. Lett. **49**, 624 (1982).

[18] D. Antreasyan et al., Phys. Rev. **D33**, 1847 (1986); Mark II, Phys. Rev. **D42**, 1350 (1990); Crystal Ball, Phys. Rev. **D41**, 3324 (1990).

[19] N. Achasov and G. Shestakov, Z. Phys. **C41**, 309 (1988); T. Truong, Proc of Hadron 89.

[20] D. Morgan and M. Pennington, Phys. Lett. **B258**, 444 (1991).

[21] Particle Data, Group, Phys. Rev. **D45**, S1 (1992).

Hadron Spectroscopy
with Instanton induced Forces

Bernard Metsch

Institut für Theoretische Kernphysik, Nußallee 14-16, W-5300 Bonn, Germany

It is shown that within the framework of a constituent quark model with string-like confinement and an instanton induced residual quark-(anti)quark interaction it is possible to account simultaneously for the baryon and the meson spectra, including the π-η-η' splitting for pseudoscalar mesons, up to energies of roughly 2 GeV.

1 The Constituent Quark Model

The most succesful description of the known baryon and meson spectra up to energies of roughly 2 GeV is the so-called Constituent Quark Model. It is based on two main ingredients, namely a (two- or three-body) confinement potential and a residual interaction which is taken to be (parts of) a non-relativistic reduction of the One Gluon Exchange. The confinement potential is understood to come ultimately from a non-perturbative treatment of QCD, whereas the OGE-interaction is based on perturbation theory and hence is highly questionable at low energies. It is therefore interesting to note that there exists another QCD based candidate for such a residual, effective interaction, computed by 't Hooft from instanton effects [1-4].

Within the framework of the constituent quark model the quark dynamics is given by the non-relativistic Hamiltonian

$$H = M + T^{rel} + V + W \tag{1}$$

where

$$M = \sum_i m_i \tag{1a}$$

is the sum of the constituent quark masses,

$$T^{rel} = \sum_i \frac{p_i^2}{2m_i} - \frac{P^2}{2M} \tag{1b}$$

the kinetic energy for the relative motion and

$$V(r_{q\bar{q}}) = a_{q\bar{q}} + b\, r_{q\bar{q}} \tag{1c}$$

is a linear confinement potential for the meson system, or

$$V(r_1, r_2, r_3) = a_3 + b\left[\min_{r_0} \sum_i |r_i - r_0|\right] \tag{1d}$$

is the three-body confinement potential for the baryon system. Confinement is thus modelled by the potential of the string configuration with minimal energy. Finally,

$$W = \frac{1}{2} \sum_{i \neq j} W_{ij} \qquad (1e)$$

is the residual quark-(anti)quark interaction to be described in more detail in section 2.

The Schrödinger equation corresponding to (1) is solved as follows: The matrix elements of the Hamiltonian (1) are calculated within a spin-flavour SU(6), O(3)-oscillator basis comprising $N = 20$ and $N = 5$ oscillator excitations for mesons and baryons, respectively. Test calculations involving up to $N = 11$ oscillator excitations for baryons have been performed in order to check the convergence with respect to the size of the model space. The hadronic spectra are obtained by the application of the variational principle, i.e. diagonalization of the Hamiltonian in the oscillator basis and minimization of the energy expectation value with respect to the oscillator length parameter. In this manner a single, consistent treatment of the two-body and the three-body problem is achieved.

2 Instanton induced Quark Interaction

The residual interaction employed here is due to 't Hooft and was originally designed to solve the U(1)-problem. An expansion of the euclidian action around the so-called single instanton solutions of the gauge fields under the assumption of zero mode dominance in the fermion sector leads to a contribution from an instanton of size ρ to the quark-Lagrangian given for three flavours by

$$
\begin{aligned}
\Delta \mathscr{L} = \int \frac{d\rho}{\rho^5} \, d_0(\rho) \Bigg\{ &\Bigg[\prod_{i=1}^{3} \left[m_i \rho - \tfrac{4}{3}\pi^2 \rho^3 (\bar{q}_{iR} q_{iL}) \right] \\
&+ \tfrac{3}{32} \left[\tfrac{4}{3}\pi^2 \rho^3 \right]^2 \Bigg[(\bar{q}_{1R} t^a q_{1L})(\bar{q}_{2R} t^a q_{2R}) - \tfrac{3}{4}(\bar{q}_{1R}\sigma_{\mu\nu} t^a q_{1L})(\bar{q}_{2R}\sigma_{\mu\nu} t^a q_{2L}) \Bigg] \\
&\hspace{4cm} \times \Bigg[m_3 \rho - \tfrac{4}{3}\pi^2 \rho^3 (\bar{q}_{3R} q_{3L}) \Bigg] \\
&+ \tfrac{9}{40} \tfrac{4}{3}\pi^2 \rho^3 \, d^{abc} \, (\bar{q}_{1R}\sigma_{\mu\nu} t^a q_{1L})(\bar{q}_{2R}\sigma_{\mu\nu} t^b q_{2L})(\bar{q}_{3R} t^c q_{3L}) \\
&\hspace{5cm} + \text{cycl. perm. of } (123) \Bigg] \\
&+ \tfrac{9}{256} i \left[\tfrac{4}{3}\pi^2 \rho^3 \right]^3 f^{abc} \, (\bar{q}_{1R}\sigma_{\mu\nu} t^a q_{1L})(\bar{q}_{2R}\sigma_{\nu\epsilon} t^b q_{2L})(\bar{q}_{3R}\sigma_{\epsilon\mu} t^c q_{3L}) \\
&+ \tfrac{9}{320} \left[\tfrac{4}{3}\pi^2 \rho^3 \right]^3 d^{abc} \, (\bar{q}_{1R} t^a q_{1L})(\bar{q}_{2R} t^b q_{2L})(\bar{q}_{3R} t^c q_{3L}) \Bigg] + (R \leftrightarrow L) \Bigg\} \qquad (2)
\end{aligned}
$$

Here, $i = 1, 2, 3$ denote the flavour degrees of freedom, m^i the corresponding current quark masses, t^a ($a = 1,....,8$) are the colour matrices and f^{abc}, d^{abc} are standard SU(3) structure constants. The instanton density for three colours and three flavours is given by

$$
d_0(\rho) = (3.63 \ 10^{-3}) \cdot \left[\frac{8\pi^2}{g^2(\rho)} \right]^6 \cdot \exp\left[- \frac{8\pi^2}{g^2(\rho)} \right] \qquad (3a)
$$

where

$$
\left[\frac{8\pi^2}{g^2(\rho)} \right] = 9 \ln\left[\frac{1}{\Lambda \rho} \right] - \frac{32}{9} \ln\left[\ln\left[\frac{1}{\Lambda \rho} \right] \right] \qquad (3b)
$$

within two loop accuracy. Here Λ is the QCD scale parameter. It should be noted that this interaction is chirally symmetric and simultaneously maximally breaks the U(1) symmetry.

After normal ordering of (2) with respect to the QCD vacuum, which exhibits non-vanishing expectation values $\langle \bar{n}n \rangle$ and $\langle \bar{s}s \rangle$ for non-strange and strange quarks, respectively and integration over ρ upto a value ρ^c for which the lnln-term of (3) is still reasonably small compared to the ln-term one obtains from $\Delta\mathscr{L}$ one obtains:

(a) An inessential constant, which renormalizes the vacuum energy.

(b) A one-body term of the form

$$\sum_{i=1}^{3} \Delta m_i : \bar{q}_i q_i:$$

(4)

where

$$\Delta m_a = \int_0^{\rho^c} \frac{d\rho}{\rho^5} d_0(\rho) \frac{4}{3}\pi^2\rho^3 \left[m_n\rho - \frac{2}{3}\pi^2\rho^3\langle \bar{n}n \rangle \right] \left[m_s\rho - \frac{2}{3}\pi^2\rho^3\langle \bar{s}s \rangle \right]$$

(4a)

and

$$\Delta m_s = \int_0^{\rho^c} \frac{d\rho}{\rho^5} d_0(\rho) \frac{4}{3}\pi^2\rho^3 \left[m_n\rho - \frac{2}{3}\pi^2\rho^3\langle \bar{n}n \rangle \right]^2$$

(4b)

which contributes to the effective (constituent) quark masses.

(c) A two-body force

$$-V = \frac{3}{16} g^{eff}(i) \ \varepsilon_{ikl} \ \varepsilon_{imn} \ q_i^\dagger q_k^\dagger (\gamma_0 \cdot \gamma_0 + \gamma_0\gamma_5 \cdot \gamma_0\gamma_5)(\mathscr{P}_6^c + 2\mathscr{P}_3^c) q_m q_n$$

(5)

with $\mathscr{P}^c$ colour projectors, which clearly exhibits the antisymmetric flavour dependence and where the coupling constants $g = 3/8 g^{eff}(s)$ and $g' = 3/8\ g^{eff}(n)$ are given by

$$g^{eff}(i) = \int_0^{\rho^c} \frac{d\rho}{\rho^5} d_0(\rho) \left[\frac{4}{3}\pi^2\rho^3\right]^2 \left[m_i\rho - \frac{2}{3}\pi^2\rho^3\langle \bar{q}_i q_i \rangle \right]$$

(5a)

(d) A three-body force W, which is obtained from (2) by dropping the mass dependent terms and normal ordering the two remaining terms. This can be shown to act only on flavour-antisymmetric, spatial-symmetric three-quark states. There are however no physical three-quark states with this property, since these states are colourfree (and thus colour-antisymmetric) and the Pauli principle can only be satisfied by spin-antisymmetric wave functions, which do not exist. Hence this three body force does not contribute in models of hadrons involving upto three constituent quarks as is the case in the present approach.

In the following the (effective) constituent quark masses and the effective coupling strength will be treated as free parameters to be adjusted to the experimental spectra. In section 4 we will check the relation to more fundamental QCD parameters as given by the expressions above. Furthermore, in order to employ 't Hoofts interaction in the (non-relativistic) constituent quark model the quark fields are expanded with respect to p/m. Keeping only the lowest order, momentum independent, terms one obtains the following non-relativistic approximation for the interaction matrix elements, e.g.

$$\langle n^2; 3,S,L,T \mid W \mid n^2; 3,S,L,T \rangle = -4.g \ \delta_{S,0}\delta_{L,0}\delta_{T,0} \ \mathscr{W}$$

(6a)

for two nonstrange quarks in a baryon being always in a colour antitriplett state and

$$\langle n\bar{n}; 1,S,L,T \mid W \mid n\bar{n}; 1,S,L,T \rangle = -8.g \ \delta_{S,0}\delta_{L,0}(\delta_{T,1}-\delta_{T,0}) \ \mathscr{W}$$

(6b)

for nonstrange quark-antiquark mesonic states. Here, $\mathcal{W}$ denotes the radial part of the matrix element. Since the present residual interaction is a pure contact force it will lead to a Hamiltonian that is unbounded. Therefore, the interaction is regularized by replacing the δ-distribution by a Gaussian

$$\delta^{(3)}(r) \rightarrow \frac{1}{\Lambda^3} \frac{1}{\pi^{3/2}} \exp(-r^2/\Lambda^2) \tag{7}$$

and Λ is interpreted as the range of the force. It is noted that a derivation of 't Hoofts force going beyond the approximation actually used, will in fact automatically yield such a finite range. Such calculations are however technically very difficult and hence do not exist. Furthermore already in this approximation the interaction has an explicit spin and flavour dependence, in contrast to the OGE, which in the same approximation would contain a central force only: the colour coulomb interaction.

The explicit flavour dependence for the meson states can be best appreciated by transforming from the isospin basis to the (u,d,s) basis. The pseudoscalar meson states with $m_T=0$ can be written as:

$$|\pi^0\rangle = \tfrac{1}{\sqrt{2}}(|u\bar{u}\rangle - |d\bar{d}\rangle) \;,\; |\eta_n^0\rangle = \tfrac{1}{\sqrt{2}}(|u\bar{u}\rangle + |d\bar{d}\rangle) \;,\; |\eta_s^0\rangle = |s\bar{s}\rangle \tag{8}$$

Then the flavour dependence of the interaction in matrix form is given by:

$$(\langle u\bar{u}| \; \langle d\bar{d}| \; \langle s\bar{s}|) \begin{bmatrix} 0 & 8g\mathcal{W} & 8g'\mathcal{W} \\ 8g\mathcal{W} & 0 & 8g'\mathcal{W} \\ 8g'\mathcal{W} & 8g'\mathcal{W} & 0 \end{bmatrix} \begin{bmatrix} |u\bar{u}\rangle \\ |d\bar{d}\rangle \\ |s\bar{s}\rangle \end{bmatrix} \tag{9}$$

which explicitly shows that the interaction is maximally (hidden) flavour mixing. In the original basis this reads

$$(\langle \pi^0| \; \langle \eta_n^0| \; \langle \eta_s^0|) \begin{bmatrix} -8g\mathcal{W} & 0 & 0 \\ 0 & 8g\mathcal{W} & 8\sqrt{2}g'\mathcal{W} \\ 0 & 8\sqrt{2}g'\mathcal{W} & 0 \end{bmatrix} \begin{bmatrix} |\pi^0\rangle \\ |\eta_n^0\rangle \\ |\eta_s^0\rangle \end{bmatrix} \tag{10}$$

Using crossing symmetry the matrix elements for strange pseudoscalar mesons are then given by

$$\langle q\bar{q}; 1,S,L,T \mid W \mid q\bar{q}; 1,S,L,T\rangle_{s^*=\pm 1} = -8.g' \, \delta_{s,0} \, \delta_{L,0} \, \mathcal{W} \tag{11}$$

In a similar fashion one obtains for the matrix elements between a strange and a non-strange quark:

$$(\langle ns| \langle sn|) \begin{bmatrix} 2g'\mathcal{W} & -2g'\mathcal{W} \\ -2g'\mathcal{W} & 2g'\mathcal{W} \end{bmatrix} \begin{bmatrix} |ns\rangle \\ |sn\rangle \end{bmatrix} \tag{12}$$

This particular flavour dependence has the following implications for the hadron spectra, as sketched in Fig. 1:
For mesons the residual interaction leads to non-vanishing contributions only in the pseudoscalar sector where it is attractive for isovector and repulsive for isoscalar states and simultaneously mixes the $n\bar{n}$- (n being a short hand notation for u and d) and $s\bar{s}$- configurations depending on the value of the same parameter g' that also

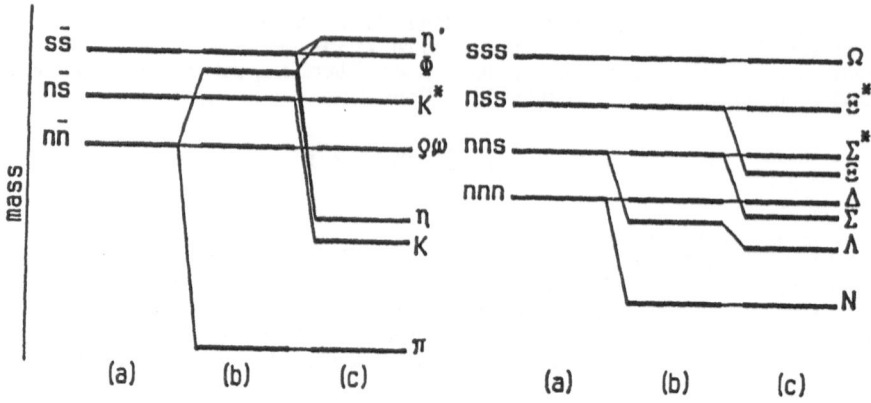

Fig. 1. Schematic mass spectra of mesons (left) and baryons (right) with (a) no residual interaction, (b) the pairing force between non-strange quarks and (c) in addition the pairing force between a non-strange and a strange quark, which mixes the isoscalar, pseudoscalar mesons and accounts for the K-K̄ and Y-Ȳ splittings. The light up and down quarks are paraphrazed by "n".

accounts for the K-K*-splitting. Thus a consistent description of the π-ρ splitting, the η-η' mixing and the K-K̄ splitting can be obtained.

In the baryonic sector the force acts whenever there is a quark pair with trivial quantum numbers thus favouring a correlation into scalar diquark configurations. Consequntly, the Δ-, Σ*-, Ξ* and Ω-states are not affected by this residual interaction and, as the non-pseudoscalar mesons, are determined by the confinement potential alone. The same parameters g and g' now determine the Δ-N-, the Σ*-Σ-Λ- and the Ξ*-Ξ-splittings, see Fig. 1.

In the following section a more detailed discusion is presented.

3 Meson and Baryon Spectra

The resulting spectra for the isovector, isoscalar and strange mesons are given in Figs. 2, 3 and 4, respectively. They show that with the confinement potential alone indeed an overall satisfactory description of the experimentally known resonances in all but the pseudoscalar channels can be obtained . A minor exception is the splitting between the singulett and triplett states for P-wave mesons. Also the splitting between the $^3P_0$-state (f0- and a0-meson) and the other ^3P-states cannot be described by the present model. It should however be noted, that there are indications that these mesons are to be understood in terms of meson-meson states rather than in terms of qq̄-configurations. The same probably also applies to higher excitions in the (isoscalar) scalar and tensor meson sector. For the pseudoscalar mesons it should be noted that in the present treatment in addition to the ground state splittings discussed in section 2 not only the π-π, but also the ρ'-π' splitting is correctly reproduced.

Concerning the baryons also the Δ- and Ω-spectra, see Figs. 6, 7, can be computed

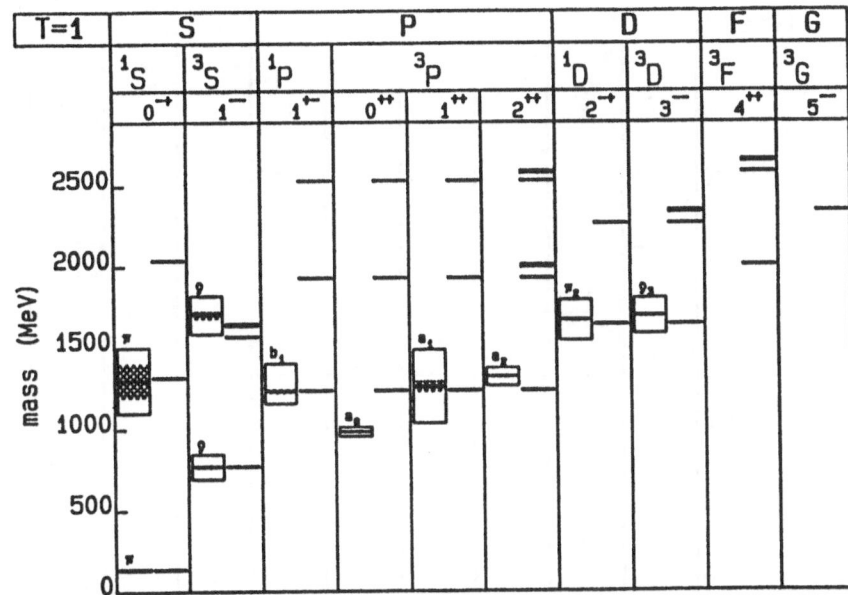

Fig. 2. Comparison of the experimental (left part of each column) and calculated (right part of each column) isovector meson spectrum. The resonances are classified by the spin, orbital angular momentum, total angular momentum, parity and charge conjugation parity. The experimental resonance position is indicated by a bar, the corresponding uncertainty by the shaded area and the width by an open rectangle. Open rectangles for the calculated triplett levels denote states with the same total angular momentum, but with a different orbital angular momentum quantum number. Experimental data are from [5].

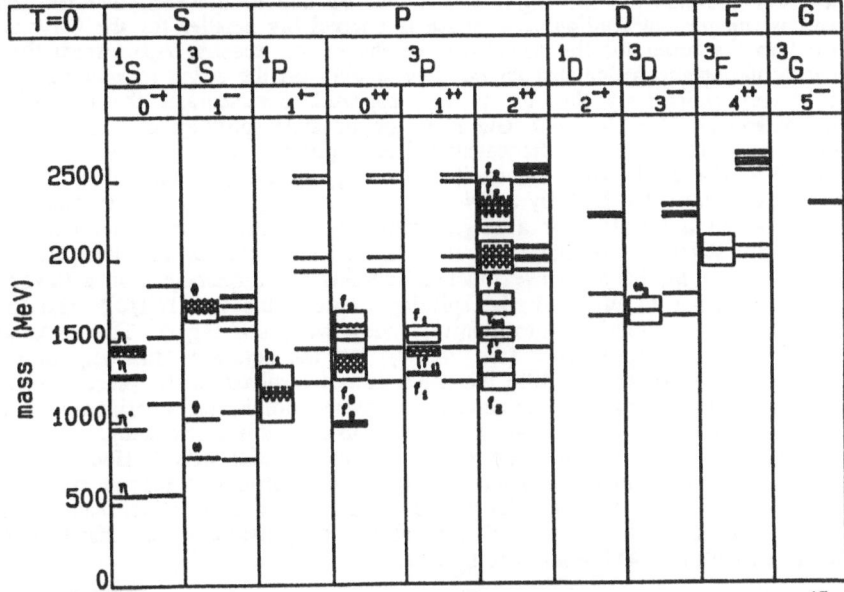

Fig. 3. Comparison of the experimental and calculated isoscalar meson spectrum. (See also caption to Fig. 2).

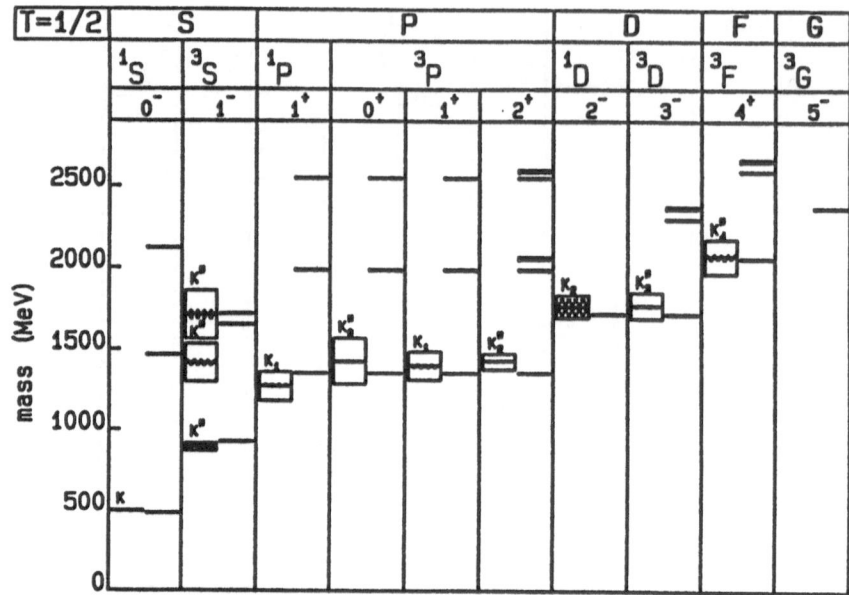

Fig. 4. Comparison of the experimental and calculated strange meson spectrum. (See also caption to Fig. 2).

from the confinement potential alone with, apart from a different energy offset $a(q^3)$, the very same parameters. The positions of the positive parity states of these resonances can be very well accounted for. Also the lowest $\Delta(1/2^-)$ and $\Delta(3/2^-)$ states are reproduced, but the calculated energies of the excited states with these quantum numbers are too high. This might be due to the fact, that for technical reasons the present configuration space is appreciably smaller for the baryons compared to that used for the calculation of the meson spectra. Apart from the quark masses the spectra calculated so far depend only on the string tension parameter b. It is very satisfactory that the present approach demonstrates the universality of this parameter. The N-Δ, Λ-Σ and Ξ-Ξ ground state splittings can be nicely reproduced by 't Hoofts force, the calculated Σ-Σ splitting being slightly too small, see Figs. 5-9. For all baryons the excitation energy of the first excited $1/2^+$-resonance state is calculated too high by a few 100 MeV, although a selective lowering (due to the instanton interaction) of a single state with respect to the other excited positive parity states can be observed. It should be remembered, that the pairing force acts whenever the baryon wave function contains a J=0 quark-pair in a flavour antisymmetric state. In this way the splitting between the two N($1/2^-$)- and the two N($3/2^-$)-states is described quantitatively very well, see Fig. 5. The same applies to the corresponding Λ-states (see Fig. 7), but the notorious difficulty in explaining the position of the $\Lambda(1405)$ resonance is not resolved in the present model. As for the a_0- and f_0-mesons, admixtures of more complicated configurations to the three-quark wave functions might be invoked to cure this discrepancy.

Our results show that the non-relativistic quark model with 't Hoofts force as a residual interaction gives an good description of meson and baryon spectra up to energies of 2 GeV. In particular 't Hoofts interaction is certainly superior to the OGE potential in the respect that it also solves naturally the standard π-η-η' puzzle without any additional assumptions.

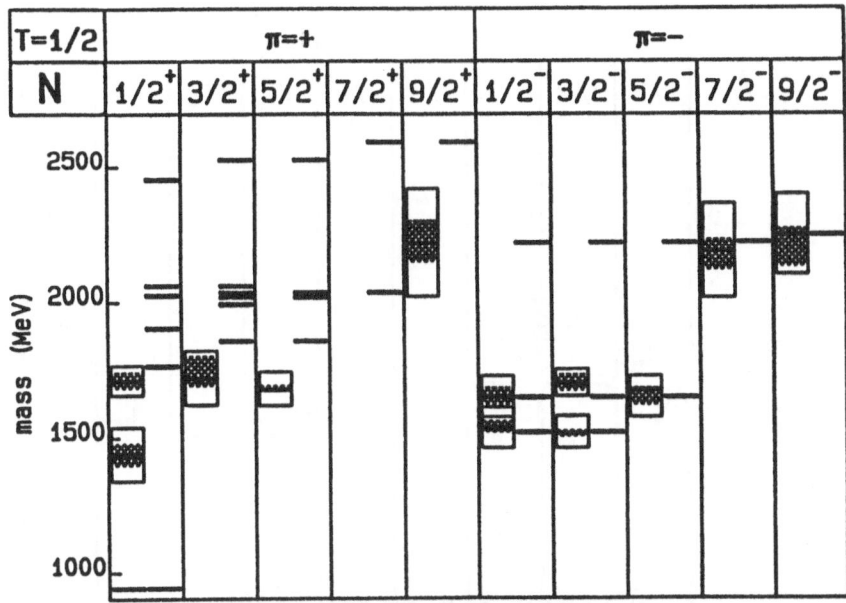

Fig. 5. Comparison of the experimental (left part of each column) and calculated (right part of each column) N-resonance spectrum. The resonances are classified by the total angular momentum and parity. The experimental resonance position is indicated by a bar, the corresponding uncertainty by the shaded area and the width by an open rectangle. Experimental data from [5].

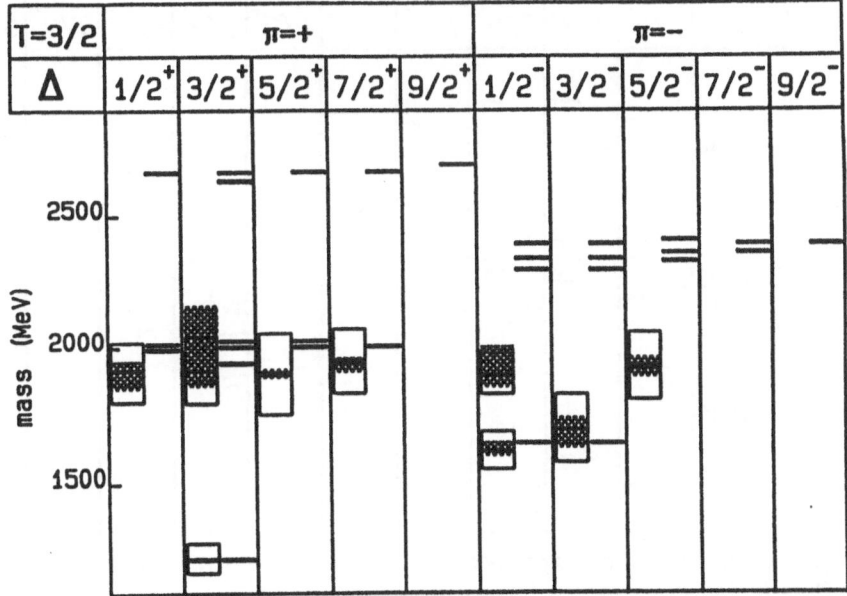

Fig. 6. Comparison of the experimental and calculated Δ-resonance spectrum. See also caption fo Fig. 5.

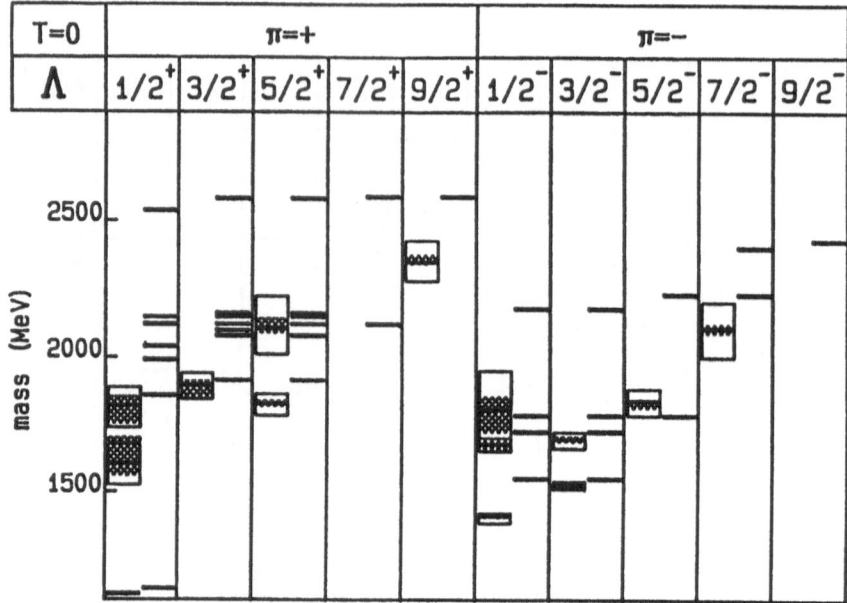

Fig. 7. Comparison of the experimental and calculated Λ-resonance spectrum. See also caption to Fig. 5.

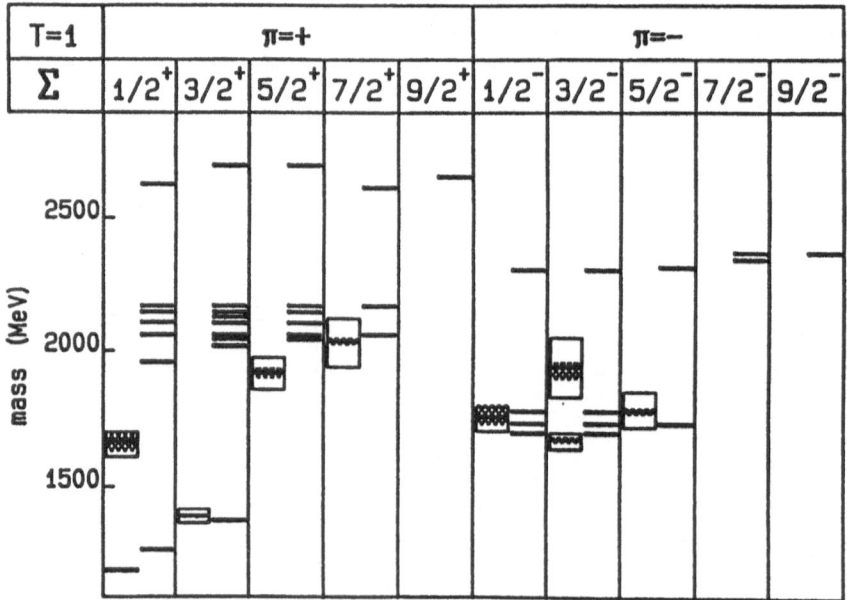

Fig. 8. Comparison of the experimental and calculated Σ-resonance spectrum. See also caption fo Fig. 5.

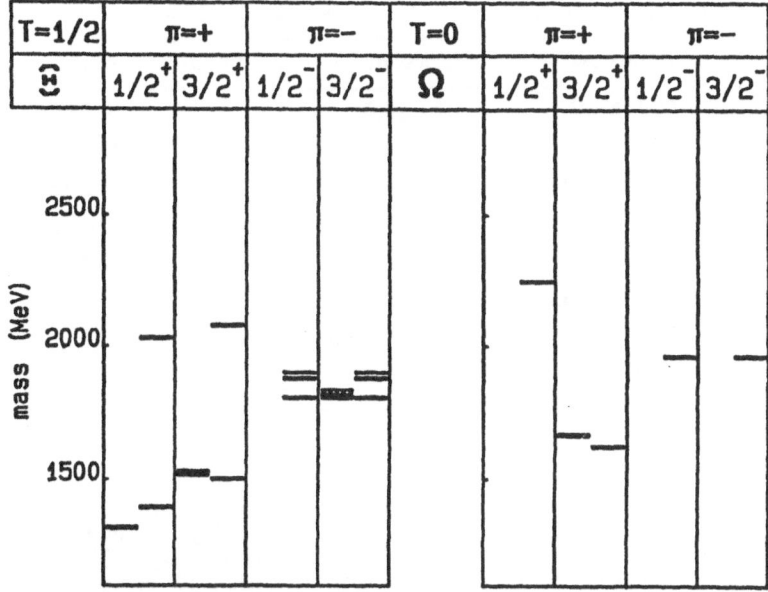

Fig. 9. Comparison of the experimental and calculated Ξ- and Ω-resonance spectra. See also caption to Fig. 5.

4 Discussion of the Parameters; Relation to QCD

The parameters entering the present calculation are given in Table 1. These parameters involve the constituent quark masses m(i) for the non-strange and strange quark, a slope parameter b that describes the string tension as well as two energy offsets a(q$\bar{q}$) and a(q^3) for mesons and baryons, respectively. It is now interesting to study to what extent the expressions for the coupling constants (5) and the constituent masses (4) are consistent with the phenomenologically determined values in our quark model. With standard values for the current masses, the quark condensates and the QCD sacle parameter given in the third column of Table 1 we obtain for an instanton size of 0.45 fm the values given in the second column. For such a value of ρ_c we found a 20% correction due to the two-loop contribution, which is quite acceptable within the present approach. Although the calculated constituent non-strange mass is reasonably close to the fitted valuethe calculated strange mass is definitely too low. In view of the various approximations made in the present phenomenological model and bearing also in mind that the instanton induced interaction is not necessarily the only source for the constituent masses, it is however hard to decide whether the difference between calculated and fitted masses actually reflects a serious discrepancy.

We want to add a remark here concerning the chiral invariance properties of 't Hoofts residual interaction: In fact this force has this desired property with the additional feature that the γ_5-U(1) invariance is maximally broken. If this interaction is used in a relativistic setup with no confining potential and zero quark

Table 1: Parameters entering the constituent quark model Hamiltonian (1).

	Emperical	Calculated	QCD-Parameters[a]	
$m_u = m_d$	300 MeV	270 MeV		9 MeV
m_s	540 MeV	330 MeV		150 MeV
b	850 MeV/fm		$\langle \bar{n}n \rangle$	$(-225 \text{ MeV})^3$
$a_{q\bar{q}}$	-892 MeV		$\langle \bar{s}s \rangle$	$0.8 \langle \bar{n}n \rangle$
a_{q^3}	-1534 MeV		Λ_{QCD}	200 MeV
g	$0.16 \ 10^{-4} \text{ MeV}^{-2}$	$0.17 \ 10^{-4} \text{ MeV}^{-2}$	ρ_o	0.45 fm
g'	$0.11 \ 10^{-4} \text{ MeV}^{-2}$	$0.09 \ 10^{-4} \text{ MeV}^{-2}$		
Λ	0.37 fm			

[a] see [6]

mass it has been shown that analogously to the Nambu Jona-Lasinio model the chiral invariance is spontaneously broken for sufficiently high coupling. As a result the quarks become massive and a massless Goldstone pion emerges. There is, however, no colour confinement at all for quark-antiquark states with energies twice this quark mass (which is roughly 300 MeV). The beautiful feature of constituent quark mass generation and the appearance of a Goldstone pion mode is thus of rather limited practical value. Nevertheless in our approach where a constituent quark mass and a confinement potential are introduced phenomenologically, a remnant of the Goldstone mechanism manifests itself in the spectra of the pseudoscalar mesons: The residual interaction has in our context the important effect that the pion mass is drastically lowered with respect to its unperturbed value (which is identical to the vector meson mass, i.e. 770 MeV) whereas the mass of the η' is pushed up by the same mechanism. Since 't Hoofts force intrinsically provides a constituent quark mass generation via the Nambu mechanism the fact that we are working directly with constituent quark masses is even natural in the present framework. Of course the subsequent non-relativistic treatment together with the phenomenological confinement potentials impedes any *analytical* computation of the desired chiral limit. Furthermore, we are facing the well-known intrinsic difficulties of energy shifts which are large compared to the constituent masses, which applies in particular to the pion. However, we feel that this price has to be paid in the present situation if one aims

Table 2: Elektro-weak decay properties of light mesons

Decay	Γ[keV]			Decay	f[MeV]		
	exp.[a]	calc.[b]			exp.[a]	calc.[b]	
		rel	non-rel.			rel	non-rel.
$\rho \to e^+ e^-$	6.77(32)	3.26	4.65	$\pi^+ \to \mu^+ \nu_\mu$	131.7(2)	465	2248
$\to \eta^0 \gamma$	56(10)	88	285	$K^+ \to \mu^+ \nu_\mu$	161(2)	347	1171
$\to \pi^0 \gamma$	119(30)	306	1120				
$\omega \to e^+ e^-$	0.60(2)	0.35	0.50				
$\phi \to e^+ e^-$	1.37(5)	1.06	1.32				

[a] see [5]   [b] see [7]

83

at a realistic unified description of the hadronic spectra up to 2 GeV.

In order to appreciate the approximations made and to illustrate the scope of the present non-relativistic treatment we present some electroweak decay properties calculated by the expectation value of free Dirac currents with the wave functions obtained by diagonalization of the Hamiltonian (1) of section 1, see also [7]. The results and those from a strictly non-relativistic treatment are compared to experimental data in Table 2.

Although for the leptonic decays the corrections are rather modest, for the pseudoscalar meson weak decay constants they are huge and can be traced back to a cancellation of contributions from the upper and lower components of the Dirac spinors. Although we recover at least the right order of magnitude, the size of the corrections makes the non-relativistic treatment of these meson states highly questionable. This applies *a forteriori* for the calculated width of the $\pi^0 \rightarrow \gamma\gamma$ decay.

As for the strong decays it should be stressed, that the non-relativistic reduction of 't Hoofts force to the order (p/m) yields in fact a vertex between three quarks and one antiquark, which can be used to compute baryon to baryon-meson and meson to meson-meson transition amplitudes, with no additional free parameters. In a pilot study of such transitions it was found that the ratio of the strong pion coupling constants for the $\Delta \rightarrow N\pi$ and $N \rightarrow N\pi$ vertices $(f_{\pi N\Delta}/f_{\pi NN})^2 = 3.4$ much better agrees with the experimental value $(f_{\pi N\Delta}/f_{\pi NN})^2 = 4$ than the SU(6) value $(f_{\pi N\Delta}/f_{\pi NN})^2 = (72/25) = 2.9$. This can be traced back to genuine two-body contributions to these kind of transitions. A systematic investigation of hadronic vertices with the instanton induced interaction is presently in progress.

We want to conclude this paper with a remark concerning the relevance of the OGE-interaction: It may be an exaggeration of the present treatment to eliminate the OGE potentials completely for light quarks. Such forces can possibly still be added, but with a much smaller strength and a short range consistent with asymptotic freedom, without essentially changing the present results. Since the instanton induced interaction should vanish for heavy quarks this means in particular that the beautiful tests of the OGE potentials in heavy quarkonia remain unaffected.

References

1. 't Hooft, G.: Phys. Rev. D 14, 3432 (1976)
2. Shifman, M.A., Vainshtein, A.I., Zakharov, V.I.: Nucl. Phys. B 163, 46 (1980)
3. Shuryak, E.V.: Nucl. Phys. B 203, 93 (1982)
4. Petry, H.R., Hofestädt, H., Merk, S., Bleuler, K., Bohr, H., Narain, K.S.: Phys. Lett. 159 B, 363 (1985)
5. Particle Data Group: Phys. Lett. 204 B, 1 (1988)
6. Reinders, L.J., Rubinstein, H., Yazaki, S.: Phys. Rep. Vol. 127, 1 (1985)
7. Huber, M.G., Metsch, B.C., Resag, J.: in: Proceedings of the Workshop on Meson Production, Interaction and Decay, Cracow 1991, Eds. Magiera, A., Oelert, W., Grosse, E., World Scientific, Singapore, 4 (1991)

STABLE MULTIQUARKS:
LESSONS FROM ATOMIC PHYSICS

Jean-Marc Richard

Institut des Sciences Nucléaires
Université Joseph Fourier–CNRS–IN2P3
53, avenue des Martyrs, F–38026 Grenoble, France

Abstract. In Atomic Physics, some configurations like $H_2^+(ppe^-)$ are particularly stable, while others, like (pe^-e^+), break into smaller subsystems. The mechanisms by which collective binding does or does not occur can tentatively be extended to hadron spectroscopy in quark models: this suggests which flavour configurations are the most likely to form stable multiquark hadrons.

Keywords. Ions. Molecules. Hadron spectroscopy. Multiquarks. Few-body systems. Quantum mechanics.

1 Introduction

Most of the hadrons which have been so far identified can be understood as quark–antiquark $(Q_1\overline{Q}_2)$ mesons or three-quark $(Q_1Q_2Q_3)$ baryons, either in the ground state or in a radially or orbitally excited state.

Potential models turn out to be rather successful, at least in the heavy-quark sector, to reproduce the salient features of the data. Most potentials incorporate the remarkable property of *flavour independence*: the same central potential $V_0(r)$ binds quarks of different masses. Some flavour dependence enter only in spin-dependent corrections of Breit–Fermi type, or in velocity-dependent corrections (Darwin-type of terms), or in mass shifts do to the coupling to decay channels.

The situation looks rather similar to the one encountered in Atomic Physics, where the same Coulomb potential binds systems as different as hydrogen (pe^-), positronium (e^+e^-), or protonium $(p\overline{p})$, the mass dependence occuring only in small corrections.

An important issue in hadron spectroscopy is whether hadrons made out of more than three quarks can exist, and in particular could be stable, i.e. lie below any threshold corresponding to a breaking into ordinary hadrons. For non-experts, the situation might at first sight look rather confusing, with many claims for multiquark hadrons which have never been confirmed. In the present survey, we shall show that some guidance is provided by Atomic Physics, to select those quark configurations which are likely to experience collective binding.

Some differences between these two fields will however be stressed in Sect. 6: spin-dependent forces are relatively more important for quarks than for electric charges, and they might give rise to new type of binding forces. Also the long-range regime of QCD, where meson exchanges take place, is a powerful tool to bind hadrons together. This is well known for nucleons, and it is regularly rediscovered that the Yukawa mechanism might also bind hyperons or mesons.

2   Flavour independence

In a given Coulomb attraction $V = -1/r$, heavy particles experience more binding than the light ones. In the 2-body case, one has

$$E(p\bar{p}) < E(pe^-) < E(e^+e^-). \tag{1}$$

In the 3-body case, one has for instance

$$E(ppe^-) < E(pe^-e^-). \tag{2}$$

Since both configurations have the same threshold energy, which corresponds to H(pe$^-$), the stability of H$^-$ implies that of H$_2^+$.

For the ground state, the energy is not only an increasing function of each inverse mass: it is also a concave function. For instance, we have the inequality

$$E_0(p\bar{p}) + E_0(e^+e^-) < E(pe^-) + E(\bar{p}e^+), \tag{3}$$

illustrating the property that heavy particles prefer to cluster together.

The same inequalities hold in heavy quark physics. For mesons, one has [1, 2]

$$Q\bar{Q} + q\bar{q} < 2Q\bar{q} \tag{4}$$

for any flavour independent potential. This can be verified for instance in the case where Q = c and q = s. Similarly, the relation

$$QQq + qqq < 2Qqq \tag{5}$$

can be derived for any "reasonable" flavour-independent potential [3]. Then the Gell-Mann–Okubo mass formula, or the equal-spacing rule of the decuplet are explained by the effect of the spin-dependent forces cancelling the convexity property (5) induced by the spin-independent forces [4].

Eqs. (4) and (5) show that one gains energy by putting heavy quarks together in the same hadron rather than apart into two different hadrons.

3   Three-unit-charge systems in Atomic Physics

Let us consider the first non-trivial charge configuration beyond binary atoms. It consists of three unit charges $q_1 = -q_2 = -q_3$, with $q_1 = \pm 1$. This includes the positronium ion, the hydrogen-molecule ion H_2^+, the hydrogen ion H^-, and several muonic ions. The stability properties of these systems have been reviewed recently by several authors [5].

Let us first stress that collective binding is not a natural trend in few-body systems. Ignoring for simplicity the virial theorem, one can say that there is a *priori* less kinetic energy in two 2-body systems than in a 4-body bound state of comparable size. The system should gain something when choosing collective binding instead of breaking into smaller subsystems. "Something" might be taking benefit of the heavy particles which experience the most attractive pockets of the potential.

The inverse masses $\alpha_i = m_i^{-1}$ are better variables than the masses themselves to express the regularities of the binding energy when one goes from one system to another. Moreover, one can use the scaling properties of the Coulomb potential to adopt the convenient normalization $\sum \alpha_i = 1$. Then each possible mass configuration can be represented, up to an overall factor, by a point inside an equilateral triangle, as shown in Fig. 1. It turns out that the region where the ion is stable consists of a narrow band

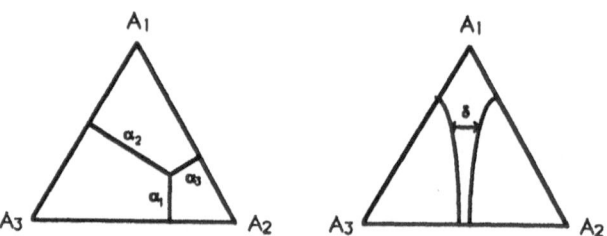

Figure 1: Stability region for three-unit-charge systems.

around the symmetry axis, where the like charges have nearly equal inverse masses. Also the largest excess of binding, as compared to the dissociation threshold, is obtained for H_2^+, i.e. with the largest possible mass asymmetry between positive and negative charges. This is illustrated in Fig. 2, where is shown the quantity

$$g(\alpha_1) = \frac{E_3\left[\alpha_1, \frac{1}{2}(1-\alpha_1), \frac{1}{2}(1-\alpha_1)\right] - E_2\left[\alpha_1, \frac{1}{2}(1-\alpha_1)\right]}{E_2\left[\alpha_1, \frac{1}{2}(1-\alpha_1)\right]} \tag{6}$$

where $E_3(\alpha_1, \alpha_2, \alpha_3)$ is the 3-body ground state energy, and E_2 its 2-body analogue, which represents the dissociation threshold. The ratio g is clearly maximal for $\alpha_1 = 1$, i.e. for $H_2^+(e^-pp)$.

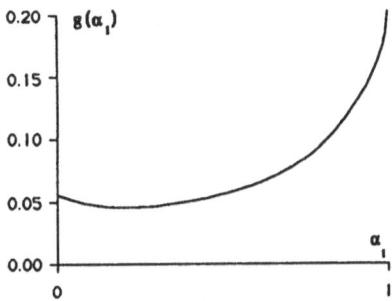

Figure 2: Relative excess of binding with respect to the threshold for the 3-unit-charge system $q_i = -1, +1, +1$, with inverse masses $\alpha_1, \frac{1}{2}(1-\alpha_1), \frac{1}{2}(1-\alpha_1)$.

4 Hydrogen-molecule-like configurations

A more subtle problem is that of four unit charges, $q_i = +1, +1, -1, -1$. A delicate variational calculation shows that the configuration with equal masses, say $(e^+e^+e^-e^-)$, is stable against dissociation into two atoms [6]. The binding energy is very small, however, since

$$g \equiv \frac{E(e^+e^+e^-e^-) - 2E(e^+e^-)}{2E(e^+e^-)} \simeq 0.03 \qquad (7)$$

It is interesting to remark that asymmetric configurations of the type (ppe^-e^-) become more and more stable as the mass ratio m_p/m_e increases [7]. One can, indeed, rewrite the Hamiltonian of the hydrogen molecule as

$$\begin{aligned} H &= \left[\left(\frac{1}{4m_p} + \frac{1}{4m_e} \right) \sum_i \vec{p}_i^2 + V \right] \\ &+ \left(\frac{1}{4m_p} - \frac{1}{4m_e} \right) \left(\vec{p}_1^2 + \vec{p}_3^2 - \vec{p}_2^2 - \vec{p}_4^2 \right) \\ &= H_S + H_A. \end{aligned} \qquad (8)$$

The part H_S, which is symmetric under charge conjugaison, is sufficient to ensure stability, by simple rescaling of the case of the positronium molecule. The antisymmetric part H_A increases the binding of the 4-body system, without changing the threshold energy. For the actual hydrogen molecule, the (relative) excess of binding is

$$g \equiv \frac{E(ppe^-e^-) - 2E(pe^-)}{2E(pe^-)} \simeq 0.17 \qquad (9)$$

Again, the heavy particles like better clustering in the same system than being separated into two atoms.

The case of $(pe^+\bar{p}e^-)$ deserves a comment: the lowest threshold consists of $(p\bar{p})$ and (e^+e^-), with the heavy particles together, so one does not gain anything in the molecule as compared to the threshold. The system $(pe^+\bar{p}e^-)$ is, indeed, unbound.

5 Stable multiquarks bound by the static potential

The simplest extension of the quark–antiquark potential $V_0(r)$ to multiquark systems is usually written as

$$V = -\frac{3}{16} \sum_{i<j} \tilde{\lambda}_i \cdot \tilde{\lambda}_j V_0(r_{ij}) \tag{10}$$

This very naive ansatz looks very similar to the electrostatic potential $\sum q_i q_j / r_{ij}$ which binds chemical molecules. Not suprisingly, when one solves the 4-quark ($qq\bar{q}\bar{q}$) problem with the above potential (10), one gets results very similar to these of Atomic Physics. They are:

  i) In the equal-mass case, one does not get any stable configuration.

  ii) For the asymmetric case $Qq\bar{Q}\bar{q}$, the instability is even worse, and the system immediately breaks into $(Q\bar{Q}) + q\bar{q}$.

  iii) On the other hand, the flavour-exotic configuration $QQ\bar{q}\bar{q}$, sometimes called "tetraquark", becomes stable, once the mass ratio M/m is large enough [8]. Unfortunately, mesons with two units of heavy flavour are very difficult to produce and to identify with the present experimental techniques, though very elaborate and powerful.

The tetraquark wave function [9] consists of a localized QQ diquark, as in ordinary baryons with double flavour (ccu or similar baryons have not yet been seen, but nobody seriously doubts about their existence). This QQ diquark, of colour $\bar{3}$, is surrounded by two light antiquarks, as in everyday antibaryons. Thus tetraquarks use binding mechanisms which are already experienced in the spectroscopy of ordinary hadrons, and well reproduced in current models. This contrasts with some speculations on multiquarks involving colour 6 or colour 8 clusters.

The tetraquark is rather unique, in the sense that it involves a heavy quark clustering which cannot exist in its possible decay products, and which does not suffer from the saturation of colour forces. Let us illustrate this remark by the example of dibaryon configurations. The system $QQqqqq$ does not gain anything, as compared to its lowest threshold $(QQq) + (qqq)$. The same is true for the triply flavoured configuration $(Q)^3(q)^3$. For $(Q)^4(q)^2$, the ground state is essentially determined by minimizing the energy of the flavoured core $(Q)^4$, and it turns out, at least in the model (10), that it is given by a $(Q)^3$ colour singlet plus an isolated Q; in other words, the $(Q)^4(q)^2$ dibaryon is unstable and breaks into two baryons, $(Q)^3 + (Qqq)$.

6 Multiquarks bound by chromomagnetic forces

To explain the hyperfine structure of mesons and baryons, one often advocate a short range interaction of the type

$$V_{SS} = -C \sum_{i<j} \frac{\tilde{\lambda}_i \cdot \tilde{\lambda}_j \vec{\sigma}_i \cdot \vec{\sigma}_j}{m_i m_j} \delta^{(3)}(r_{ij}) \tag{11}$$

Spin–orbit and tensor forces are also used for $\ell > 0$ partial waves. Eq. (11) is inspired by the one-gluon-exchange model, but is slightly more general.

If taken seriously (this implies in particular colour 6 or 8 states in some sub-clusters), the chromomagnetic interaction (11) induces striking coherences in some particular configurations. The most famous is the H or "Hexaquark" proposed by Jaffe in 1977 [10]. It is a (ssuudd) configuration with spin $J = 0$.

More recently the "Pentaquark" (P) was proposed in the baryon number $B = 1$ sector [11]. It consists of an heavy antiquark surronded by four ordinary or strange quarks in a flavour triplet (of the old SU(3) group of flavour) and spin $s = 0$ state. An exemple is ($\bar{c}$suud).

Several calculations have been done for studying the stability of Hexaquark or Pentaquark states. Unfortunately, the results depend on the detailed assumptions made on confining forces.

One shoud also stress that one is dealing here with a delicate 5-body or 6-body problem. The wave function "hesitates" between a collective binding and a molecular structure, of the type baryon–baryon for H, and meson-baryon for P. A serious variational calculation should incorporate these two aspects of the internal dynamics.

7 Hadron–hadron molecules

The long range part of the nucleon–nucleon potential is mediated by exchange of mesons. Although one does not understand very well how this Yukawa mechanism is derived from basic QCD, it gives rise to a reasonable phenomenology in low-energy Nuclear Physics.

An obvious question is whether meson-exchange can produce bound states involving other hadrons. We already know examples of "hypernuclei", with a Λ hyperon attached to a nuclear core, and one suspects the existence of a 2-body bound state in the $\Sigma - N$ sector [12].

More recently, Törnqvist has proposed to explain some of the recently discovered resonances as meson–meson states bound by pion exchange [13]. Note that this is a rather recurrent idea: for instance, some puzzling states of the Charmonium spectrum were sometimes interpreted as $D^*\overline{D}^*$ molecules; the data in fact are better explained in terms of the more conventional $c\bar{c}$ model [14].

Here, one should recall some elementary quantum mechanics. First, a Yukawa potential,

$$Y(r) = -g\frac{\exp(-\mu r)}{r} \tag{12}$$

associated with a reduced mass m in the 2-body problem, can be rescaled in terms of a simplified problem with $\mu = m = 1$. Second, the potential does not support bound states unless the strength is large enough, say $g \geq g_2$. Finally, the minimal coupling for getting 3-body bound states in the potential $\sum Y(r_{ij})$ is lower than for 2-body bound states, say $g \geq g_3$, with $g_3 < g_2$ (in practice $g_3 \simeq 0.85g_2$). In the actual situation, there are spin and isospin factors associated with pion exchange, and the multimeson potential energy is sligthly more involved than $\sum Y(r_{ij})$. Anyhow, one should retain that if meson–meson states exist, multimeson molecules might well be present.

These multihadrons states bound by pion exchange differ from the $K\overline{K}$ molecules proposed by Isgur, Barnes and others, and reviewed by Close at this Workshop [15]. The binding of $K\overline{K}$ molecules is due to chromomagnetic forces between quarks of both kaons.

8 Conclusions

Let us briefly summarize. In Atomic Physics, the stability of a system of given charges $q_i = +1, -1, -1$ or $q_i = +1, +1, -1, -1$ requires very particular distributions of the constituent masses m_i. If the conditions on the m_i are not fulfilled, the system spontaneously breaks into smaller atoms or ions. Explicit few-body calculations in simple quark models show the same pattern: with a purely central and flavour-independent potential, one does not get stable multiquark states, except for peculiar combinations of heavy and light quarks or antiquarks. Chromomagnetic forces offer interesting possibilities of coherent attraction, but the existence of stable multiquarks such as Hexaquark and Pentaquark is rather model dependent. Finally, new branches of Nuclear Physics await development, with the recent emphasis on the long-range attractive forces between hadrons, leading to the prediction of multihadron clusters. Needless to say that on needs a better understanding of the multiquark potential in QCD before taking too seriously the few-body calculations which have been discussed in this review.

Acknowledgments

I would like to thank the organizers, P. Kroll, K. Goeke and H.R. Petry, for the enjoyable and stimulating atmosphere of this Workshop.

References

[1] R.A. Bertlmann and A. Martin, *Nucl. Phys.* **B168** (1980) 111.

[2] S. Nussinov, *Phys. Rev. Lett.* **51** (1983) 2081.

[3] S. Nussinov, *Phys. Rev. Lett.* **52** (1984) 966;
J.-M. Richard and P. Taxil, *Phys. Rev. Lett.* **54** (1985) 847;
E. Lieb, *Phys. Rev. Lett.* **54** (1985) 1987;
A. Martin, J.-M. Richard and P. Taxil, *Phys. Lett.* **176B** (1986) 224.

[4] J.-M. Richard, *Phys. Rep.* **212** (1992) 1.

[5] A.M. Frolov and D.M. Bishop, *Phys. Rev.* **A45** (1992) 6236;
A.V. Gur'yanov and T.K. Rebane, *Sov. Phys. JETP* **71** (1990) 1;
E.A.G. Armour and W.B. Brown, *The stability of three-body Coulombic systems*,
to be published;
A. Martin, J.-M. Richard and T.T. Wu, *Phys. Rev.* **A** , in press.

[6] E.A. Hylleraas and A. Ore, *Phys. Rev.* **71** (1947) 493.

[7] M.A. Abdel-Raouf, in *Few-Body Problems in Particle, Nuclear, Atomic, and
Molecular Physics*, Proc. 11th Eur. Conf. Few-Body Phys., Fontevraud, France,
1987, ed. J.-L. Ballot and M. Fabre de la Ripelle, *Few-Body Systems Suppl.* **2** ;
T.K. Rebane, *Sov. J. Nucl. Phys.* **50** (1989) 746;
J.-M. Richard, in preparation.

[8] J.-P. Ader, J.-M. Richard and P. Taxil, *Phys. Rev.* **D25** (1982) 2370.

[9] S. Zouzou et al., *Z. Phys.* **32** (1982) 427;
L. Heller, in *The Elementary Structure of Matter*, Proc. Les Houches Workshop,
1987, ed. J.-M. Richard et al. (Springer–Verlag, Berlin, 1988);
H.J. Lipkin, *Phys. Lett.* **B172** (1986) 242.

[10] R.L. Jaffe, *Phys. Rev. Lett.* **38** (1977) 195.

[11] H.J. Lipkin, *Phys. Lett.* **195B** (1987) 484;
C. Gignoux, B. Silvestre–Brac and J.-M. Richard, *Phys. Lett.* **193B** (1987) 323.

[12] See, for instance, C.B. Dover and G.E. Walker, *Phys. Rep.* **89** (1982) 1;
C.B. Dover, D.J. Millener and A. Gal, *Phys. Rep.* **14** (1989) 1.

[13] N.A. Törnqvist, *Phys. Rev. Lett.* **67** (1991) 556.

[14] A. Le Yaouanc, L. Oliver, O. Pène and J.C. Raynal, *Phys. Lett.* **71B** (1977) 397;
72B (1977) 57.

[15] F.C. Close, Contribution to this Workshop.

Dynamical Supersymmetry for Mesons and Baryons

D. B. Lichtenberg

Physics Department, Indiana University, Bloomington, IN 47405, USA

Abstract. An approximate dynamical supersymmetry exists between mesons and baryons. This supersymmetry arises in QCD dynamics because both diquarks and antiquarks belong to $\bar{3}$ multiplets of color SU(3). The supersymmetry is broken by mass and size differences between diquarks and antiquarks and also by spin-dependent QCD forces. I discuss a method to use this supersymmetry to obtain the masses of some baryons from the masses of the corresponding mesons.

Keywords. Quark, diquark, meson, baryon, supersymmetry

1 Introduction

To my knowledge, Miyazawa [1,2] was the first person to apply the idea of supersymmetry to particle physics. As early as 1966 he developed a supersymmetric algebra in order to classify mesons and baryons in a single multiplet. He did this with the aim of obtaining relations between properties of baryons and mesons.

The supersymmetry between mesons and baryons is badly broken. If it were not, then, because a pion and a nucleon belong to the same supermultiplet, they would have the same mass. An important question is whether one can make use of the supersymmetry to obtain baryon properties from meson properties despite the symmetry breaking. In this talk, I take a modest step toward giving a positive answer to that question.

Miyazawa's supersymmetry is very different from the fundamental supersymmetry proposed in the 1970's, which requires that there exist *elementary* supersymmetric partners to the elementary fermions and bosons of the standard model.

Miyazawa did his work before the invention of QCD, and he did not use the word "supersymmetry." I believe that Catto and Gürsey [3,4] were the first physicists to bring the work of Miyazawa into the framework of QCD. Previously, Gao and Ho [5,6] independently developed similar ideas, but their work is not widely known. A few years ago I suggested [7] that the meson-baryon symmetry might be more applicable to hadrons containing one heavy quark than to hadrons containing only light quarks. A brief discussion of meson-baryon supersymmetry has appeared in a recent review of diquarks [8].

As modernized by Catto and Gürsey, Miyazawa's idea is as follows: Consider an approximation in which a diquark can be considered as a single particle. The diquark in a baryon, like an antiquark in a meson, belongs to a $\bar{3}$ multiplet of color-SU(3). In each case, the particle is bound to a quark to form a composite color-singlet hadron. According to QCD, it follows that the interaction in a baryon between a diquark and a quark is approximately the same as the interaction in a meson between an antiquark and a quark. The similarity of the quark-antiquark and quark-diquark interactions gives rise to an approximate supersymmetry between mesons and baryons.

One can see from the work of Catto and Gürsey that anyone who treats diquarks and antiquarks on a similar footing is, in effect, making use of supersymmetry even when not mentioning the word explicitly. I cannot quote the many papers that fall in this category. But I do wish to point out that one of the most important manifestations of meson-baryon supersymmetry is that the Regge trajectories of mesons and baryons have similar slopes [9, 10]. I also want to point out a contribution in this category by Reinhardt [11] at this meeting.

Miyazawa [1,2] and Catto and Gürsey [3,4] have discussed the algebraic and group-theoretical aspects of meson-baryon supersymmetry in great detail. In particular, they discussed the superalgebras SU(3/3) and SU(6/21), both of which allow certain mesons and baryons to belong to a single multiplet. I shall not discuss the group theory of SU(3/3) or SU(6/21) in any detail here, as the mathematics has been well treated by Catto and Gürsey.

A fundamental multiplet of SU(3/3) contains two SU(3) flavor antitriplets. These are the spin-1/2 antiquarks:

$$\bar{s}, \quad \bar{d}, \quad \bar{u} \tag{1}$$

and the spin-zero diquarks:

$$(ud - du)/\sqrt{2}, \quad (us - su)/\sqrt{2}, \quad (ds - sd)/\sqrt{2}, \tag{2}$$

where the symbol for a quark denotes its flavor wave function. A fundamental multiplet of SU(6/21) contains the above spin-1/2 antiquarks and spin-zero diquarks and in addition contains a flavor sextet of spin-one diquarks:

$$uu, \quad (ud + du)/\sqrt{2}, \quad dd, \quad (us + su)/\sqrt{2}, \quad (ds + sd)/\sqrt{2}, \quad ss. \tag{3}$$

The algebra SU(3/3) allows members of the pseudoscalar meson flavor octet and spin-1/2 baryon flavor octet to belong to the same supermultiplet. The algebra of SU(6/21) allows the vector meson nonet and the spin-3/2 baryon decuplet also to belong in the multiplet with the pseudoscalar mesons and spin-1/2 baryons. It is not clear, however, that either of these classification schemes is useful, because, as I have already noted, the pion and nucleon belong to the same supermultiplet but have very different properties.

Because the supersymmetry is so badly broken for hadrons containing light quarks, I suggested [7] that the method be applied to hadrons containing at least one heavy quark, which is treated as a singlet of SU(3/3). It turns out that the symmetry breaking is smaller in this case, because some of the main characteristics of a hadron containing a heavy quark derive principally from the properties of the heavy quark alone. For example, consider mesons and baryons belonging to a supermultiplet of SU(3/3) containing a b quark and light quarks and/or antiquarks. The B meson and Λ_b baryon, among other particles, belong to this supermultiplet. To a first approximation, the mass of the hadron is determined by the mass of the b quark. The B meson has a mass of 5.28 GeV, while the Λ_b baryon has a mass of about 5.64 GeV, only 7% heavier. (In contrast, the nucleon is is almost 7 *times* as heavy as the pion.) Also, the B and Λ_b probably have very similar lifetimes, as the decay in each case arises principally from the decay of the b quark (spectator model).

But the fact that supersymmetry is rather good for hadrons containing a b quark follows not so much from diquark–antiquark supersymmetry as from the fact that many of the properties of such hadrons depend only slightly on the properties of the light constituents. Here, I want to make use of supersymmetry in a more essential way.

2. Diquark clustering

Meson-baryon supersymmetry depends to some degree on the existence of two-quark (diquark) correlations in baryons, although there should exist remnants of supersymmetry even if no diquark correlations occur. I should like to review briefly some cases in which diquark clustering in baryons ought to exist.

1) If the quarks in a baryon have unequal masses, then the two heavier quarks ought to cluster into a diquark [12]. If the two lightest quarks have the same mass, there should still be diquark clustering of one of the light ones with the heavy one. (The wave function should be antisymmetric with respect to the interchange of the two light quarks.)

2) If one quark is radially or orbitally excited, then the two remaining quarks should cluster into a diquark [9, 10].

3) If three light quarks in a spin-1/2 baryon are in their ground state, then spin-dependent forces should cause diquark clustering [12].

The above three cases of diquark clustering can be demonstrated in potential models, although the amount of clustering depends on details. Potential models usually fail to lead to diquark clustering only in the case of a spin-3/2 baryon containing three quarks of the same mass. In this connection it is a good approximation to neglect the mass difference between the u and the d quark.

Dziembowski [13], in a contribution to this meeting, showed that in his model diquark correlations in the nucleon are necessary to get reasonable agreement with both the static nucleon properties and results of deep inelastic lepton-nucleon scattering.

3. Supersymmetry breaking

Next I shall consider various causes of supersymmetry breaking arising from differences between an antiquark and a diquark. I identify the following differences:

1) A diquark and an antiquark have different masses. Although the static potential between two colored particles is supposed to be independent of their masses, nonstatic corrections to the potential do depend on mass. Furthermore, even in a nonrelativistic treatment with an interaction which is

independent of mass, the particle masses appear in the Schrödinger equation and therefore affect the wave function. Because bound-state energy levels and decay rates depend on the wave function, they are mass dependent. Of course, decay rates depend also on phase space, which is a function of hadron masses.

2) A diquark and an antiquark have different spins. According to QCD, spin-dependent terms contribute to the interaction of colored particles. These spin-dependent terms are also mass dependent, vanishing in the limit of infinite mass. However, for particles of finite mass, the spin-dependent interactions exist and are different for antiquark and diquark, both because of their different spins and different masses.

3) A diquark and an antiquark are expected to have different sizes. The static interaction between two colored particles is obtained from QCD under the assumption that the particles are point-like. A diquark is certainly an extended object, being a state of two quarks considered collectively. Although a *current* quark is supposed to be point-like, a *constituent* quark, which is the kind of quark most appropriate to use in the calculation of hadron masses, should have a size greater than zero. Nevertheless, I think that a diquark is larger than a constituent quark.

4) A meson containing a quark and antiquark of the same flavor can annihilate strongly if its mass is sufficiently high, but a quark and a diquark cannot annihilate. However, if the quark and antiquark in a meson have different flavors, neither strong nor electromagnetic annihilation can occur, and there is only the possibility of weak quark-antiquark annihilation in mesons. The supersymmetry breaking arising from weak annihilation should be small.

4. Properties of constituent quarks and diquarks

In calculating static properties of a hadron in the quark model, it turns out to be a good approximation to consider only the hadron's valence quarks, neglecting gluons and quark-antiquark pairs of the sea. However, in so doing, one should use the constituent mass of a quark, which is about 300 MeV larger than the mass of a current quark. I like to think of a constituent quark as a current quark clothed with gluons and quark pairs. The reason for the heavier constituent quark mass is that, when observed at low momentum transfer, the quark apparently drags its clothing along with it.

Although, as far as is known, a bare or current quark is a point particle, a constituent or clothed quark should have a radius bigger than zero. Calculations of baryon properties in quark models and quark-diquark models support this idea. For example, in most potential models using constituent quarks, the size of the quark is neglected. This neglect leads [14] to a calculated proton radius of around 0.5 fm, in contrast to the experimental value of 0.8 fm. Within the framework of the constituent-quark model, the only way I know to reconcile this discrepancy is to assume that quarks have radii greater than zero, perhaps of order 0.5 fm. (If one includes sea-quark or pion degrees of freedom explicitly to enlarge the proton radius, one is going beyond the constituent-quark model.) In the model of Ref. [14], a diquark can be as large as a hadron, so at least in that model a diquark is larger than a constituent quark.

The mass of a constituent quark is not necessarily a quantity which is a constant independent of the hadron in which the quark finds itself. However, if one treats the constituent quark masses as parameters which can vary from hadron to hadron, one has too many parameters to be able to say anything useful. I shall therefore adopt a single set of constituent-quark masses. From my experience with potential models and semi-empirical mass formulas, I believe that the following quark masses are reasonable (I neglect the u, d mass difference and give masses in MeV):

$$m_u = m_d = 320, \quad m_s = 520, \quad m_c = 1760 \quad m_b = 5140. \tag{4}$$

One can in principle use a potential model and these quark masses to calculate the masses of mesons and diquarks [14]. However, such calculations are quite model dependent, especially in the case of light quarks, where relativistic effects are large. In this talk, I do not use a potential model, but rather use meson masses as input and use a simple semi-empirical formula to obtain diquark masses in terms of meson masses.

To simplify matters I confine myself to ground-state mesons and neglect spin-dependent forces. (Later, I shall put back approximate effects of such forces.) Without spin-dependent forces, the appropriate empirical meson mass to use is the spin-averaged meson mass. To obtain the spin-averaged masses M_{12} of mesons containing quarks q_1 and q_2, I use the standard formula

$$M_{12} = \frac{3}{4}V_{12} + \frac{1}{4}P_{12}, \tag{5}$$

where V_{12} and P_{12} are the masses of the vector and pseudoscalar mesons respectively as given in the tables of the Particle Data Group [15] and in the recent work of Yanagisawa et al. [16]. The values of the spin-averaged masses are given in Table 1. Because I am uncertain about the quark content of the η and η' mesons, the spin-averaged value of the $s\bar{s}$ meson is my estimate. Likewise, because the η_b has not yet been seen, the spin-averaged mass of the $b\bar{b}$ meson is my estimate.

The mass of a meson can be written in terms of its constituent quark masses m_1 and m_2 and an energy eigenvalue term E_{12} as

$$M_{12} = m_1 + m_2 + E_{12}. \tag{6}$$

Using the assumed quark masses from (4) and the values of M_{12} from (5), I can easily compute the values of E_{12}, and these are also shown in Table 1. The spin-averaged meson masses in Table 1 are similar to to those calculated many times previously; see, e.g., Ref. [17].

I can plot the values of E_{12} against the reduced masses μ_{12} of the constituents If I connect the points by eye with a curve, I get the curve shown in Fig. 1.

Table 1. Spin-averaged meson masses from Eq. (5) and energy eigenvalues from Eq. (6) in MeV. The symbol q stands for either a u or d quark. The reduced mass of the two quarks is also given.

Quark content	Mass M_{12}	Energy E_{12}	Reduced mass μ_{12}
$q\bar{q}$	612	-28	160
$q\bar{s}$	793	-47	198
$s\bar{s}$	936	-104	260
$q\bar{c}$	1974	-106	271
$q\bar{b}$	5313	-147	301
$s\bar{c}$	2075	-205	401
$s\bar{b}$	5397	-263	472
$c\bar{c}$	3068	-452	880
$b\bar{b}$	9446	-834	2570

One can prove rigorously from the Feynman-Hellman theorem that the curve should be smooth and monotonically decreasing, provided the Hamiltonian is nonrelativistic and the potential is independent of quark flavor. In my opinion, relativistic effects will not disturb this behavior very much, although, strictly speaking, in a relativist treatment, E_{12} depends on m_1 and m_2 individually rather than just on the reduced mass μ_{12}. The fact that the curve of Fig. 1 is montonically decreasing and reasonably smooth reinforces my belief that I have taken reasonable values for the quark masses, or at least, for the quark mass differences.

I now need to determine the diquark masses D_{12}. I begin by making the simple ansatz

$$D_{12} = M_{12} + a + b\mu_{12}, \tag{7}$$

where a and b are constants to be determined. I obtain these constants from the experimental masses [15] of the N, Λ, and Σ belonging to the baryon octet and the Δ and Σ^* belonging to the baryon decuplet. From these baryon masses, the spin-averaged baryon masses M_{qqq} and M_{qqs} have been computed [17]. The result is

$$M_{qqq} = 1086, \quad M_{qqs} = 1270, \tag{8}$$

where the masses are in MeV. Next, I assume that a spin-averaged baryon mass can be written as a sum of a diquark mass, a quark mass, and an energy eigenvalue E_{123}:

$$M_{123} = D_{12} + m_3 + E_{123}. \tag{9}$$

In this formula I assume that if the quark masses are unequal, the diquark is made of the two heaviest quarks, as these are likely to be closest together on the average.

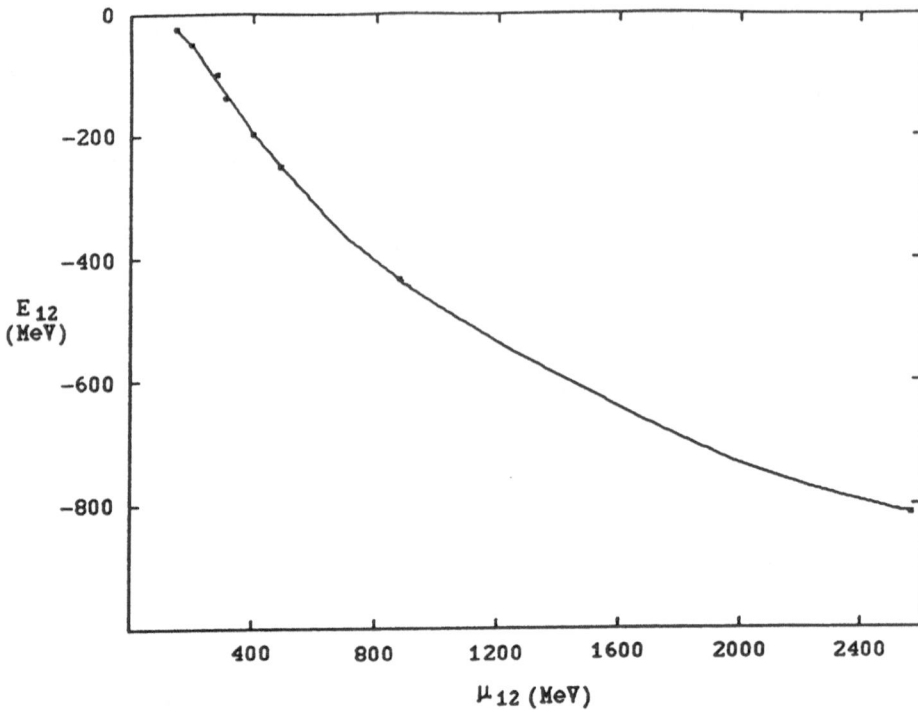

Fig. 1. Energies (excluding quark rest energies) of mesons vs reduced mass of constituent quarks. The dots are calculated from the data of Refs. [15] and [16].

The essential use of supersymmetry comes about as follows: I assume that the eigenvalue E_{123} is the *same function* of the reduced mass of the diquark and quark as E_{12} is of the reduced mass of quark and antiquark. Thus, I can use the baryon data of (8) together with the curve in Fig. 1 to determine the diquark masses D_{qq} and D_{qs}. I find I need a diquark mass $D_{qq} = 835$ MeV in order to get a baryon mass $M_{qqq} = 1086$ MeV, and I need a diquark mass $D_{qs} = 1030$ MeV in order to get a baryon mass $M_{qqs} = 1270$ MeV. These diquark masses are the first two entries in Table 2, column 2. These two diquark masses enable me to use (7) to obtain the parameters a and b. Then I use these parameters to obtain the remaining entries in Table 2.

Unfortunately, it turns out that the mass of a diquark containing two given quarks depends on the flavor of the third quark. In the case that the third quark is a u or d, the values of a and b turn out to be

$$a_q = 164 \text{ MeV}, \quad b_q = 0.368, \tag{10}$$

where I have put subscripts on the parameters to make explicit their dependence on the mass of the third quark.

Another expression for a diquark mass is

$$D_{12} = m_1 + m_2 + \epsilon_{12}, \tag{11}$$

where ϵ_{12} is the energy eigenvalue. The diquark masses D_{12} and values of ϵ_{12} are given in Table 2. The entries in this table apply only if the third quark is u or d. The value of the mass of the Ω (somewhat reduced to reflect the omission of spin-dependent forces) is not enough to allow me to determine a_s and b_s separately. Tentatively, I assume

$$a_s = a_q + 50 \text{ MeV}, \quad b_s = b_q = b. \tag{12}$$

Table 2. Diquark ground-state masses and energy eigenvalues in MeV. Spin-dependent forces are neglected.

Quark content	Mass D_{12}	Energy ϵ_{12}
qq	835	195
qs	1030	190
ss	1196	156
qc	2238	158
qb	5588	128
sc	2387	107
sb	5735	75

For the most part, the diquark energy ϵ_{12} is a monotonically decreasing function of the reduced mass, but there is a single reversal: ϵ_{qc} should be slightly smaller than ϵ_{ss}, whereas it turns out 2 MeV larger. I believe that this problem would go away if I increased the mass of the c quark by a few MeV. However, I shall live with this reversal and proceed in the next section to calculate some baryon masses.

5. Baryon masses from meson masses

It is now straightforward to calculate the mass of any ground-state baryon containing a diquark of mass given in Table 2 and a third quark q (u or d), provided spin is neglected. I simply calculate the reduced mass of the diquark and quark, read off the energy eigenvalue from Fig. 1, and use (9). Here again, the essential use of supersymmetry is the use of the curve in Fig. 1 for baryons, although it was obtained from meson data. To obtain the mass of a baryon in which the lightest quark is an s quark, I follow the same procedure, except that I increase the diquark masses of Table 2 by 50 MeV in accordance with (12). The results for baryon masses neglecting spin effects are given in Table 3. Entries for baryons containing the quark content qqq, qqs, and sss are omitted from Table 3 because the masses were used as input data. Also, entries for baryons containing two heavy quarks are omitted, as such baryons are unlikely to be observed for some time. However, it is straightforward to predict their masses.

The masses of the excited states of the baryons of Table 3 can be obtained by requiring the baryons to lie on Regge trajectories with the same slopes as the trajectories of their meson partners. In assuming the slopes are the same for baryons and corresponding mesons, I am again making use of supersymmetry. For brevity, I do not include the predicted masses here.

Table 3. Predicted ground-state baryon masses in MeV. Spin-dependent forces are omitted. The diquark is composed of the two heaviest quarks.

Quark content	Mass M_{123}
qss	1430
qqc	2440
qsc	2590
qqb	5770
qsb	5920
ssc	2730
ssb	6040

The last step is to add spin-dependent forces as a perturbation so as to get predictions for actual baryon masses. To do this I make use of a semi-empirical mass formula [18] for the spin-dependent forces in ground-state baryons. The formula of Ref. [18] is within the framework of the three-quark model, not the quark-diquark model. But this is just what is needed, for the following reason: The only spin-dependent force contributing to the ground state energy in lowest-order perturbation theory is the color-hyperfine (spin-spin) interaction. This interaction is extremely short range (a δ-function interaction in one approximation). The quark-diquark approximation breaks down when the third quark is very close to one of the quarks of the diquark, so one should use the three-quark model for this term.

Including spin-dependent forces, I obtain the predictions for baryon masses shown in Table 4. Table 4 also gives the experimental values [15,19] where known. The N, Λ, Σ, Δ, Σ^*, and Ω baryons have been omitted from the table, as their masses have been used as input in order to obtain the three parameters of the model: a_q, b, and a_s.

The entries in Tables 3 and 4 are given to the nearest 10 MeV because the method has a precision which is no better than that. Of course, additional errors arise from deficiencies in the model.

The method I have discussed makes essential use of supersymmetry in treating a diquark of a certain mass like a fictitious antiquark of the same mass. Spin effects are added later as perturbations. I believe that there is a tendency of

Table 4. Predicted values of some ground-state baryon masses in MeV compared to the experimental values [15,19] where known.

Baryon name	Quark content	Predicted mass	Experimental mass
Ξ	qss	1310	1318
Ξ^*	qss	1530	1533
Λ_c	qqc	2280	2285
Σ_c	qqc	2450	2453
Σ_c^*	qqc	2520	?
Ξ_c	qsc	2470	2470
Ξ_c'	qsc	2580	?
Ξ_c^*	qsc	2660	?
Ω_c	ssc	2710	2716 ± 6
Ω_c^*	ssc	2790	?
Λ_b	qqb	5600	5641 ± 50
Σ_b	qqb	5810	?
Σ_b^*	qqb	5840	?

the method to underestimate the masses of some baryons because the method neglects the size difference between diquarks and antiquarks. A larger particle (a diquark) does not bind as strongly to a quark as a smaller particle (an antiquark) does. But I am encouraged by the generally good agreement of the predictions in Table 4 with experiment, and I look forward to future measurements of the masses of baryons containing a c or b quark for further testing of broken meson-baryon supersymmetry.

Acknowledgements. I am grateful to the organizers of this conference for their partial support. Additional support has come from the U.S. Department of Energy.

References

[1] H. Miyazawa, Prog. Theor. Phys. 36 (1966) 1266.
[2] H. Miyazawa, Phys. Rev. 170 (1968) 1586.
[3] S. Catto and F. Gürsey, Nuovo Cimento 86 (1985) 201.
[4] S. Catto and F. Gürsey, Nuovo Cimento 99 (1988) 685.
[5] C.-s. Gao and T.-h. Ho, 1982 Commun. Theor. Phys. (Beijing) 1 (1982) 761.
[6] C.-s. Gao and T.-h. Ho, 1982 Commun. Theor. Phys. (Beijing) 2 (1983) 1045.
[7] D. B. Lichtenberg, J. Phys. G: Nucl. Part. Phys. 16 (1990) 1599.

[8] M. Anselmino, E. Predazzi, S. Ekelin, S. Frederksson, and D. B. Lichtenberg, Lulea University report TULEA 1992:05 (unpublished).

[9] T. Eguchi, Phys. Lett. B 59 (1975) 457.

[10] A. Martin, Z. Phys. C 32 (1986) 359; In *Workshop on Diquarks*, ed. by M. Anselmino and E. Predazzi (World Scientific, Singapore, 1989), p. 70.

[11] H. Reinhardt, contribution to this conference.

[12] S. Fleck, B. Silvestre-Brac, and J.M. Richard, Phys. Rev. D 38 (1988) 1519.

[13] Z. Dziembowski, contribution to this conference.

[14] D. B. Lichtenberg, W. Namgung, J. G. Wills, and E. Predazzi, Z. Phys. C – Particles and Fields 19 (1983) 19.

[15] K. Hikasa et al., Phys. Rev. D 45 (1992) S1.

[16] C. Yanagisawa et al., Phys. Rev. Lett. 66 (1991) 2436.

[17] M. Anselmino, D. B. Lichtenberg, and E. Predazzi, Z. Phys. C – Particles and Fields 48 (1990) 605.

[18] Y. Wang and D. B. Lichtenberg, Phys. Rev. D 42 (1990) 2404.

[19] K. Schubert, contribution to this conference.

K$\bar{\text{K}}$ molecules

B.C. Pearce

Institut für Kernphysik, Forschungszentrum Jülich, D-5170 Jülich, Fed. Rep. Germany

Abstract. I present a review of the description of the coupled $\pi\pi - K\bar{K}$ system within the meson exchange framework developed recently by the Jülich group. In this model the $f_0(975)$ emerges as a $K\bar{K}$ bound state. I also show results for the $K\pi$ system.

1 Introduction

Despite it being generally accepted as the correct theory of the strong interactions, its non-perturbative nature has meant that QCD has still failed to provide us with detailed descriptions of phenomena in the low energy (~ 1 GeV) region. Hence the abundance of models of strong interaction physics. However, there are indications[2] that most of the dynamics in this regime can be understood in terms of colour-neutral objects, namely nucleons, mesons and isobars. This is the basis of the success of the meson exchange models of medium energy nuclear physics.

An important advantage of meson exchange models is that they enable the (possibly strong) effects of the meson cloud to be analysed. In many models based on quarks and gluons these effects are ignored. For example, the quark model[3] describes states only in terms of valence quarks. In that model, the ρ meson is entirely $q\bar{q}$, yet presumably in nature it must have a component that is $qq\bar{q}\bar{q}$. In the effective meson picture, this component arises from the two pion self energy correction to the ρ and gives rise to its width for decay to two pions. Calculations of the decay width within the quark model exist but the effect of the self energy diagrams on the real part of the mass is ignored on the assumption that it can be accounted for by a renormalisation of the parameters of the model. It is interesting that, to describe the $f_0(975)$ and $a_0(980)$ mesons within the quark model, it is necessary to first project out an effective meson type potential and then proceed in a similar vein to the meson exchange picture[4].

In other words, we assume that in nature any meson state m emerges as a combination of a "bare" meson m_0 (the $q\bar{q}$ or valence quark component) and two-meson contributions such as $\pi\pi$ and $K\bar{K}$ (the $qq\bar{q}\bar{q}$ component clustered as two mesons). That is, the meson wave function can be expressed as

$$|m\rangle = \alpha|m_0\rangle + \beta|\pi\pi\rangle + \gamma|K\bar{K}\rangle + \dots \tag{1}$$

The quark model[3] puts a lot of effort into the $|m_0\rangle$ component but ignores the remaining components. Here we adopt the opposite philosophy. We assume that $|m_0\rangle$ can be parametrized simply in terms of a bare mass. We then put our major effort into describing the 2-meson components as well as possible. Clearly the ideal approach would incorporate both ideas in a unified manner. However, the present approach plays an important role in describing cases where the quark model fails. A case in point is the $f_0(975)$, which we find can be described with no contribution from $|m_0\rangle$, a small but non-negligible component from $|\pi\pi\rangle$ and the major part from $|K\bar{K}\rangle$. That is, it emerges as a $K\bar{K}$ bound state or molecule.

2 $\pi\pi$ scattering

The starting point for our calculation is an effective meson Lagrangian such as that of Refs. [5, 6, 7] The exact solution to the field theory would then be obtained by solving the Bethe-Salpeter equation with a complete kernel (i.e., including *all* two-particle irreducible diagrams in step (c) above). This is of course impossible in practice so we make two approximations. First, the infinite set of two-particle irreducible diagrams constituting the potential is truncated to include only t- and s-channel meson exchanges. Secondly, we utilise either the Time Ordered Perturbation Theory (TOPT) or Blankenbeclar-Sugar (BbS) approaches to reduce the dimensionality of the integral equation from four to three, which makes it more amenable to solution[1].

The integral equation that we solve for the t-matrix, T, is expressed in operator form as

$$T = V + VGT, \qquad (2)$$

where G is the two-body propagator. This is to be understood as a 2×2 matrix equation to account for couplings between the $\pi\pi$ and $K\bar{K}$ channels.

The relevant t-channel meson exchange diagrams that will contribute to the $\pi\pi$ potential V are (a) ρ exchange between $\pi\pi$ and $\pi\pi$ states, (b) $K^*(890)$ exchange between $\pi\pi$ and $K\bar{K}$ states, and (c) ρ, ω and ϕ exchange between $K\bar{K}$ and $K\bar{K}$ states. These are illustrated in Figs. 1(a)–1(c). To these we must also add

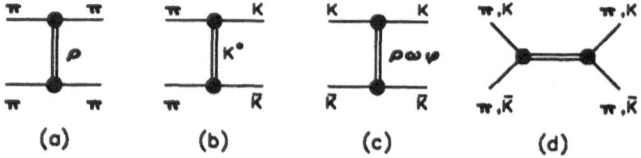

(a)   (b)   (c)   (d)

Figure 1. Meson exchange contributions to the coupled $\pi\pi - K\bar{K}$ system. (a)–(c) t-channel. (d) s-channel.

s-channel pole diagrams as illustrated in Fig. 1(d). These correspond to the genuine $q\bar{q}$ states while the t-channel exchanges provide the (possibly strong) background. In solving the integral equation, many diagrams are generated that renormalise both the mass and coupling constants of these s-channel poles (this

---

[1]Reference [1] utilised the TOPT approach. However, in order to obtain the correct analytic behaviour of the scalar form factor in the unphysical region[8], it was necessary to redo the calculation using BbS. Both methods gave similar results in the physical region after some readjustment of cutoff parameters. In particular, the conclusions regarding the nature of the $f_0(975)$ were unchanged. Here the $\pi\pi$ results use BbS and the $K\pi$ results use TOPT.

can be seen by iterating Eq. (2) with the potential of Fig. 1). One example of such a diagram is given in Fig. 2. Hence, the physically observed state is a combination

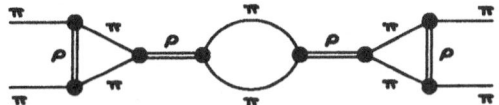

Figure 2. An example of a diagram generated by solving Eq. (2) that renormalises Fig. 1(d)

of bare $q\bar{q}$ state and two-meson dressing. Rather than explicitly calculating the mass shift and change in the coupling constant due to renormalisation, we introduce the bare mass and bare coupling constant as free parameters which are adjusted to fit the resonant phases.

We consider first the scalar-isoscalar channel with $I^G(J^{PC}) = 0^+(0^{++})$. This channel contains a narrow resonance, the $f_0(975)$ which was once identified as a member of the scalar $q\bar{q}$ nonet, but it is now generally agreed that its mass and decay properties are inconsistent with that assignment.

The phase shifts and inelasticities resulting from the meson-exchange coupled channel calculation for this partial wave are shown in the solid curves of Fig. 3. To obtain complete agreement with the empirical $\pi\pi$ phase shifts, it is necessary to include an s-channel pole corresponding to the broad $f_0(1400)$. The bare mass and coupling constant are taken as free parameters which are adjusted to fit the data. If we calculate $K\bar{K}$ scattering with the potential given only by Fig. 1(c) (i.e., no coupling to the $\pi\pi$ channel) then we find typical bound state phase shifts. This confirms that the structure observed in Fig. 3 is due to a bound $K\bar{K}$ state. It comes about because the ρ, ω and ϕ exchange contributions of Fig. 1(c) are each attractive, providing enough attraction to bind. The dashed curve is the result of turning off the diagram of Fig. 1(c), which, since Fig. 1(c) is the mechanism for creating the bound state, illustrates the strong background to the $f_0(975)$ provided by t-channel exchanges and the $f_0(1400)$.

The partial waves with isospin 2 and spin 0 and 2 ($2^+(0^{++})$ and $2^+(2^{++})$) provide a useful check on the model. There are no experimentally observed resonances in these channels, so the phase shifts are completely determined within the model by the t-channel processes, whose parameters have already been constrained by the $0^+(0^{++})$ channel. Also, isospin 2 means there is no coupling to the $K\bar{K}$ channel. The results in Fig. 4 show excellent agreement with the data.

Results for the partial waves $1^+(1^{--})$ and $0^+(2^{++})$ are shown in Fig. 5. These waves contain the ρ and the $f_2(1270)$ respectively, which both decay predominantly to $\pi\pi$. In both cases it was necessary to include an s-channel pole (corresponding to a genuine $q\bar{q}$ state) in order to reproduce the data. The results of just the t-channel processes are indicated in the figures by the dashed lines.

3 $K\pi$ scattering

It is straightforward to extend the model of the $\pi\pi$ system to the $K\pi$ interaction. Since all of the t-channel exchange processes are completely determined by the $\pi\pi$ system, it provides a useful consistency check of the model. Only the bare masses and coupling constants of the s-channel poles remain to be adjusted in the cases where they are necessary. This system is somewhat simpler than the

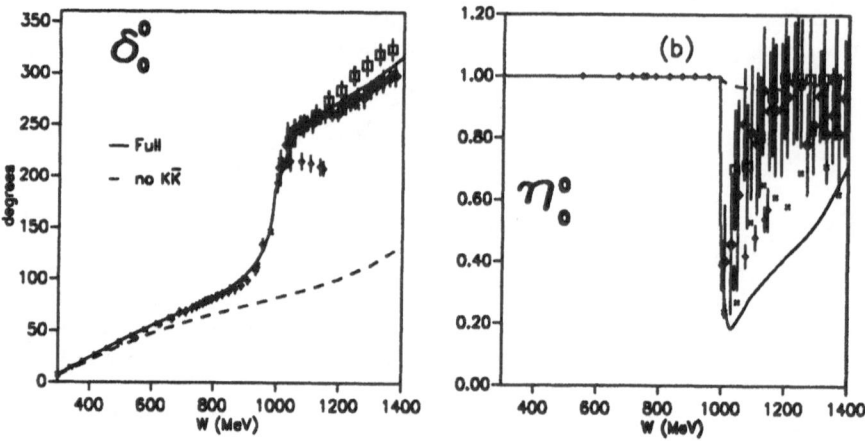

Figure 3. Results for the $\pi\pi$ phase shifts (a) and inelasticity (b) in the $I^G(J^{PC}) = 0^+(0^{++})$ partial wave. For references to the data, see Ref. [1].

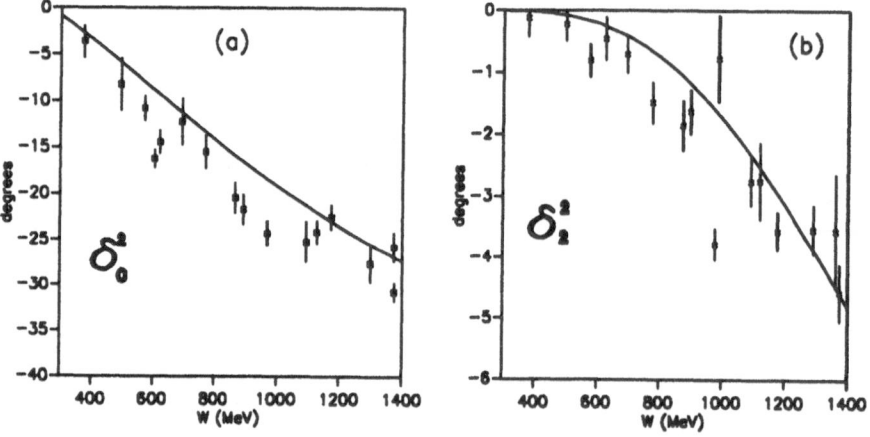

Figure 4. Results for the non-resonant $\pi\pi$, $T = 2$ partial waves. (a) $I^G(J^{PC}) = 2^+(0^{++})$ and (b) $2^+(2^{++})$. For references to the data, see Ref. [1].

$\pi\pi$ case since there are no coupled channels to consider in the energy region of interest.

In Fig. 6 we show the results for the resonant partial waves, $I(J^P) = \frac{1}{2}(0^+)$ and $\frac{1}{2}(1^-)$. These correspond respectively to the $K_0^*(1430)$ and $K^*(892)$ states listed by the Particle Data Group. The dashed curves are the results using only the t-channel exchange driving terms while the solid curves include s-channel states (with bare masses and couplings adjusted to reproduce the data as usual). The $K_0^*(1430)$ is of special interest since it is a member of the lowest lying scalar nonet which we have just demonstrated should not contain the $f_0(975)$. In fact, as can be seen, the model requires a genuine $q\bar{q}$ state in order to agree with experiment. However, the t-channel exchanges provide a strong, non-negligible

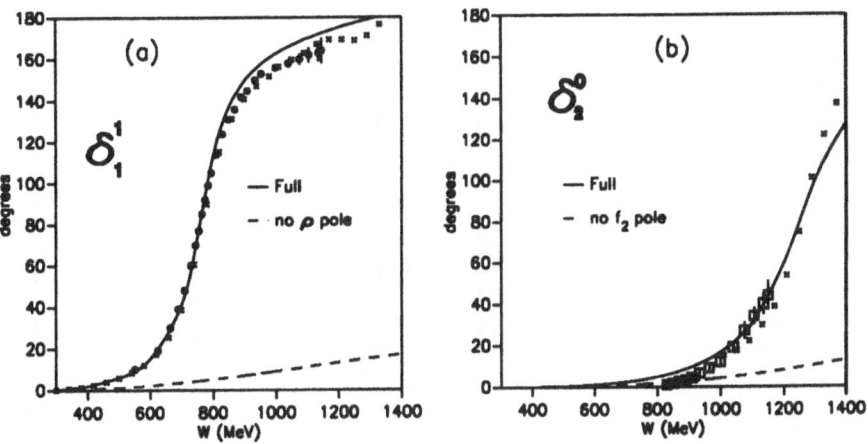

Figure 5. Results for the resonant $\pi\pi$ (a) $I^G(J^{PC}) = 1^+(1^{--})$ and (b) $0^+(2^{++})$ partial waves. For references to the data, see Ref. [1].

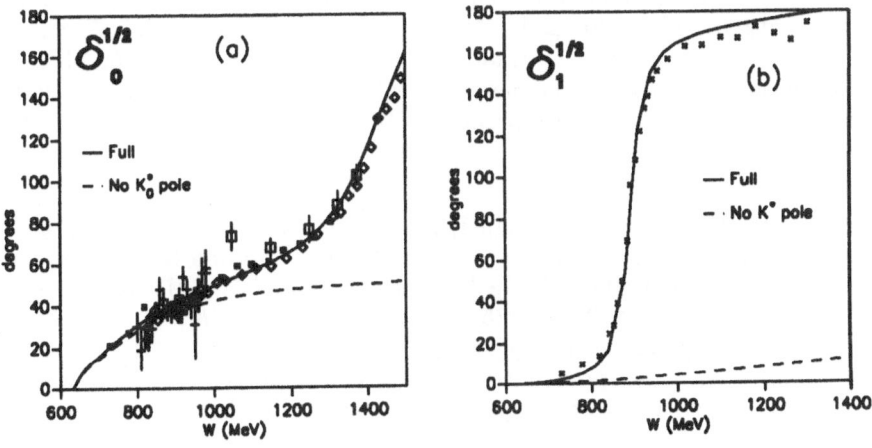

Figure 6. Results for the resonant (a) $I(J^P) = \frac{1}{2}(0^+)$ and (b) $\frac{1}{2}(1^-)$ partial waves of the $K\pi$ system. For references to the data, see Ref. [1].

background. Similarly, the $\frac{1}{2}(1^-)$ requires an s-channel pole term in order to reproduce the vector $K^*(892)$, although the background is much less in that case.

Once again, the non-resonant partial waves provide a useful check of the self-consistency of the model. In this case, all the parameters are determined in the $\pi\pi$ sector. As can be seen in Fig. 7, the results are very reassuring.

It is clear that the model described is very successful at describing the low-lying states that couple strongly to $\pi\pi$ and $K\pi$. The $f_0(975)$, about which there is some controversy, emerges as a $K\bar{K}$ bound state (or molecule). This interpretation agrees in spirit (although not in the details) with the non-relativistic quark model calculation of Weinstein and Isgur[4] but is at variance with the

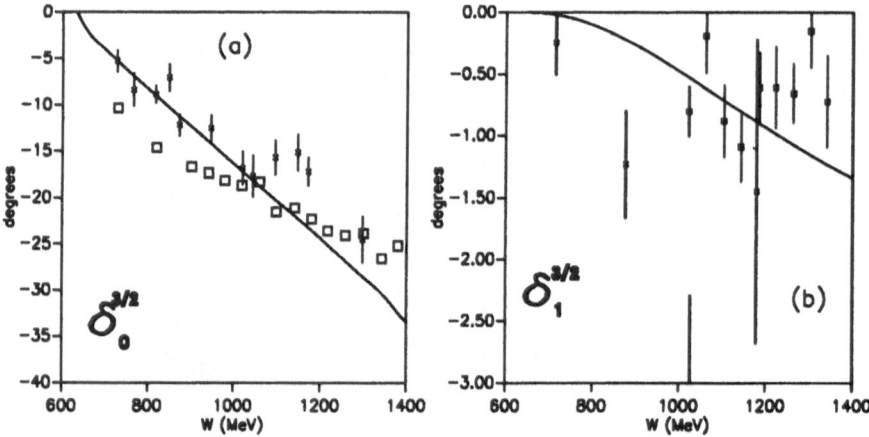

Figure 7. Results for the non-resonant (a) $I(J^P) = \frac{3}{2}(0^+)$ and (b) $\frac{3}{2}(1^-)$ partial waves of the $K\pi$ system. For references to the data, see Ref. [1].

analysis of Morgan and Pennington[9]. All other resonance states examined ($\rho(770)$, $f_2(1270)$, $K_0^*(1430)$ and $K^*(892)$) required the inclusion of s-channel poles. It was also necessary to include a state corresponding to the $f_0(1400)$. To extend the calculation to higher energies it will be necessary to include coupling to other channels (for example, $\rho\rho$ and $N\bar{N}$).

4 Conclusions

The model of the $\pi\pi - K\bar{K}$ system[1] I have reviewed is driven by s- and t-channel meson exchanges with the vertices obtained from an SU(3) symmetric Lagrangian. Only the well-established mesons (ρ, ω, ϕ, $K^*(892)$) are used in the t-channel driving terms. The $f_0(975)$ then emerges as a $K\bar{K}$ bound state. All other resonances considered (ρ, $f_2(1270)$, $K^*(892)$ and $K_0^*(1430)$) required the introduction of s-channel poles corresponding to genuine $q\bar{q}$ states. It was also necessary to include a state corresponding to the $f_0(1400)$.

References

[1] D. Lohse, J.W. Durso, K. Holinde and J. Speth, Nucl. Phys. **A516** (1990) 513.

[2] G. t'Hooft, Nucl. Phys. **B72** (1974) 461; E. Witten, Nucl. Phys. **B160** (1979) 57.

[3] S. Godfrey and N. Isgur, Phys. Rev. D **32** (1985) 189.

[4] J. Weinstein and N. Isgur, Phys. Rev. D **41** (1990) 2236.

[5] M. Bando, T Kugo, S. Uehara, K. Yamawaki and T. Yanagida, Phys. Rev. Lett. **54** (1985) 1215.

[6] S. Weinberg, Phys. Rev. **166** (1968) 1568.

[7] J. Schwinger, Phys. Lett. **24B** (1967) 473.

[8] B.C. Pearce, K. Holinde and J. Speth, Nucl. Phys. **A541** (1992) 663.

[9] D. Morgan and M.R. Pennington, Phys. Lett. **B258** (1991) 444.

[10] J.F. Donoghue, J. Gasser and H. Leutwyler, Nucl. Phys. **B343** (1990) 341.

[11] J. Gasser, H. Leutwyler and M.E. Sainio, Phys. Lett. **B253** (1991) 260.

Production of $Q\bar{Q}g$ hybrid mesons in $N\bar{N}$ annihilation

T. GUTSCHE

Institut für Theoretische Physik, Universität Tübingen,
Auf der Morgenstelle 14, 7400 Tübingen, F.R. Germany

Abstract: Nucleon-antinucleon $(N\bar{N})$ annihilation reactions provide a copious source for the production of gluons (g) through the destruction of quark-antiquark $(Q\bar{Q})$ pairs in the initial state. Therefore, in addition to the production of the usual $Q\bar{Q}$ meson states, $N\bar{N}$ annihilation can serve as an entrance channel to mesonic resonances containing dynamical excitations of the gluon field, for instance hybrids $(Q\bar{Q}g)$. We discuss the main features of the spectrum and decay modes for these hybrid mesons. Particularly, we focus on the production mechanism in $N\bar{N}$ annihilation, thereby employing a microscopic model with constituent quarks and gluons. The interpretation of the E(1420), observed in the reaction $p\bar{p} \rightarrow \pi^+\pi^- E$ at rest, as a hybrid meson is investigated.

1. Introduction:

Our understanding of hadronic structure is to a large extent based on the constituent quark model where mesons are made up of a valence quark-antiquark pair $(Q\bar{Q})$ and baryons are made up of three valence quarks. However, a fundamental aspect of QCD is that coloured gluons exist in addition to the well-established coloured quarks. If quarks are permanently confined to colour-singlet clusters, then it is to be expected that gluons will also be confined. As a result, new types of hadronic matter are predicted, among them the glueballs (no valence quarks) and the hybrids with a constituent quark-antiquark pair and an excited glue degree of freedom. Therefore, the conventional spectrum of $Q\bar{Q}$ mesons is expected to be supplemented by new objects containing dynamical excitations of the gluon field g. Already there are a number of experimental candidates, which do not fit into the usual SU(3) $Q\bar{Q}$ nonets. For a review of this situation see ref. [1]. In the following we discuss the potentialities of nucleon-antinucleon $(N\bar{N})$ annihilation as means of producing $Q\bar{Q}g$ hybrid states.

2. Spectrum of $Q\bar{Q}g$ hybrids:

Estimates for the mass spectrum of hybrids $(Q\bar{Q}g)$ have been carried out in a variety of models, for example the bag model [2], the flux tube model [3], QCD sum rules [4] and lattice QCD [5]. For a recent review concerning the hybrid spectrum see [6].

In the Bag Model [2] the lowest lying hybrid states correspond to the coupling of a colour octet $Q\bar{Q}$ pair ($^{2S+1}L_J = {}^1S_0, {}^3S_1$) to a transverse electric (TE) mode of the constituent gluon ($J^{\pi C} = 1^{+-}$). Hybrid states involving the lowest transverse magnetic (TM) mode of the gluon lie approximately 500 MeV higher. The states formed by coupling a $Q\bar{Q}$ pair to a TE gluon are

$$[(Q\bar{Q})_{^3S_1} \otimes TE]_{J^{\pi C}=0-+,1-+,2-+} \quad and \quad [(Q\bar{Q})_{^1S_0} \otimes TE]_{J^{\pi C}=1--} \tag{1}$$

with each state coming in four flavour combinations (according to the convention of the Particle Data Group [23] these states are denoted by $\hat{\rho}$, $\hat{\omega}$, $\hat{K}^*$, $\hat{\phi}$ for the $J^{\pi C}$ multiplets of $Q\bar{Q}(^3S_1) \otimes TE$, correspondingly for the multiplets of $Q\bar{Q}(^3S_1) \otimes TE$).

The principal signatures of the hybrid mass spectrum are that for a given combination of $J^{\pi C}$, $\hat{\rho}$ lies lowest with an isospin splitting $M(\hat{\omega}) - M(\hat{\rho}) \approx 100 - 150\ MeV$, while the $\hat{\omega}$ and the $\hat{K}^*$ are nearly degenerate. The lowest lying hybrids (with $J^{\pi C} = 0^{-+}, 1^{-+}$) are predicted to lie in a mass range of 1.3-1.6 GeV/c^2. The overall mass scale depends sensitively on the self-energy of the TE mode, however not the mass splittings between the flavor combinations. Hybrids with $J^{\pi C} = 1^{-+}$ have true "exotic" quantum numbers, i.e. quantum numbers which can not be formed in the usual $Q\bar{Q}$ systems. An experimental observation of such exotic quantum numbers would be a signal for the occurrence of mesonic resonances beyond the $Q\bar{Q}$ system. Particularly hybrid states with exotic quantum numbers are expected to occur in a mass range below 2 GeV/c^2, whereas other possible exotic systems (four quark $Q^2\bar{Q}^2$ states and glueballs for instance) are predicted to lie much higher [7].

Alde et al. [8] have presented first evidence for such a meson with $J^{\pi C} = 1^{-+}$ near 1.4 GeV/c^2 in the $\pi^- p \to \pi^0 \eta n$ reaction at 100 GeV. See however ref. [9] for a critical discussion of the data. The possible interpretation of this state as a $\hat{\rho}(1^{-+})$ hybrid was discussed by Iddir et al. [10].

3. Decays of $Q\overline{Q}g$ hybrids:

Branching ratios for the decay of hybrids have been estimated in the constituent gluon picture [10,11] and the flux tube model [3]. In the constituent quark-gluon picture the decay of a $Q\bar{Q}g$ hybrid is provided through destruction of glue in the TE mode into a $Q\bar{Q}$ pair as indicated in the effective diagram of fig. 1a. The transition amplitude for the process of fig. 1a is defined as

$$T = <\Psi_{M_1 M_2}|\mathcal{O}_{^3S_1}|\Psi_{Q\bar{Q}g}> \tag{2}$$

with the quark-antiquark-gluon transition operator given by

$$\mathcal{O}_{^3S_1} = g\vec{\sigma}^{Q\bar{Q}} \cdot \vec{\epsilon}_g \delta^3(\vec{k}_Q + \vec{k}_{\bar{Q}} - \vec{k}_g) \tag{3}$$

where $\vec{\epsilon}_g$ is the polarization vector of the gluon. The strength of the transition is given by g. The transition operator corresponds to the nonrelativistic quark-

Fig.1: Mechanisms for the decay of a $Q\bar{Q}g$ hybrid or a $Q\bar{Q}$ meson are shown in (a) and (b). The production mechanisms for $Q\bar{Q}g + Q\bar{Q}$ or $Q\bar{Q} + Q\bar{Q}$ in $N\bar{N}$ annihilation are indicated in (c) and (d). The strenghts of the $^3P_0$ $Q\bar{Q}$ and the $^3S_1$ $Q\bar{Q}g$ vertices, indicated by dots, are denoted by λ and g, respectively.

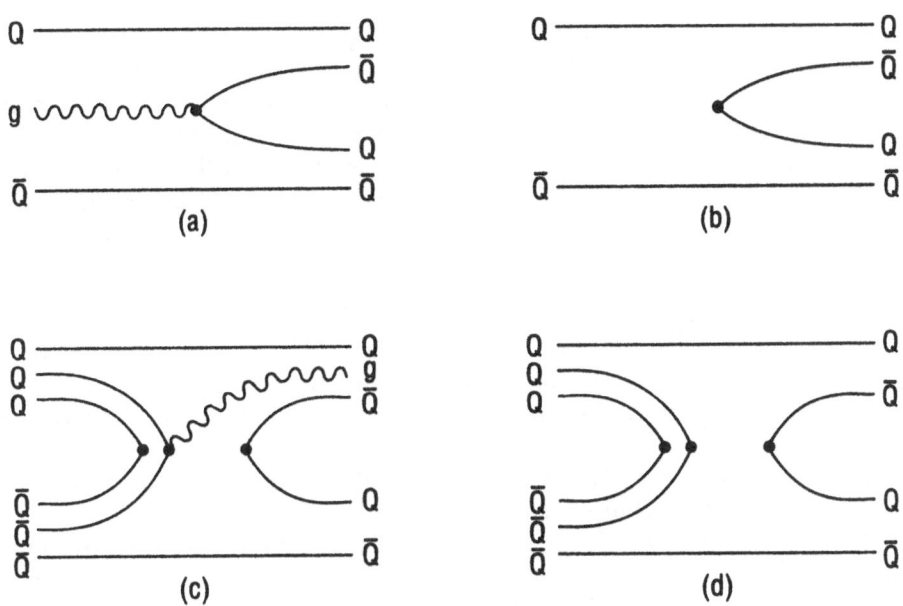

antiquark-gluon interaction of lowest order QCD, where the created $Q\bar{Q}$ pair carries the quantum numbers $^{2S+1}L_J = {}^3S_1$. The final two-meson state is described by SU(6) quark cluster wave functions, whereas a three-body wave function is used for the $Q\bar{Q}g$ system. The nonrelativistic gluon wave function is in simple correspondence to the relativistic bag-model one, thereby taking care that the gluon has the correct physical polarization according to the boundary condition. For details in constructing nonrelativistic gluon wave functions see ref. [11] and the articles referenced therein. For $Q\bar{Q}g$ hybrids involving a 1^{+-} TE gluon, we obtain the approximate selection rule:

$$(Q\bar{Q})_{\ell=0} \otimes TE \rightarrow sp(\ell_f = 0)$$

$$\nrightarrow sp(\ell_f = 2)$$

$$\nrightarrow ss(\ell_f = 1) \tag{4}$$

where s denotes a s-wave meson $(Q\bar{Q})_{\ell=0}$, p a p-wave meson $(Q\bar{Q})_{\ell=1}$ and ℓ_f is the relative orbital angular momentum between the final mesons. The orbital angular momentum p-wave excitation of the gluon is transferred as an internal orbital excitation of a $Q\bar{Q}$ meson instead of a relative p-wave between two s-wave mesons. The above selection rules in the decay of hybrid mesons both occur in the constituent gluon picture as well as in the flux tube picture. Therefore the favoured final states in the decay of hybrids all contain broad p-wave mesons.

3. The E(1420) as a $Q\overline{Q}g$ hybrid ?

Recent experimental results of the ASTERIX group [12] confirm the earlier established [13] resonance of the E(1420) meson. They investigated the reaction $p\bar{p} \rightarrow \pi^+\pi^- K^\pm\pi^\mp K^0$ with a gaseous hydrogen target (mixture of initial atomic S and P states), and observe a peak in the $K^\pm\pi^\mp K^0$ invariant mass spectrum corresponding to the E(1420). By comparing with the earlier bubble chamber experiment [13] (dominantly atomic S-wave annihilation), Duch et al. [12] assign the quantum numbers $J^{\pi C}(I^G) = 0^{-+}(0^+)$ to the E(1420), with the dominant transition

$$p\bar{p}(^{11}S_0) \rightarrow [\sigma \otimes E]_{\ell=0} \tag{5}$$

corresponding to s-waves in the final state ($\sigma \equiv (\pi\pi)_{\ell=0,I=0}$). In the decay analysis of the E(1420) it is suggested that $E \rightarrow \pi a_0(980)$ is the dominant $K^\pm\pi^\mp K^0$ decay mode, although $\bar{K}K^\star$ is not ruled out at the 20-30% level. The E(1420) carries the quantum numbers of the η or η', however it is too low in mass to be a radial excitation. Therefore, the E(1420) is suggested to be a dominant candidate for a non-$Q\bar{Q}$ meson.

In radiative J/ψ decays [14] a resonance, called $\eta(1440)$, with the same quantum numbers, with dominant $K\bar{K}\pi$ decay mode and approximately the same mass and width has been seen. However, it is not evident if the $\eta(1440)$ seen in J/ψ decays is the same resonance seen in $N\bar{N}$ annihilation [1]. The $\eta(1440)$ is often considered to be a glueball candidate [15].

The two-body decay modes of the E(1420) allowed by general $J^{\pi C}(I^G)$ conservation are

$$E \rightarrow \rho\rho(\ell_f = 1), \omega\omega(\ell_f = 1), \bar{K}K^\star + c.c.(\ell_f = 1), \pi a_0(\ell_f = 0),$$

$$\pi a_2(\ell_f = 2), \eta\sigma(\ell_f = 0). \tag{6}$$

In the observed $K\bar{K}\pi$ mode, both πa_0 and $\bar{K}K^\star$ compete. The experimentally observed preference for $E \rightarrow \pi a_0$ over $\bar{K}K^\star + c.c.$ is naturally explained by the decay mechanism of fig. 1a in accordance with the approximate selection rules of eq. (4). Due to the experimentally observed strong selection rule in the decay mechanism it seems reasonable to interpret the E(1420) as a hybrid meson of the type $\hat{\omega}(0^{-+}) = [\omega \otimes TE]_{0-+}$. The observed mass of the E is in accordance with the range of predictions as derived in the Bag Model [2]. For the proposed decay mechanism of fig. 1a we obtain a relative branching ratio [16]

$$B(E \rightarrow \pi a_0)/B(E \rightarrow \eta\sigma) \approx 8 , \tag{7}$$

while the contribution of the other modes of eq. (6) are zero for equal meson radii. This large ratio can also explain why the decay $E \rightarrow \eta\sigma \rightarrow \eta\pi\pi$ was not seen in the experiment by Ando et al. [17].

The branching ratio for the production of the E measured in the bubble chamber experiment [13] (dominantly S-wave (L=0) annihilation) is

$$B(p\bar{p}(L=0) \to \pi^+\pi^- E) \cdot B(E \to K_s K^\pm \pi^\mp) \approx (7.1 \pm 0.35) \times 10^{-4} . \qquad (8)$$

Duch et al. [12] give a branching ratio of $(3 \pm 0.9) \times 10^{-4}$ for a gaseous hydrogen target (mixed L=0,1 initial state, $\sim 60\%$ $L = 1$, $\sim 40\%$ $L = 0$). This is consistent with a small production rate of the E from an L=1 $N\bar{N}$ state. Using the estimate $B(E \to \pi a_0) \approx 0.9$ (as derived from eq. (7)) we obtain $B(E \to K_s K^\pm \pi^\mp) \approx 0.1$. Taking this number at face value and combining it with the experimental result of the bubble chamber experiment eq. (8) we obtain a branching ratio for the E meson production via $p\bar{p}$ annihilation of

$$B(p\bar{p}(L=0) \to (\pi^+\pi^-)_{\ell=0}E) \approx 7 \times 10^{-3}. \qquad (9)$$

under the assumption that $(\pi^+\pi^-)_{\ell \geq 1}E$ is negligible. E and η carry the same quantum numbers, therefore it is reasonable to compare the branching ratio of eq. (9) to the branching ratio of ordinary meson production $(\pi^+\pi^-)_{\ell=0}\eta$. The latter branching ratio is obtained by subtracting the $\rho\eta$ piece from the total $\pi^+\pi^-\eta$ rate and results in the experimental value [18]

$$B(p\bar{p}(L=0) \to (\pi^+\pi^-)_{\ell=0}\eta) \approx 6 \times 10^{-3} . \qquad (10)$$

Hence we obtain the experimental ratio

$$R_{exp} = \frac{B(p\bar{p}(L=0) \to (\pi^+\pi^-)_{\ell=0}E)}{B(p\bar{p}(L=0) \to (\pi^+\pi^-)_{\ell=0}\eta)} \approx 1 . \qquad (11)$$

To investigate the theroretical interpretation of the E as a $\hat{\omega}(0^{-+})$ hybrid [16], we estimated the ratio (11). In $N\bar{N}$ annihilation gluons can be produced through the destruction of $Q\bar{Q}$ pairs in the initial state. The effective gluon produced can be in its ground state (the $0^{++}(0^+)$ condensate, corresponding to the $^3P_0$ vertex) or in an excited state (the TE mode, for instance, corresponding to the $^3S_1$ vertex). Meson decays [19], corresponding to fig. 1b, and $N\bar{N}$ annihilation into two mesons (the A2 model of ref. [20]), corresponding to fig. 1d, are well described phenomenologically using the effective $^3P_0$ vertex; this approximation has been given a basis in strong-coupling QCD [3,21]. In correspondence to the usual meson decay and $N\bar{N}$ annihilation into mesons we consider analogous graphs for the hybrid decay (fig. 1a) and the hybrid production (fig. 1c), replacing the $^3P_0$ $Q\bar{Q}$ vertex by the $Q\bar{Q} \to TE$ $^3S_1$ vertex. The strength g of the $Q\bar{Q} \to TE$ vertex is adjusted to fit the experimental total width $\Gamma \approx 62 MeV$ of the E meson; the strength λ of the $^3P_0$ $Q\bar{Q}$ vertex is chosen such as to reproduce the $f_2(1270) \to \pi\pi$ decay width. The same strength parameters are used, when estimating the rates for $(\pi\pi)_{\ell=0}E = \sigma E$ and $\sigma\eta$ production in $N\bar{N}$ annihilation (σ is described by a $(Q\bar{Q})_{\ell=1}$ pair with vacuum quantum numbers). Therefore our calculation contains no free parameters. Although the Born approximation of figs. 1c, 1d is not valid in obtaining absolute values for the branching ratios, it is reasonable to use it for estimating the ratio of rates involving the same inital and final state quantum numbers, thus cancelling out the effect of initial state interaction. Depending sensitively on the choice of radial parameters, we obtain values for R which are very small

$$R_{theor} = 10^{-8} - 10^{-6} . \tag{12}$$

The small values of eq. (12) arise from a suppression of the spatial overlap and of the spin-flavor-color recoupling coefficients for σE production relative to that for $\sigma \eta$.

Although the dominance of the πa_0 over the $K \bar{K}^\star$ decay mode is consistent with the identification of the E as a $\hat{\omega}(0^{-+}(0^+))$ hybrid meson, the theoretical estimate for the production branching ratio of the E, relative to the η, is much smaller than the experimental value of eq. (11). Therefore, the interpretation of the E as a hybrid meson appears rather dubious.

4. Production of $Q\overline{Q}g$ hybrids in $N\overline{N}$ annihilation:

The theoretical result for the production rate of the E meson, when interpreting it as a hybrid meson, might suggest that the production of hybrid mesons in $N\bar{N}$ annihilation is a truely suppressed process. However the ratio of $Q\bar{Q}g$ to ordinary $Q\bar{Q}$ production is rather sensitive to the quantum numbers of the mesons and the initial $N\bar{N}$ partial wave involved.

The transition amplitude T describing the hybrid production in $N\bar{N}$ annihilation corresponding to fig. 1c is given by

$$T = < \Psi_{(Q\bar{Q})(Q\bar{Q}g)}|\mathcal{O}_{^3S_1}\mathcal{O}_{^3P_0}\mathcal{O}_{^3P_0}\mathcal{O}_{^3P_0}|\Psi_{N\bar{N}} > \tag{13}$$

where the $^3P_0$ operators $\mathcal{O}_{^3P_0}$ and the $^3S_1$ operator $\mathcal{O}_{^3S_1}$ act on the respective quark, antiquark and gluon components in the initial and final state. The $Q\bar{Q}$ $^3P_0$ vertex is defined in momentum space as

$$\mathcal{O}_{^3P_0} = \lambda \vec{\sigma}^{Q\bar{Q}} \cdot (\vec{k}_Q - \vec{k}_{\bar{Q}})\delta^3(\vec{k}_Q + \vec{k}_{\bar{Q}}) \tag{14}$$

Again we assume SU(6) wave functions for N, $\bar{N}$, $Q\bar{Q}$ and $Q\bar{Q}g$. For each transition $N\bar{N} \rightarrow (Q\bar{Q}) + (Q\bar{Q}g)$ with final c.m. momentum q, we obtain for the production rate Γ

$$\Gamma = C_q F(q) SFC[(Q\bar{Q})(Q\bar{Q}g)] \tag{15}$$

where F(q) is a form factor proportional to $|T|^2$ and C is a constant depending on the quantum numbers of the initial $N\bar{N}$ state. SFC is a weight factor depending on the spin-flavor-color quantum numbers of the initial and final state. Details of this calculations will be given in ref. [22]. The values of the recoupling coefficients SFC, assuming the dynamics of fig. 1c, display a significant dependence on L, the $N\bar{N}$ relative orbital angular momentum, spin S and isospin I and the final state $(Q\bar{Q})(Q\bar{Q}g)$ combination. In tables 1 and 2 we list transitions of the

Table 1: Dynamical selection rules for TE hybrid production in $N\bar{N}(L = 0)$ annihilation. The subscript at $[Q\bar{Q} \otimes Q\bar{Q}g]$ denotes the spin coupling in the final state. The final state relative orbital angular momentum $l_f = 1$ is determined by parity conservation.

Dynamically forbidden channels in the transition $N\bar{N}(L = 0) \rightarrow (Q\bar{Q})_{\ell=0} + [(Q\bar{Q})_{\ell=0} \otimes TE]\ (l_f = 1)$
$^{31}S_0 \not\rightarrow [\pi^\pm \otimes \hat{\pi}^\mp(1^{--})]_1$
$^{13}S_1 \not\rightarrow [\omega \otimes \hat{\omega}(1^{-+})]_{0,1},\ \ [\rho \otimes \hat{\rho}(1^{-+})]_{0,1}$
$^{33}S_1 \not\rightarrow [\rho^0 \otimes \hat{\omega}(1^{-+})]_0,\ \ [\omega \otimes \hat{\rho}^0(1^{-+})]_0$

Table 2: Dynamical selection rules for TE hybrid production in $N\bar{N}(L = 1)$ annihilation. The final state relative orbital angular momentum $l_f = 0$ is determined by parity conservation.

Dynamically forbidden channels in the transition $N\bar{N}(L = 1) \rightarrow (Q\bar{Q})_{\ell=0} + [(Q\bar{Q})_{\ell=0} \otimes TE]\ (l_f = 0)$
$^{11}P_1 \not\rightarrow \eta\,\hat{\eta}(1^{--}),\ \pi\,\hat{\pi}(1^{--})$
$^{31}P_1 \not\rightarrow \pi^0\,\hat{\eta}(1^{--}),\ \eta\,\hat{\pi}^0(1^{--})$
$^{13}P_1 \not\rightarrow \omega\,\hat{\eta}(1^{--}),\ \rho\,\hat{\pi}(1^{--}),\ \eta\,\hat{\omega}(1^{-+}),\ \pi\,\hat{\rho}(1^{-+})$
$^{33}P_0 \not\rightarrow \rho^\pm\,\hat{\rho}^\mp(1^{-+})$

type $N\bar{N}(L = 0,1) \rightarrow (Q\bar{Q})_{\ell=0} + [Q\bar{Q}_{\ell=0} \otimes TE]$ forbidden by the dynamics of fig. 1c, which are otherwise allowed by conservation of $J^{\pi C}(I^G)$. Due to phase space considerations, the relevant channels in hybrid production from the $p\bar{p}$ atom, as studied by the Crystal Barrel collaboration at LEAR [24], are of the type $N\bar{N}(L = 0,1) \rightarrow \pi + (Q\bar{Q}g)$. Tables 3 and 4 give a list of the weight factors SFC for all such transitions allowed by general conservation of $J^{\pi C}(I^G)$.

Out of the possible transitions in $N\bar{N}(L = 0)$ annihilation of table 3 we considered the interesting case of exotic $\hat{\rho}^0(1^{-+})$ production [16]. For the $\hat{\rho}$ we take a mass of 1.4 GeV/c^2 from Alde et al.[8]. In analogy to the estimate for the E

Table 3: Spin-flavor-color weight factors SFC for the transition $N\bar{N}(L = 0) \rightarrow \pi + (Q\bar{Q}g)$. Note that the zero in the entry of SFC corresponds to the forbidden transition of table 1.

Transition	SFC
$^{11}S_0 \rightarrow \pi^0 \hat{\rho}^0(1^{-+})$	1.00
$^{31}S_0 \rightarrow \pi^\pm \hat{\pi}^\mp(1^{--})$	0
$\pi^0 \hat{\omega}(1^{-+})$	2.78
$^{13}S_1 \rightarrow \pi^0 \hat{\pi}^0(1^{--})$	1.57
$^{33}S_1 \rightarrow \pi^\pm \hat{\rho}^\mp(0^{-+})$	0.25
$\pi^\pm \hat{\rho}^\mp(1^{-+})$	0.67
$\pi^\pm \hat{\rho}^\mp(2^{-+})$	0.73
$\pi^0 \hat{\eta}(1^{--})$	0.54

Table 4: Spin-flavor-color weight factors SFC for the transition $N\bar{N}(L = 1) \rightarrow \pi + (Q\bar{Q}g)$. Note that the zeros in the entries of SFC correspond to the forbidden transitions of table 2. No charge labels indicate, that all charged modes are allowed.

Transition	SFC
$^{13}P_0 \rightarrow \pi^0 \hat{\rho}^0(0^{-+})$	1.00
$^{33}P_0 \rightarrow \pi^0 \hat{\omega}(0^{-+})$	0.003
$^{31}P_1 \rightarrow \pi^\pm \hat{\rho}^\mp(1^{-+})$	0.052
$\pi^0 \hat{\eta}(1^{--})$	0
$^{13}P_1 \rightarrow \pi \hat{\rho}(1^{-+})$	0
$^{33}P_1 \rightarrow \pi^0 \hat{\omega}(1^{-+})$	0.037
$\pi^\pm \hat{\pi}^\mp(1^{--})$	0.079
$^{11}P_1 \rightarrow \pi \hat{\pi}(1^{--})$	0
$^{13}P_2 \rightarrow \pi^0 \hat{\rho}^0(2^{-+})$	0.040
$^{33}P_2 \rightarrow \pi^0 \hat{\omega}(2^{-+})$	0.016

meson production of section 3, we applied the model of figs. 1c, 1d to obtain the ratio

$$\frac{B(p\bar{p}(L=0) \to \pi^0 \; \hat{\rho}^0)}{B(p\bar{p}(L=0) \to \eta \; \sigma)} \approx 0.02 - 0.1 \; . \tag{16}$$

The initial $^{11}S_0$ state is the same for both final states of eq. (16), therefore effects of initial state interaction tend to cancel. Applying eq. (10), the $\pi^0\hat{\rho}^0(1^{-+})$ branching ratio is of the order $(2-9)\times 10^{-4}$, which still could be measured at LEAR. The allowed nonstrange decay channels of the $\hat{\rho}(1^{-+})$ are

$$\hat{\rho}(1^{-+}) \quad \to \quad \pi^0\eta(l_f = 1), \; \pi^{\pm}\rho^{\mp}(l_f = 1), \; \rho^0\omega(l_f = 1,3), \; \pi^0 f_1(1285)(l_f = 0,2),$$

$$\pi^0 f_2(1270)(l_f = 2), \; \pi^{\pm} b_1^{\mp}(1235)(l_f = 0,2) \tag{17}$$

The decay modes $\pi^0\eta$ and $\pi^0 f_2(1270)$ result in 4γ and 6γ final states, respectively, suitable for measurement by Crystal Barrel. If the selection rules of eq. (4) indeed apply, then the decay channels $\pi^0 f_1(l_f = 0)$ and $\pi^{\pm} b_1^{\mp}(l_f = 0)$ are the dominant ones, leading to 10γ (with $f_1 \to \pi^0 a_0$, $a_0 \to \pi^0\eta$) or $\pi^+\pi^- + 5\gamma$ (with $b_1^{\pm} \to \pi^{\pm}\omega$, $\omega \to \pi^0\gamma$) final states.

From tables 3 and 4 it is evident that other final state channels in the production of $Q\bar{Q}g$ hybrids are probably more dominant, assuming favourable phase space. Particularly, the production of the exotic $\hat{\omega}(1^{-+})$ in $N\bar{N}(L=0)$ annihilation and of the nonexotic $\hat{\rho}^0(0^{-+})$ in $N\bar{N}(L=1)$ annihilation are the most favoured cases. For the $\hat{\omega}(1^{-+})$ the allowed nonstrange decay channels are

$$\hat{\omega}(1^{-+}) \quad \to \quad \pi\pi(l_f = 1), \; \eta\eta(l_f = 1), \; \rho\rho(l_f = 1,3), \; \omega\omega(l_f = 1,3),$$

$$\pi a_1(1260)(l_f = 0,2), \; \pi a_2(1320)(l_f = 2), \; a_0 a_0(980)(l_f = 1) \; . \tag{18}$$

Assuming the selection rules of eq. (4) the dominant decay mode is the channel $\pi a_1(l_f = 0)$ leading to a 4π final state for the decay of the $\hat{\omega}(1^{-+})$ (with $a_1 \to \rho\pi$).

Note that the production of $\pi^0\hat{\rho}^0(1^{-+})$ as calculated for S-wave annihilation expressed in eq.(16) is forbidden in P-wave annihilation assuming the model of fig. 1c. Calculations for estimating relative branching ratios analogous to eqs. (12) and (16) are in progress [22].

All transitions display a strong dependence on spin-flavor quantum numbers and on the orbital angular momentum L of the the $N\bar{N}$ state. The strong dependence on L(=0,1) should be explored at LEAR, thus displaying a striking signature for the production of hybrids. The search for hybrids in $N\bar{N}$ annihilation processes posts one of the most important challenges in the current round of experiments at LEAR.

Acknowledgement: This work was supported by a grant from the Deutsche Forschungsgemeinschaft (Fa 67/10-5).

5. References

1. L.G. Landsberg, IHEP report 89-54 (Serpukhov, 1989);
   S.U. Chung, Z. Phys. C 46 (1990) S111;
   N.N. Achasov, Nucl. Phys. B (Proc. Suppl.) 21 (1991) 189;
   T.H. Burnett and S.R. Sharpe, Ann. Rev. Nucl. Part. Sci. 40 (1990) 327.
2. T. Barnes, F.E. Close and F. deViron, Nucl. Phys. B 224 (1983) 241;
   M. Chanowitz and S.R. Sharpe, Nucl. Phys. B 222 (1983) 211.
3. N. Isgur, R. Kokoski and J. Paton, Phys. Rev. Lett. 54 (1985) 869;
   N. Isgur and J. Paton, Phys. Rev. D 31 (1985) 2910.
4. I.I. Balitsky, D. Dyakanov and A.V. Yung, Z.Phys. C 33 (1986) 265;
   J.I. Latorre, P. Pascual and S. Narison, Z. Phys. C 34 (1987) 347.
5. N.A. Campbell, A. Huntly and C. Michael, Nucl. Phys. B 306 (1988) 51.
6. S. Godfrey, in: Glueballs, Hybrids and Exotic Hadrons, AIP Conf. Proc.
   No. 185, ed. S.U. Chung, New York (1989), p. 373.
7. C.B. Dover, in: Glueballs, hybrids and exotic hadrons, AIP Conf.
   Proc. No. 185, ed. S.U. Chung (AIP, New York, 1989), p. 251.
8. D. Alde et al., Phys. Lett. B 205 (1988) 397.
9. S.F. Tuan, T. Ferbel and R.H. Dalitz, Phys. Lett. B 213 (1988) 537.
10. F. Iddir et al., Phys. Lett. B 205 (1988) 564.
11. A. Le Yaouanc et al., Z. Phys. C 28 (1985) 315.
12. K.D. Duch et al., Z. Phys. C 45 (1989) 223.
13. P. Baillon et al., Nuovo Cim. 50 A (1967) 393.
14. L. Köpke and N. Wermes, Phys. Rep. 174 (1989) 67.
15. M. Chanowitz, Phys. Lett. B 187 (1987) 409.
16. C.B. Dover, T. Gutsche, A. Faessler and R. Vinh Mau, Phys. Lett. B 277
   (1992) 23.
17. A. Ando et al., Phys. Rev. Lett. 57 (1986) 1296.
18. C. Amsler, Nucl. Phys. A 508 (1987) 501;
   P. Weidenauer et al., Z. Phys. C 47 (1990) 353.
19. A. Le Yaouanc et al., Phys. Rev. D 8 (1973) 2223;
   D 9 (1974) 1415; D 11 (1975) 1272.
20. M. Maruyama, S. Furui and A. Faessler, Nucl. Phys. A 472 (1987) 643.
21. H.G. Dosch and D. Gromes, Phys. Rev. D 33 (1986) 1378.
22. C.B. Dover, T. Gutsche, and A. Faessler, in preparation.
23. Particle Data Group, J.J. Hernandez et al., Review of particle properties,
   Phys. Lett. B 239 (1990) p. I.7.
24. H. Koch, Nucl. Phys. B (Proc. Suppl.) 21 (1991) 87.

Higher Twist Effects in Deep Inelastic Scattering

U. Landgraf[1]

[1]Fakultät für Physik der Universität Freiburg, Hermann-Herder-Str. 3,
D-7800 Freiburg i.Br., Germany.

Abstract. Measurements of nucleon structure functions in deep inelastic lepton nucleon scattering have now reached a precision where a separation of the Q^2-dependence predicted by QCD and additional higher twist contributions becomes feasible. Analyses of the structure functions F_2 of the proton and the deuteron and of the quantity $R = \sigma_L/\sigma_T$ are presented.

1 Introduction

Ever since scaling violations in deep inelastic scattering have been detected, people have been worried to what extent they are caused by QCD processes or are a trivial consequence of multi-parton correlations, called higher twist effects. Until quite recently data from deep inelastic lepton-nucleon scattering experiments were too unprecise, however, to allow a reliable separation of these two contributions.

The nucleon structure function F_2 can be derived from the inclusive scattering cross section in charged lepton-nucleon scattering in the one-photon exchange approximation

$$\frac{d^2\sigma}{dx\,dQ^2} = \frac{4\pi\alpha^2}{Q^4}\frac{F_2(x,Q^2)}{x}\left(1 - y - \frac{xyM}{2E} + \frac{y^2}{2}\frac{1 + \frac{4M^2x^2}{Q^2}}{1 + R(x,Q^2)}\right), \qquad (1)$$

where the cross section is meant to be corrected for higher order QED processes. Here, E is the energy of the incident lepton, Q^2 is the negative of the four-momentum squared of the exchanged virtual photon, M is the nucleon mass, $x = Q^2/2M\nu$ is called the Bjorken scaling variable, ν is the energy of the virtual photon in the laboratory system, and $y = \nu/E$ is the fractional energy transfer from the lepton to the nucleon. In place of the second structure function F_1 the quantity $R(x,Q^2) = \sigma_L/\sigma_T$, the ratio of the photoabsorption cross sections for longitudinally and transversely polarized virtual photons, has been used in this formula. From helicity considerations R is exacly zero for non-interacting massless spin-1/2 quarks. It is measured to be small and the influence of the uncertainties of this measurement on the extracted value of F_2 can be neglected.

In the derivation of the above cross section formula the electromagnetic current-current interaction leads to the product of two tensors

$$\frac{d^2\sigma}{dx\, dQ^2} = \frac{2\pi\alpha^2}{Q^4}\frac{y^2}{Q^2} L_{\mu\nu} W^{\mu\nu}(p, q). \tag{2}$$

The leptonic Tensor $L_{\mu\nu}$ can be calculated in QED and is given by

$$L_{\mu\nu} = \sum_{s\, s'} \overline{u}(p'_\mu, s')\gamma_\mu u(p_\mu, s)\, \overline{u}(p_\mu, s)\gamma_\nu u(p'_\mu, s'), \tag{3}$$

whereas the hadronic tensor

$$W^{\mu\nu} = \frac{1}{4\pi}\int d^4x\, e^{iq\cdot x}\langle p|[J^\mu(x), J^\nu(0)]|p\rangle \tag{4}$$

can only be calculated if the proton wave function $|p\rangle$ is known. The most general form of the hadronic tensor is

$$W^{\mu\nu} = F_1\left(-g^{\mu\nu} + \frac{q^\mu q^\nu}{q^2}\right) + \frac{F_2}{p\cdot q}\left(p^\mu - \frac{p\cdot q\, q^\mu}{q^2}\right)\left(p^\nu - \frac{p\cdot q\, q^\nu}{q^2}\right). \tag{5}$$

with two structure functions F_1 and F_2. Together with the relation

$$R = \frac{F_L}{2xF_1} = \frac{\left(1 + 4x^2\frac{M^2}{Q^2}\right)F_2 - 2xF_1}{2xF_1} \tag{6}$$

between R and F_1 this leads to eq. (1).

2 The origin of higher twist terms[1]

If one is interested in the behaviour of the hadronic tensor $W^{\mu\nu}$ at "large" momentum transfer q where only a small region in space contributes to the Fourier integeral in (4), one is allowed to expand the product of the current operators around $x = 0$. The general form for the small distance expansion of a product of operators is given by

$$\lim_{x\to 0} O_a(x)O_b(0) = \sum_k \hat{c}_{abk}(x)O_k(0). \tag{7}$$

This operator product expansion (OPE) is very useful since on the right-hand side only the coefficient functions $\hat{c}_{abk}(x)$ are dependent on the space coordinates and the sum runs only over *local* operators. In momentum space this formula reads

$$\lim_{q\to\infty}\int d^4x\, e^{iq\cdot x} O_a(x)O_b(0) = \sum_k c_{abk}(q)O_k(x=0) \tag{8}$$

---

[1] This chapter is based on lecture notes from A.V.Manohar, preprint UCSD/PTH 92-10.

where $c_{abk}(q)$ are the Fourier transforms of the coefficient functions $\hat{c}_{abk}(q)$.

As will be scetched here, dimensional analysis can be used to obtain the momentum dependence of the hadronic tensor, since the matrix elements of the local operators between the unknown proton states are independend of q and all momentum dependence is contained in the coefficient functions.

All local operators possibly entering the sum of eq. (8) can be classified according to their mass (or energy) dimension and their spin. As an example, the operator for the quark current $\bar{\psi}\gamma^\mu\psi$ has spin 1 (since it involves two spinors of spin-1/2 quarks) and dimension 3 (as can be seen e.g. from the mass term $M\bar{\psi}\psi$ in the Lagrange density which has mass dimension 4). An operator of dimension d with spin n has n Lorentz indices $\mu_1\mu_2\ldots\mu_n$, so the local operators in (8) can be written as

$$O_{d,n}^{\mu_1\mu_2\cdots\mu_n} \tag{9}$$

The matrix elements of these operators between hadronic states depend only on one Lorentz vector, the four-momentum of the hadron. So the only possibility to produce the required number of Lorentz indices is

$$\langle p|O_{d,n}^{\mu_1\mu_2\cdots\mu_n}|p\rangle = \alpha_1\, p^{\mu_1}p^{\mu_2}\ldots p^{\mu_n} + \alpha_2\, g^{\mu_1\mu_2}p^{\mu_3}\ldots p^{\mu_n} + \ldots \tag{10}$$

where in the second and following terms one or more pairs of momenta are replaced by the metric tensor $g^{\mu\nu}$.

Since the mass dimension of hadronic states in the conventional relativistic normalisation is minus one, dimensional analysis of this equation leads to

$$\underbrace{\langle p|O_{d,n}^{\mu_1\mu_2\cdots\mu_n}|p\rangle}_{\dim d\,-\,2} = \alpha_1\,\underbrace{p^{\mu_1}p^{\mu_2}\ldots p^{\mu_n}}_{\dim n} + \alpha_2\,\underbrace{g^{\mu_1\mu_2}p^{\mu_3}\ldots p^{\mu_n}}_{\dim n\,-\,2} + \ldots. \tag{11}$$

The mass dimensions of the reduced matrix elements α_1, α_2 etc. can be made explicit by a power of the hadron mass M:

$$\begin{aligned}
\langle p|O_{d,n}^{\mu_1\mu_2\cdots\mu_n}|p\rangle &= \beta_1\, M^{d-n-2}\, p^{\mu_1}p^{\mu_2}\ldots p^{\mu_n} \\
&\quad + \beta_2\, M^{d-n}\, g^{\mu_1\mu_2}p^{\mu_3}\ldots p^{\mu_n} + \ldots
\end{aligned} \tag{12}$$

where β_1, β_2 etc. are now dimensionless.

Applying the OPE (8) to the hadronic current (4) one obtains

$$\underbrace{W^{\mu\nu}}_{\dim 0} \propto \sum \underbrace{c_{\mu_1\ldots\mu_n}(q)}_{\dim 2\,-\,d}\, \underbrace{\langle p|O_{d,n}^{\mu_1\mu_2\cdots\mu_n}|p\rangle}_{\dim d\,-\,2} \tag{13}$$

where the sum runs over a complete set of operators of all possible dimensions and all possible spins. For the dimensional analysis it has been used that $W_{\mu\nu}$ is dimensionless by definition, as can be seen e.g. from (5), to obtain the mass dimensions of the coefficient functions.

The dimension of the coefficient functions can now be used with the same arguments as above to construct their momentum dependence, i.e.

$$c_{\mu_1\ldots\mu_n}(q) \propto \frac{q_{\mu_1}}{Q}\ldots\frac{q_{\mu_n}}{Q}Q^{2-d}. \tag{14}$$

Note that the coefficient functions $c_{\mu_1 \ldots \mu_n}$ have two spin indices less than the operators $O_{d,n}^{\mu_1 \ldots \mu_n}$ in order to leave the two indices μ and ν uncontracted. Inserting (14) and (12) into (13) yields

$$W^{\mu\nu} \propto \sum \beta_1^{d,n} \; p^\mu p^\nu \frac{(p \cdot q)^{n-2}}{Q^{n-2}} Q^{2-d} M^{d-n-2} \tag{15}$$

Only the first term of eq. (12) has been kept here since it produces the leading power of Q^2 of each coefficient.

From the definition of x_{Bj} the product $p \cdot q$ is equal to $Q^2/2x_{Bj}$. The remaining Lorentz indices μ and ν are to be contracted with the leptonic tensor $L_{\mu\nu}$ of eq. (3) and will produce products $p \cdot k$ and $p \cdot k'$ that are also of the order Q^2. This finally leads to

$$L_{\mu\nu} W^{\mu\nu} = \sum g_{d,n} \left(\frac{Q}{M} \right)^{2+n-d} \tag{16}$$

where all neglected factors are absorbed in the dimensionless functions $g_{d,n}$.

It is apparent from the last equation that the momentum dependence of the contributions of the individual operators to the hadronic tensor depends on the combination

$$t = d - n \tag{17}$$

which has been named *twist*.

Of course, this dimensional analysis can give only the *leading* power contribution in the free field approximation. In QCD the coefficient functions and their Q^2-dependence can be calculated, however, to any order in perturbation theory.

The simplest operator one can have is the quark current $\overline{\psi}\gamma^\mu\psi$. It corresponds to the contribution of the quark density function to the structure functions. Since it has dimension 3 and spin 1 it is of twist 2. From (16) one can see that this leading twist contribution in the case of a noninteracting quark field is independent of Q^2 – it exhibits the scaling behaviour of the naive quark parton model.

Since quark and gluon operators always appear in pairs in the OPE, the first "higher" twist contribution is of twist 4 and corresponds to a two-particle correlation function. Its leading Q^2-behaviour is given by M^2/Q^2. Similarly, the next term in the expansion has twist 6, corresponds to three-particle correlations and contributes like M^4/Q^4.

In a theory with interactions there will be Q^2-dependend corrections to the simple power law behaviour of the free theory. In QCD these corrections produce the famous logarithmic scaling violations and have in principle to be applied to each term in the twist expansion (16) separately. However, the calculations for higher twist terms are more involved than for the leading twist contribution since the number of correlation functions is much higher than the number of distribution functions. The mixing between the corrections for quarks and gluons adds another complication.

3   Determination of higher twist terms from structure function data

Because the exact determination of the higher twist terms with their proper Q^2-dependence is currently not possible, the following parametrisation is used:

$$F_2(x, Q^2) = F_2^{LT}(x, Q^2)\left(1 + \frac{C^1(x)}{Q^2} + \frac{C^2(x)}{Q^4} + \cdots\right). \qquad (18)$$

It contains the assumption that the logarithmic Q^2-dependence of the leading twist term $F_2^{LT}(x, Q^2)$ and the higher twist terms is the same. This is certainly an approximation, but the limited amount of experimental information available does not allow a more involved analysis.

In order to extract the higher twist functions $C^1(x), C^2(x) \ldots$ from the measured structure functions one has to analyse the Q^2-dependence and to separate the $(1/Q^2)$-behaviour of the higher twist terms from the $(\log Q^2)$-dependence of the scaling violations. To make this separation possible, a rather precise measurement of F_2 over a wide range in Q^2 is needed. In general the data of one experiment are not sufficient for such an analysis, and in the combination of data from different experiments the relative normalisation of the measurements can play an important role.

The other difficulty is the large number of unknown parameters that have to be determined at the same time: the scaling violations of the leading twist term depend mainly on the QCD parameter Λ, but also on the initial parametrisations of the quark and gluon distributions at some starting value Q_0^2. In the analysis presented below, 16 parameters are needed to describe $F_2^{LT}(x, Q^2)$. The parameters describing the unknown functions $C^1(x)$ etc. have then to be added to this number.

It is common in this kind of analysis to correct for "target mass effects". The origin of these corrections are the neglected terms the expression for the matrix elements of the local operators in eq. (12). Although these so called "trace terms" belong to the operator of a given twist, they nevertheless are proportional to higher powers of M^2/Q^2. If they are not to be neglected one has to reorder the summation of eq. (16) which leads to modified coefficient functions. Such a calculation results in the prescription that the structure functions should not be analysed in the usual variable x, but rather the Nachtmann variable

$$\xi = \frac{2x}{1 + \sqrt{1 + 4x^2 \frac{M^2}{Q^2}}} \qquad (19)$$

should be used, and some extra factors are to be applied to the QCD evolution equations [1].

Higher twist terms also contribute to the second structure function F_1, or the ratio $R(x, Q^2)$. In principle, R should be more suited than F_2 to look for higher twist effects since it vanishes to lowest order (in leading twist and without QCD). Moreover, the QCD-corrections to R can be calculated without additional input once the QCD parameter Λ and the quark and gluon distribution functions have been obtained from a F_2-analysis. Unfortunately the contribution of R to the cross section is rather small and therefore measurements of R are not very precise, even if a large data sample has been used.

4 Higher twist analyis from absolute structure functions F_2

Until recently the experimental situation of F_2-measurements was rather confusing as can be seen from figure 1. Published data from the two muon experiments at CERN disagreed by more than 10% at low x and neither of the data sets matched well with the low energy measurements from SLAC. These experimental problems spoiled any analysis of higher twist contributions.

After intense work in the last two years the situation has greatly improved and is shown in figure 2. The low Q^2 data from SLAC have been reanalysed with an improved radiative correction program and with a better internal normalisation. It has been found that the change in slope between the data of SLAC and BCDMS at high values of x can be understood as a systematic effect from the calibration of the energy measurement in the BCDMS experiment [2]. In addition new and preliminary data from the New Muon Collaboration that overlap the acceptance of both experiments and extend the measured region to lower

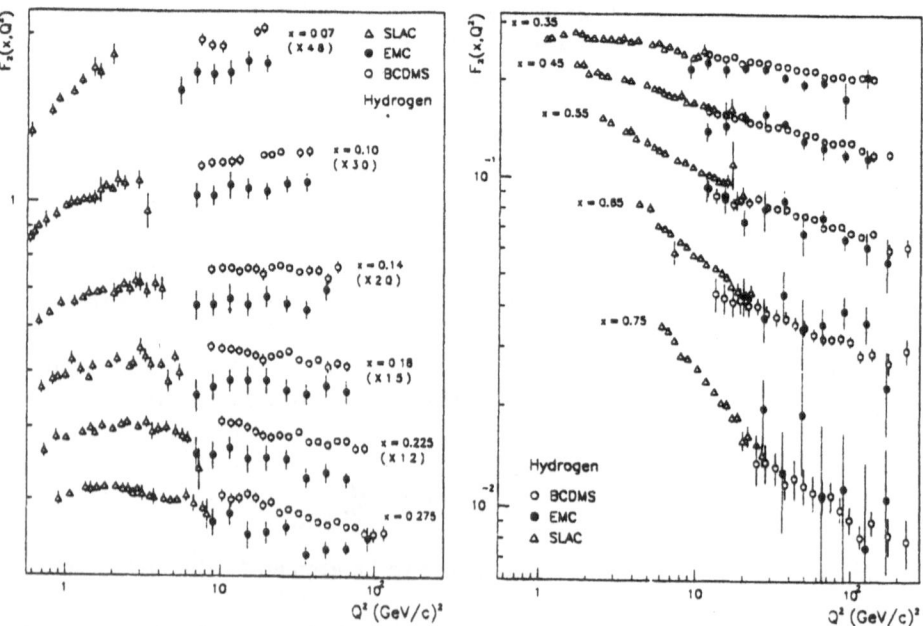

Figure 1: Comparison of the previously published F_2-data sets of SLAC, EMC and BCDMS on hydrogen (from [2]).

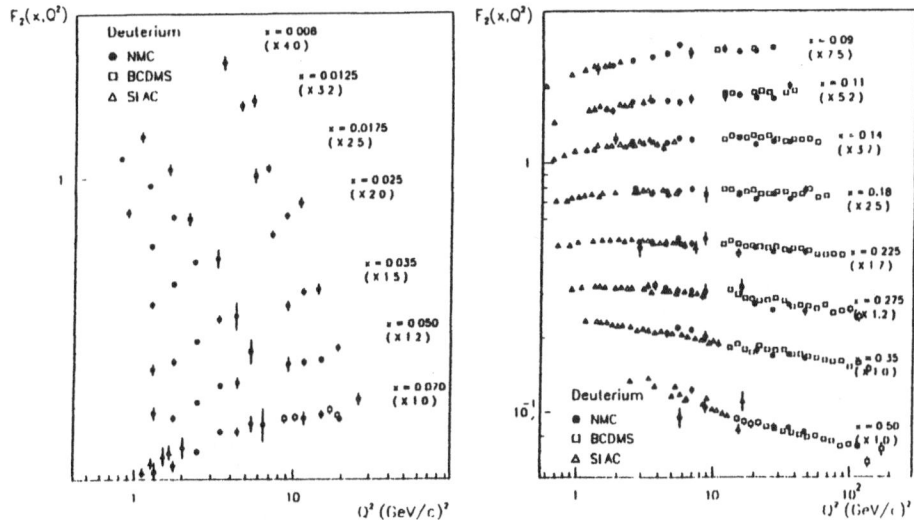

Figure 2: Comparison of the revised SLAC and BCDMS data set on deuterium with preliminary data from the NMC.

values of x are in excellent agreement with the revised data sets of SLAC and BCDMS. The EMC data, reevaluated with the same function for R as the other data sets and renormalised to give the best fit, still disagree to some extent with the data of the other three experiments shown in figure 2.

The high statistics SLAC/BCDMS data set has now been used to perform fits with the parametrisation of eq. (18) [3]. Only twist-4 coefficients have been determined since the addition of a $1/Q^4$-term did not significantly improve the quality of the fit. Figure 3 shows the result of this fit as a solid line. As described above target mass corrections have been included in the leading twist structure functions. For each x-bin one additional parameter was included to describe the higher twist contribution. The difference between the solid and the dashed line indicates the size of the twist-4 contribution. It is observed that the contribution of the higher twist terms looks simliar in hydrogen and deuterium, being negative at low x and getting positiv and larger at high x.

In figure 4 the higher twist coefficients are plotted as a function of x. The size of these coefficients at low x is rather small, although their contribution to the structure function is significant since the data are at rather small Q^2. But because at small x a similar contribution can also be produced by a change in the gluon distribution, these measurements have a large error that makes them compatible with zero. The twist-4 contribution at $x > 0.4$ is significant, however, and it seems that all values are higher for deuterium than for hydrogen.

The difference in the higher twist terms for neutron and proton has been analysed by the NMC collaboration in more detail [4]. They used a complementary arrangement of two hydrogen and two deuterium targets to measure the ratio F_2^n/F_2^p with very small systematic uncertainty. From eq. (18) the

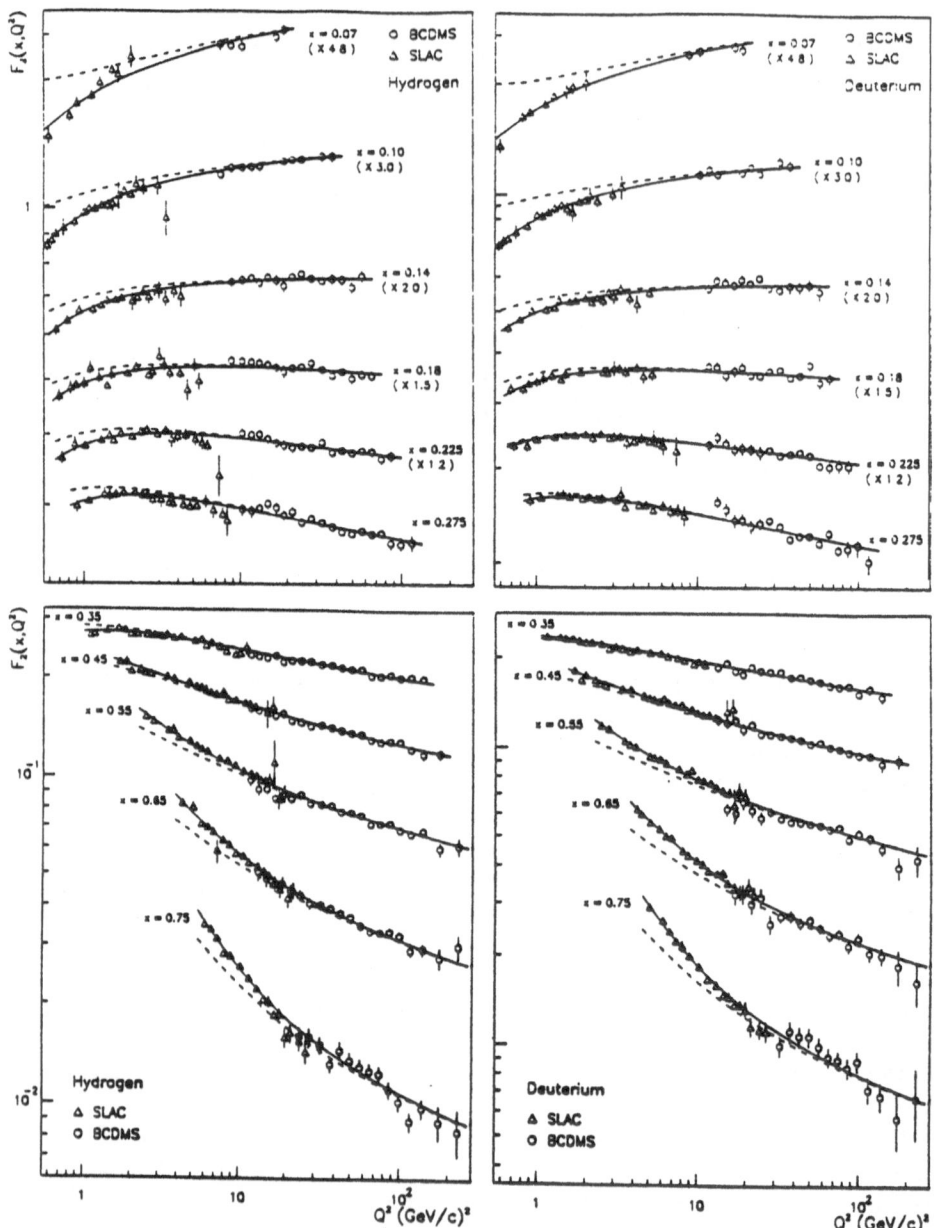

Figure 3: Fits of next to leading order QCD with target mass corrections and higher twist contributions to the revised SLAC-BCDMS data set.

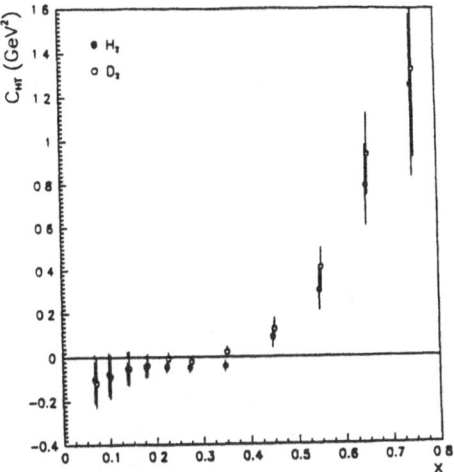

Figure 4: Twist-4 coefficients obtained from an analysis of the revised SLAC-BCDMS F_2 data sets.

Q^2-dependence of this ratio can be parametrized as

$$\frac{F_2^n}{F_2^p} = \left(\frac{F_2^n}{F_2^p}\right)^{LT} \cdot \left(1 - \frac{C^p - C^n}{Q^2}\right) \qquad (20)$$

since the correction to the leading twist is small. In figure 5 the measurements of this ratio by NMC, SLAC and BCDMS are plotted for selected x-bins. The NMC data have been used to renormalize the SLAC and BCDMS ratios that are

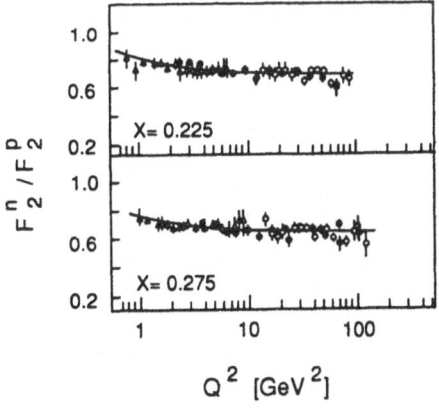

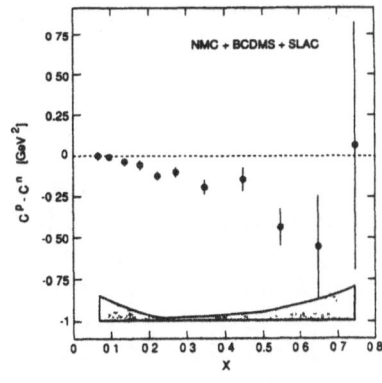

Figure 5: The Q^2-dependence of the ratio F_2^n/F_2^p in selected x-bins. Data are from NMC (full circles), SLAC (triangles) and BCDMS (open circles). The curves are fits to eq. (20).

Figure 6: The difference of the twist-four coefficients for proton and neutron obtained from fits like the ones in the previous figure.

obtained from independend measurements on hydrogen and deuterium targets and have correspondingly larger normalisation errors. No binding corrections have been applied to extract the neutron structure function from the deuterium measurement. Also shown is the fit to eq. (20) which determines the higher twist coefficients $C^p - C^n$. The resulting set of coefficients for all x-bins is plotted in figure 6. The analysis clearly shows that the twist-4 coefficient in the neutron is smaller than in the proton.

5 Higher twist analysis from $R(x, Q^2)$

At SLAC a dedicated high statistics measurement of the quantity R has been performed [5]. The resulting data are displayed in figure 7 together with data from the CERN muon and neutrino experiments. The solid line which gives a good description of all data points is the best fit to a phaenomenological parametrisation of the data, known as R^{1990}. It is this parametrisation that has been used by most collaborations to extract the structure function from the measured cross section (eq. (1)). Also shown are different predictions for R starting from the quark and gluon parametrisations obtained by the CDHS collaboration. The dotted line is the prediction from perturbative QCD alone. It is much lower than the data. If target mass corrections are included the prediction comes closer to describe the data (dashed curve) but falls still significantly below. It is believed that this discrepancy is due to higher twist effects.

An earlier subset of these data has been used to extract the higher twist contributions [6]. The longitudinal structure function F_L defined in eq. (6) has to lowest order only one twist-4 contribution, namely the one from the correlation between a quark and a gluon with opposite helicity. It has been assumed in this analysis that $T_-(x)$, the probability for a quark and a gluon to carry the momentum fraction x, is in the limit of large x equal to the quark distribution function itself. Then

$$F_L^{\text{twist-4}} = \frac{8\kappa^2}{Q^2} T_-(x) \tag{21}$$

where the twist scale κ^2 is the only free parameter of the theory. The fit to the early SLAC data set resulted in a value of $\kappa^2 \simeq 0.05$ GeV^2.

The prediction for R from this analysis is compared with the new R-measurement in figure 8 where the data point at each value of x has been obtained from the data in figure 7 by averaging over the interval $5 \leq Q^2 \leq 10$ GeV^2. Again, the solid line is the R^{1990} parametrisation, the dotted line is the QCD prediction without and the dashed line with target mass corrections. The curve with two short and one long dashes gives the result of the higher twist model. It fits the data as well as the R^{1990} parametrisation. The fifth curve in this figure, which is in total contradiction to everything else, is the prediction of a model with spin-0 diquarks [7].

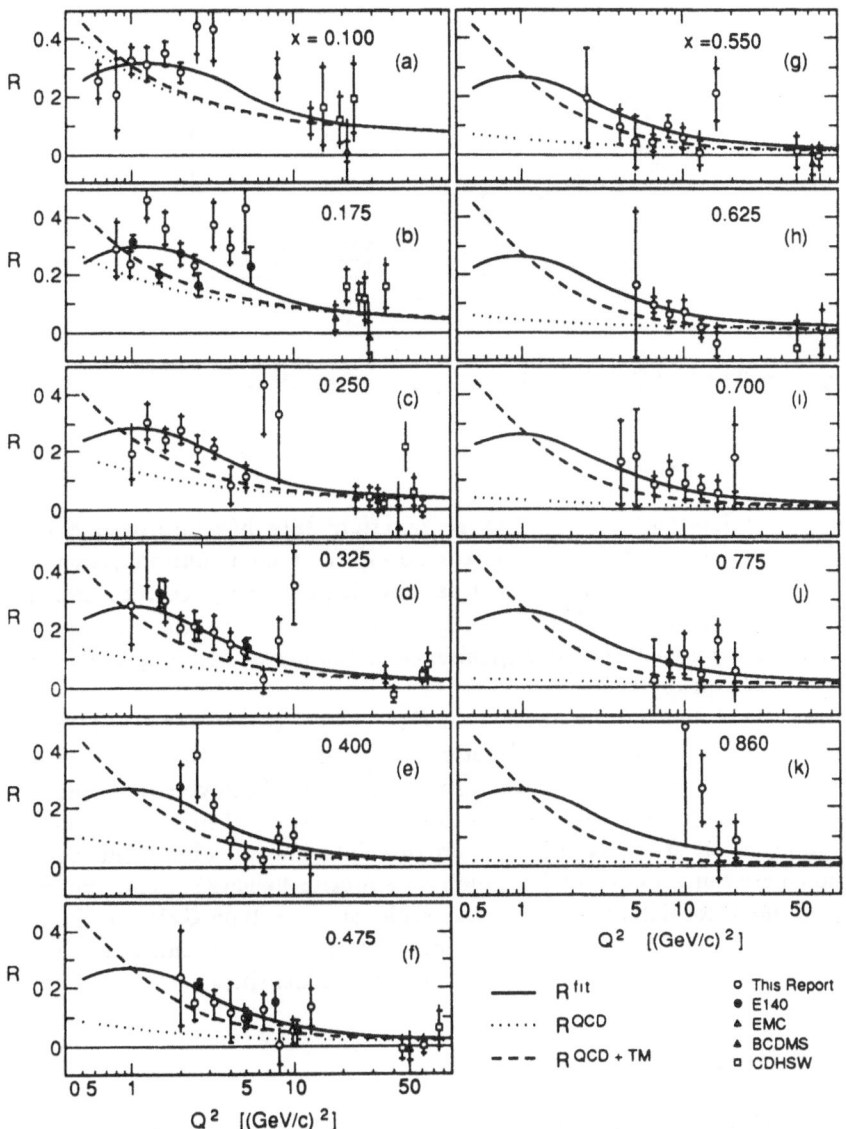

Figure 7: $R(x, Q^2)$ **as measured at SLAC. Also shown are data from** μ-N **and** ν-N **scattering.**

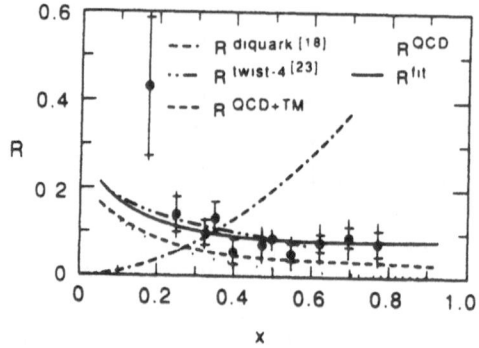

Figure 8: The SLAC data on R averaged over the Q^2-range from 5 to 10 GeV^2.

6 Conclusions

Since the understanding of structure function data has improved a lot, it is possible now to extract the next to leading twist contribution (twist-4) from charged lepton-nucleon scattering data. Limitations of the current analysis are, that the Q^2-dependence of the leading twist term is calculated to second order in QCD only and that the Q^2-dependence of the next-to-leading twist term can not be treated properly at all.

Within these limitations there is a clear evidence for nonvanishing twist-4 contributions to the structure function which increase with raising x and are larger for the neutron than for the proton. For QCD analyses it should be kept in mind that they are significant up to $Q^2 = 10$ GeV^2 at high values of x.

An analysis of $R = \sigma_L/\sigma_T$ can provide information about the correlation function between a quark and a gluon of opposite helicities. The experimental information is consistent with a twist scale of $\kappa^2 \simeq 0.05$ GeV^2, although this analysis was carried out on a subsample of the data only and did not use the most recent parametrisations of quark and gluon densities.

References

[1] O. Nachtmann, Nucl. Phys. **B163**, 237 (1973).

[2] A. Milsztajn et al., Z.Phys. **C49**, 527 (1991).

[3] M. Virchaux and A. Milsztajn, SACLAY preprint DPhPE 91-08, submitted to Phys.Lett. **B**.

[4] P. Amaudruz et al., Nucl.Phys **B371**, 3 (1992).

[5] L.W. Whitlow et al., Phys.Lett. **B250**, 193 (1990).

[6] J.L. Miramontes et al., Phys.Rev. **D40**, 2184 (1989).

[7] S. Ekelin and S. Fredriksson, Phys.Lett **162B**, 373 (1985).

The Internal Structure of Constituent Quarks[*),+)]

Harald Fritzsch

Sektion Physik der Universität München
and
Max–Planck–Institut für Physik und Astrophysik
Werner Heisenberg Institut für Physik — München, Germany

Abstract. Constituent quarks have a nontrivial spin structure, due to the gluonic anomaly of QCD and the related breaking of the chiral symmetry in the U(1) channel.

Deep inelastic scattering reveals that the nucleon is a rather complicated object consisting of an infinite number of quarks, antiquarks and gluons. Although there is only scarce information about the internal structure of the other strongly interacting particles, nobody doubts that the same is true for all mesons and baryons. Nevertheless it seems that under certain circumstances they behave as if they were composed of a single constituent quark and another constituent antiquark or three constituent quarks. Examples are the magnetic moments of the baryons, the spectroscopy of mesons and baryons, the meson–baryon couplings, the ratios of total cross sections like $\sigma(\pi N)/\sigma(NN)$ etc. . Thus it seems to make sense to slice the proton into three pieces, into three constituent quarks called U or D . A proton would have the composition (UUD). The constituent quarks would carry the internal quantum numbers of the nucleon.

In deep inelastic scattering one observes that a nucleon has the composition $|u\ u\ d\ \bar{q}q...g...>$ (g: gluon, $q = u,d,s$), i.e. the quark density functions (which are scale dependent) are described by a valence quark and an essentially infinite number of quark–antiquark pairs. One might be tempted to identify the valence quark, defined by the corresponding quark density function, with a constituent quark. This identification would imply that the three quark picture denoted above is nothing but a very rough approximation, and both $\bar{q}q$-pairs and gluons need to be added to the picture. However in this case one would not be able to understand why the picture of a baryon consisting of three constituent quarks works so well in many circumstances.

It seems much more likely to us that a constituent quark is a quasiparticle which has a nontrivial internal structure on its own, i.e. consisting of a valence quark, of many $\bar{q}q$-pairs and of gluons — in short, it looks like one third of a proton, including all the complications provided by the internal structure of the proton. Thus a constituent quark has an effective mass, an internal size, etc.

[*)] Supported in part by DFG contract Nr. F412/12–2 and C.E.C. project SCI–CT91-0729

[+)] Invited talk given at the 99. WE–Heraeus–Seminar on "Quark Cluster Dynamics", Physikzentrum Bad Honnef, Germany, June 1992, to be published in the Proceedings.

This interpretation of a constituent quark is not new; it was first discussed already about 20 years ago[1]. Nevertheless it is still unclear to what extent it can be derived from the basic laws of QCD, since it is deeply related to nonperturbative aspects of QCD, like the confinement problem.

One way to gain deeper insights into the internal structure of the constituent quark is to consider spin problems, which is the topic of this talk. In the constituent quark picture it is, of course, assumed that the nucleon spin is provided by the combination of the spins of the three constituent quarks. If the latter have a nontrivial internal structure, the question arises whether the spin structure of the constituent quarks is a complex phenomenon, as it seems to be for the nucleon, or not.

A simple model for the spin structure would be to assume that the spin of, say, a constituent U quark is provided by the valence u–quark inside it, and the $\bar{q}q$–cloud and the gluonic cloud do not contribute to the spin. We shall conclude that this "naive" picture is not correct.

Before we discuss the constituent quarks, let us summarize the results about the spin structure of the proton. As usual we define the distribution functions of the quarks of flavor q and helicity + 1/2 (−1/2) by q_+ (q_-). The lowest moment of the structure function g_1, measured in the deep inelastic scattering of polarized leptons off hadronic targets, is given by the moments of these quark densities Δq:

$$\int_0^1 g_1 dx = \frac{1}{2}(\frac{4}{9}\Delta u + \frac{1}{9}\Delta d + \frac{1}{9}\Delta s), \tag{1}$$

$$(\Delta u = \int_0^1 dx\,(u_+ + \bar{u}_+ - u_- - \bar{u}_-)\ etc.)$$

The spin density moments Δq are determined by the nucleon matrix elements of the associated axial vector currents (s_μ: spin vector):

$$\Delta q \cdot s_\mu = <p,s\,|\bar{q}\gamma_\mu\gamma_5 q|p, s> \tag{2}$$

The experimental data give[2]:

$$g_1 dx = 0.114 \pm 0.012 \pm 0.026 \tag{3}$$

The Bjorken sum rule which follows from the algebra of currents in QCD, relates the difference of the u/d — moments to the axial vector coupling constant measured in β–decay:

$$\Delta u - \Delta d = g_A \tag{4}$$

(we neglect radiative corrections of the order of α_s/π).

Using SU_3 one finds:

$$g_A = F + D, \quad \Delta u + \Delta d - 2\Delta s = 3F - D \tag{5}$$

Here F and D are defined by the axial vector matrix elements of the members of the baryon octet. An analysis of the hyperon decays gives:[2]

$$F = 0.47 \pm 0.04 \quad D = 0.81 \pm 0.03$$
$$\Delta u = 0.78 \pm 0.06 \quad \Delta d = -0.48 \pm 0.06$$
$$\Delta s = -0.19 \pm 0.06 \tag{6}$$
$$\Delta\Sigma = \Delta u + \Delta d + \Delta s = 0.10 \pm 0.17$$

An essential feature of the data is that the "sea" of the $\bar{s}s$–pairs in the nucleon appears to be highly polarized; it contributes significantly to the axial singlet charge. This implies that the

"Zweig" rule does not seem to work for the matrix elements of the axialvector currents.

The quantity $\Delta\Sigma$ is the nucleon matrix element of the axial baryonic current and can be defined as the axial baryon charge of the nucleon:

$$\Delta\Sigma \cdot s_\mu = <p|\bar{u}\gamma_\mu\gamma_5 u + \bar{d}\gamma_\mu\gamma_5 d + \bar{s}\gamma_\mu\gamma_5 s|p> = <p|j_\mu^{05}|p> \qquad (7)$$

In a naive wave function picture of the nucleon the axial baryon number corresponds to the portion of the nucleon spin carried by the quarks. Independent of a specific wave function model we can define $\Delta\Sigma$ as the relative amount of the nucleon spin carried by the intrinsic spins of the quarks. In the simplest SU_6 type model of the nucleon this quantity is one. In reality it may depart

significantly from one, due to the contributions of orbital momenta and of the $\bar{q}q$–pairs or the gluons to the nucleon spin. Nevertheless it is surprising to observe that $\Delta\Sigma$ seems to be small compared to one. However we emphasize that the experiments give solely an information about the axial baryonic charge of the nucleon, and not about the spin. Only in a nonrelativistic $SU(6)$ type model, in which the quarks move in an s–wave, the axial baryonic charge and the spin of the nucleon, multiplied by two, are both equal to one. There is no reason why $\Delta\Sigma$ could not be much less than one, or even zero, if we doubt the validity of the "naive" $SU(6)$ model.

Of course, a possible vanishing of the axial singlet nucleon charge must be discussed in view of the fact that the octet axial charges are, of course, different from zero. Nevertheless they depart substantially from the values one obtains in a nonrelativistic $SU(6)$ approach, which, for example, predicts $g_A / g_V = 5/3$, while in reality one has $g_A / g_V \approx 1.27$.

Furthermore the octet charges obey the Goldberger–Treiman relations, which relate the mass of the nucleon and the axial charges to the coupling and decay constants of the pseudoscalar mesons. The latter act as massless Nambu–Goldstone bosons in the chiral limit of QCD[4]. This suggests that also the value of the singlet axial charge is not unrelated to the chiral symmetry of QCD and its dynamical breaking. For this reason it is useful to examine the nucleon matrix element of the axial singlet current in this respect. First we consider it in the chiral limit of $SU(3)_L \times SU(3)_R$, in which $m_u = m_d = m_s = 0$. In this limit the octet of axial vector currents is conserved, while the singlet current is not conserved due to the gluonic anomaly:

$$\partial^\mu j_\mu^{i5} = 0 \quad (i = 1,2, ...8) \qquad (8)$$

$$\partial^\mu j_\mu^{05} = 3 \cdot \frac{\alpha_s}{2\pi} \cdot \text{tr } G\tilde{G} = a$$

It is known that this limit, in which the masses of the three light quark flavors are neglected, is not far away from the real world of hadrons. In the limit there exist eight massless pseudoscalar mesons, serving as the Goldstone bosons. However the ninth pseudoscalar, the η' – meson, remains massive and has a mass not far from its physical mass, i.e. about 900 MeV. The axialvector charges of the baryons are related to the coupling constants of the pseudoscalar mesons with the baryons by the Goldberger – Treiman relations, e.g. those for the pions (f_π: pion decay constant, M: nucleon mass):

$$2 M g_A = 2 f_\pi g_{\pi NN} \qquad (9)$$

We remind the reader how these relations are obtained. The matrix element of the axial vector current in the octet channel can be described by two form factors:

$$\bar{u}(p)[G_1(q^2) \gamma_\mu\gamma_5 - G_2(q^2) q_\mu\gamma_5] u(p') \qquad q = p - p' \qquad (10)$$

The induced pseudoscalar formfactor G_2 acquires a pole at $q^2 = 0$, since the pion mass vanishes in the chiral limit:

$$G_2(q^2) = \frac{2f_\pi g_{\pi NN}}{q^2} \tag{11}$$

Due to the conservation of the current one finds $2M \cdot G_1(0) = 2M\, g_A = 2f_\pi \cdot g_{\pi NN}$. We stress that this relation follows as the result of an interplay between the axial vector form factors G_1 and G_2. It is the latter, which contains the pion pole. But the conservation of the current leads to the constraint about G_1, i.e. to a condition about the axial charge – the Goldberger – Treiman relation. In other words: the chiral symmetry allows us to convert a statement about the divergence of the axial vector current into a statement about the matrix element of the current. Due to the pole in G_2 one finds a nonzero matrix element, even though the current is conserved. In the absence of the pole the chiral symmetry would be trivially fulfilled – the nucleon mass would have to vanish.

Let us consider the nucleon matrix element of the axial baryonic current in the chiral limit:

$$<p|j_\mu^{05}|p'> = \bar{u}(p)\,(G_1^0\,\gamma_\mu\gamma_5 - G_2^0\,q_\mu\,\gamma_5)\,u(p') \tag{12}$$

Here the induced pseudoscalar form factor does not have a Goldstone pole at $q^2 = 0$. Instead of the Goldberger Treiman relation one finds after taking the divergence and setting $q = 0$[5,6]:

$$G_1^0(0) = \Delta\Sigma = A(0) \tag{13}$$

where A is the formfactor of the anomalous divergence:

$$<p|\, 3 \cdot \frac{\alpha_s}{2\pi}\, tr\, G\tilde{G}|p'> = 2\,M\,A(q^2)\bar{u}(p)\,i\gamma_5 u(p') \tag{14}$$

We conclude: The axial baryonic charge $\Delta\Sigma$ ("the spin of the nucleon") is nothing but the nucleon matrix element of the anomalous divergence, i.e. a purely gluonic quantity. This quantity is essentially unknown. Eventually it may be calculated in the future within the lattice approach to QCD.

It is interesting to note that the fact that the singlet quantity $\Delta\Sigma$ is a gluonic quantity while the octet spin densities, e.g. $\Delta u + \Delta d - 2\,\Delta s$, are determined by the nucleon matrix elements of quark bilinears, indicates a substantial violation of the "Zweig rule" for the axial vector nonet. The latter would imply $\Delta s = 0$, and we would have $\Delta\Sigma = \Delta u + \Delta d + \Delta s = \Delta u + \Delta d - 2\Delta s$. Thus the matrix element of the anomalous divergence, a gluonic quantity, would have to be equal to the matrix element of the eighth component of the axial vector octet, a quark bilinear. There is no reason why this should be the case.

We conclude: The violation of the "Zweig rule" in the pseudoscalar channel, which is well–known and caused by the QCD anomaly, implies via the mechanism of spontaneous symmetry breaking another violation of this rule for the nucleon matrix elements of the axial vector current. The strength of this violation is given by the magnitude of the spin density moment Δs. Therefore it is not surprising that in particular this spin density moment appears to be large.

Apparently the violation of the "Zweig" rule is such that the axial singlet charge $\Delta\Sigma$ is rather small, perhaps even zero. Thus the constituent quark model needs a revision which must take into account this effect, being a consequence of the dynamics of chiral symmetry and its breaking. Below we shall discuss such a revision, which is able to combine both chiral dynamics and the "naive" constituent quark model[7,8,9].

First we consider a simplified case, namely the one of QCD with the two flavors u and d only. The strange quarks and the "heavy" flavors c, b and t are disregarded. Furthermore we assume $m_u = m_d = 0$, i.e. the chiral symmetry $SU(2)_L \times SU(2)_R$ is exactly fulfilled. The pions are massless.

Due to the QCD anomaly the singlet pseudoscalar η (quark composition $(\bar{u}u + \bar{d}d)/\sqrt{2}$) has a mass of the order of the nucleon mass M. The Goldberger – Treiman relation is exactly valid:

$$2 M_n g_A = 2 F_\pi g_{\pi NN}. \tag{15}$$

In the SU(6) type constituent quark model the axialvector coupling constant g_A is given by the nucleon expectation value of the quark spin operator $\frac{1}{2} \sigma_z$:

$$g_A = <\sigma_z(u)> - <\sigma_z(d)> = 5/3 \tag{16}$$

where one has:

$$1/2 \; \sigma_z(u) = 2/3 \qquad 1/2 \; \sigma_z(d) = -1/6, \tag{17}$$

$$1/2 \; (\sigma_z(u) + \sigma_z(d)) = 1/2 \; (= \text{nucleon spin}).$$

In reality g_A is not equal to 5/3, but about 1.27, i.e. the prediction of the "constituent model" is violated by about 24 %. This violation can be understood without giving up the simple ideas of the constituent quark model, as an effect due to orbital motions and relativistic effects. Thus in the isovector channel both the chiral dynamics and the constituent quark model do not in contradict, but rather supplement each other. This observation encourages us to consider the "constituent quarks" as separate entities. In a "Gedankenexperiment" we could consider a polarized "constituent quark" Q (Q = U,D) and study its coupling constants. They would also obey a Goldberger – Treiman relation[10]:

$$2 M_q \tilde{g}_A = 2 F_\pi g_{\pi QQ} \tag{18}$$

(M_q: constituent quark mass, $\tilde{g}_A$ axialvector coupling constant of the constituent quark, $g_{\pi QQ}$: pion – quark coupling constant).

Suppose we consider the corresponding matrix elements of the vector and axialvector currents and relate them to the various moments of the quark density functions. One finds naively:

$$<U|\bar{u}\gamma_\mu u|U> = p_\mu/M_U = \left[\frac{p_\mu}{M_U} \right] \int_0^1 (u_+ + u_- - \bar{u}_+ - \bar{u}_-) \, dx \tag{19}$$

$$<U|\bar{d}\gamma_\mu d|U> = 0$$

$$<U|\bar{u}\gamma_\mu \gamma_5 u|U> = s_\mu \cdot \Delta u = s_\mu \cdot \int_0^1 (u_+ + \bar{u}_+ - u_- - \bar{u}_-) dx$$

$$= s_\mu \cdot 1$$

$$<U| \bar{d}\gamma_\mu \gamma_5 d|U> = 0$$

(s_μ: spin vector, p_μ: four–momentum, the quark density functions refer to the U–quark and should carry an index u, which is not explicitly denoted here.) These relations reflect the expectation that in a constituent U–quark the quark density functions must be arranged such that the correct flavor structure is obtained and that its total spin is carried by the u–flavor. The d–flavor is not supposed to contribute to the spin.

We could go further and be more specific about the structure of the quark density functions. The success of the "Zweig" rule relies on the assumption that $(\bar{q}q)$–pairs contribute very little to the hadronic wave functions. Correspondingly we could consider a limit in which the $(\bar{q}q)$–pairs are neglected ("valence quark dominance"). In this limit we find for a U—quark:

$$\bar{u}_+ = \bar{u}_- = u_- = 0 \tag{20}$$

$$d_+ = d_- = \bar{d}_+ = 0.$$

Only the density function u_+ is different from zero. This is easily understood if we consider the free quark model, in which the "constituent quarks" and the "current quarks" are identical and we have not only the relations (20), but in addition the function u_+ is known: $u_+ = \delta(x-1)$.

Thus the essential difference between a "constituent quark" inside a hadron and a free quark lies in the shape of the density function u_+. The confinement forces merely cause this function to depart from a δ—function and to spread out over the available x—range.

It turns out that the picture of a constituent quark described above is not consistent with the constraints given by the chiral symmetry. In the constituent model we have $^C\bar{d}_\pm{}^\Im = 0$. This implies for a U—quark that both the isoscalar and the isovector combinations of the spin density moments are equal to one:
$\Delta u - \Delta d = \Delta u + \Delta d = 1$. The isovector part is determined by the pion pole. If the isosinglet η—meson would also be a Goldstone particle, the associated coupling constants would conspire such that the isovector and isoscalar spin density moments would be equal, and the results of the "naive" constituent model would be obtained. However due to the QCD anomaly the isosinglet spin density function does not receive a Goldstone pole contribution. Instead it is given by the constituent quark matrix element of the anomalous divergence:

$$\Delta u + \Delta d = A$$

$$<u|\, 2 \cdot \frac{\alpha_s}{2\pi} \mathrm{tr} \; G \, \bar{G}|\, u> = 2\, M_u \, A \, \bar{u} \, i\gamma_5 u \tag{21}$$

$$\Delta u = (1 + A) \,/\, 2 \qquad \Delta d = (A{-}1) \,/\, 2$$

There is no dynamical reason why A should be equal to one. If it were, the spin density moments would indeed reproduce the constituent quark model result. In particular Δd would vanish. This is not ruled out a prori, but if it were, it would be a miraculous coincidence.

For all other values of A the spin density moment Δd does not vanish. We conclude that for $A \neq 1$ the constituent U quark must contain $(\bar{q}q)$—pairs. Thus a violation of the "Zweig rule" is automatically implied. It is interesting that these pairs are generated by the same nonperturbative mechanism due to the gluon anomaly which causes the η—meson to acquire a mass and not to act as a Goldstone particle in the chiral limit of $SU(2)_L \times SU(2)_R$.

Intuitively one can understand the violation of the "Zweig rule" discussed above as follows. The chiral dynamics of a "constituent quark" would obey the "Zweig rule" if it were surrounded by a cloud of π and η Goldstone bosons. The Goldstone poles of the axialvector current matrix elements would imply, via the Goldberger — Treiman relations in the isovector and isoscalar channel, that the matrix elements obey the constraints given by the "Zweig rule" (in particular $\Delta d = 0$ for a u—quark etc.). However the QCD anomaly causes the η—pole at $q^2 = 0$ to disappear. As a result the "Goldstone cloud" of an U—quark consists only of π—mesons. Thus the dynamical structure of the constituent quark is drastically changed. In particular $(\bar{u}u)$ and $(\bar{d}d)$ pairs are generated, which modify the spin structure.

We note that these pairs cannot simply be regarded as the pairs inside virtual π—mesons. Their presence is caused by the chiral dynamics, in particular by the Goldberger — Treiman relations for the axialvector matrix elements. Their appearance is a nonperturbative phenomenon just like the generation of the η—mass due to the gluonic anomaly.

In ref. (2) the author has argued that the nucleon matrix element of the anomalous divergence should be very small. The argument is based on the observation that the "constituent quark model" requires the quarks in a nucleon to be in an s—wave. In particular the gluonic components of the wave function should also be dominantly in an s—wave. This implies that a pseudoscalar density like $\alpha_s \cdot G_{\mu\nu} G^{\mu\nu}$ is not expected to have a sizeable matrix element, in accordance with the experimental results.

It is easy to apply similar considerations to the constituent quarks. Also for them we should have A = 0, at least to a good approximation. Thus we find for a constituent U quark that the "Zweig rule" for the density moments is maximally violated:

$$\Delta u = 1/2 \qquad\qquad \Delta d = -1/2. \tag{22}$$

We can go further and specify the various density moments. If the "Zweig rule" were valid (both π and η Goldstone modes present), we would have

$$\int u_+ dx = 1, \quad u_- = \bar{u}_+ = \bar{u}_- = 0 \quad d_+ = d_- = \bar{d}_+ = \bar{d}_- = 0 \tag{23}$$

Such a constraint which is not invariant under the renormalization group can only be imposed for a particular value of the energy scale μ, which is expected to be the characteristic hadronic energy scale. The removal of the η Goldstone pole causes a shift in the density moments, which we can parametrize by two functions h_+ and h_-:

$$u_+ = u_+^v + h_+ \qquad \bar{u}_+ = d_+ = \bar{d}_+ = h_+ \tag{24}$$

$$u_- = \bar{u}_- = d_- = \bar{d}_- = h_-.$$

(u_+^v: intrinsic density function of U—quark in the absence of the anomaly, $\int u_+^v dx = 1$). We find:

$$\Delta\Sigma = \Delta u + \Delta d = 1 + 4 \int_o^1 (h_+ - h_-) dx = 0 \tag{25}$$

$$\Delta u - \Delta d = 1$$

It follows:

$$\int_o^1 (h_+ - h_-) \, d_+ = -1/4. \tag{26}$$

We observe that $\Delta\Sigma$ vanishes because the constituent U quark contribution to $\Delta\Sigma$ is cancelled by the pairs. A cancellation is only possible, if the density function h_- is different from zero. On the other hand h_+ can be zero, in accordance with the sum rule (26). The simplest model obeying the constraints discussed above is one in which we have

$$h_+ = 0, \qquad\qquad \int h_- dx = 1/4,$$

$$u_+ = u_+^v, \qquad\qquad \int u_- dx = \int \bar{u}_- dx = \int d_- dx = \int \bar{d}_- dx = 1/4$$

$$d_+ = \bar{d}_+ = \bar{u}_+ = 0 \tag{27}$$

Thus we obtain in the case A = 0 the following picture of a polarized constituent U quark in the $SU(2)_L \times SU(2)_R$ limit: The density function u_+, which describes the density of u—quarks polarized in the same direction as the U—quark, is unaffected by the QCD anomaly. The latter causes a large violation of the "Zweig rule" in the sense that $(\bar{q}q)$—pairs are generated. We shall refer to this "cloud" of $(\bar{q}q)$—pairs as the "anomaly cloud". The density functions $\bar{u}_-$ and $\bar{d}_-$ are different from zero, i.e. the pairs are polarized oppositely to the original constituent quark. The sum of all (anti) quark spins is zero. Thus for A = 0 the quarks do not contribute to the spin of the constituent quark. The latter is provided by the orbital angular momentum of the pairs. This can be seen as follows. If we would turn off the QCD anomaly (e.g. formally by setting $n_c = \infty$), the "naive" picture should hold, i.e. the spin of the U–quark is carried by the valence quark u_v. Once the

anomaly is introduced, the u valence quark continues to contribute its spin, but the $(\bar{q}q)$ pairs cancel the latter. Their total angular momentum must be zero. Otherwise the introduction of the anomaly would violate the conservation of angular momentum. Thus we have:

$$J_z(U) = + 1/2 = J_z(u_v) + J_z(\text{cloud}) + L_z(\text{cloud}) \tag{28}$$

$$= + 1/2 + (-1/2) + (+1/2)$$

In the case A ≠ 0 the cancellation between the spin of the valence quark and the spins of the "anomaly cloud" would not be complete, but the sum of the spins and of the orbited angular momenta of the pairs in the "anomaly cloud" would still be zero.

Thus far we have disregarded the polarization effects due to gluons. We find it unlikely, but not impossible that polarized gluons contribute also to the total angular momentum of the anomaly cloud, and the gluonic term would appear also in eq. (28).

Finally we consider the case of the three light flavors u,d,s. In the chiral limit of $SU(3)_L$ x $SU(3)_R$ we obtain for a constituent U quark in analogy to eq. (27):

$$\Delta U = 2/3 \qquad \Delta d = -1/3 \qquad \Delta s = -1/3. \tag{29}$$

In the symmetry limit the "anomaly cloud" is, of course, SU(3) symmetric. In reality symmetry breaking will be present. The result will be that the effects of the $(\bar{s}s)$ pairs are somewhat reduced compared to those of the $(\bar{u}u)$ and $(\bar{d}d)$ pairs. For example, in a U–quark we expect: $\Delta d > \Delta s$. The actual spin density momenta of the U,D constituent quarks will lie between the extreme case of SU(2) x SU(2) ($\Delta d = -1/2$ for a U–quark) and of SU(3) x SU(3) ($\Delta d = -1/3$ for a U–quark). However we note that for A = 0 the limit of $SU(2)_L$ x $SU(2)_R$ is not ruled out experimentally. It would imply for a proton:

$\Delta u + \Delta d = 0$
$\Delta u - \Delta d = |g_A/g_V| = 1.26$ $\qquad$ (30)
$\Delta u = 0.63 \qquad \Delta d = -0.63$

$\int_0^1 g_1 \, dx = 1/2 \, (4/9 \, \Delta u + 1/9 \, \Delta d) = 0.105,$

while the experimental value for this integral is

$$\int_0^1 g_1 \, dx = 0.114 \pm 0.038 \tag{31}$$

In this limiting case the SU(3) symmetry for the axialvector currents would be very badly broken. It is known that the SU(3) breaking in the axialvector channel is sizeable, but certainly not as large as to allow the case $\Delta s = 0$. Details of the symmetry breaking cannot be presented here.

Recently it was argued that the anomaly could contribute to the axial singlet charge if gluons are highly polarized in a polarized nucleon. In this case their contribution to the singlet charge could be calculated perturbatively[12,13]. In our approach we see no reason for a large gluonic polarization. Thus the effect discussed in ref. (12, 13) would be negligible in comparison to the nonperturbative phenomenon discussed here.

The smallness of the axialsinglet charge, parametrized above by the parameter A, follows also within the Skyrme type model, discussed in ref. (14). The connection of this model to the scheme discussed here remains unclear, although some common features exist. In our approach we would also expect that in the case of one flavor the spin of a constituent quark is cancelled partially or fully by the "anomaly cloud". Thus we see no qualitative difference between the cases of one or two (three,...) flavors. On the other hand in the Skyrme model the case of one flavor is not defined.

The picture of "constituent quarks", carrying a polarized "anomaly cloud", described here implies that many aspects of hadronic physics, especially those in which polarization and spin aspects are relevant, must be reconsidered. Among them are the magnetic moments of the baryons, the polarization phenomena of hyperons in hadronic processes and the spin asymmetries observed in strong interaction processes. Many further tests of the ideas presented here can be envisaged, once spin asymmetries can be measured in electroweak lepton–hadron reactions at high energies. The

generation of a cloud of ($\bar{q}q$)–pairs by the QCD anomaly reminds us of the "Cooper pairs" in the BCS–theory of superconductivity. Indeed there are some analogies between superconductivity and hadronic physics in the chiral limit, e.g. the appearance of the mass gap, which in QCD is related to the anomaly as well as to the dynamical breaking of scale invariance and the chiral symmetry, and the presence of pairing forces, which in QCD are responsible for the removal of the Goldstone pole in the singlet axialvector channel.

In this lecture I have described why polarized constituent quarks should be surrounded by a cloud of polarized quark–antiquark pairs. Our reasoning was entirely based on phenomenological arguments. It would be interesting to see how these polarized pairs are generated dynamically, via those nonperturbative effects, due to instantons etc., which are responsible also for the QCD mass gap and the breaking of the chiral symmetry in the axial singlet channel. An explicit dynamical model along these lines is not yet available..

References

1. H. Fritzsch and M. Gell–Mann, Proc. of XVI. Int. Conference on High Energy Physics, Vol. 2, p. 135 (Chicago, 1972)
2. G. Baum et al., Phys. Rev. Lett. 51 (1983) 1135
    J. Ashman et al., Phys. Lett. B 206 (1988) 364
    V.W. Hughes et al., Phys. Lett. B 212, (1988) 511
3. M. Bourquin et al., Zeitschrift für Physik C 21 (1983) 27
    R.L. Jaffe and A. Manohar, MIT–preprint (1989)
4. See e.g.: H. Pagels, Phys. Rep. 16 (1975) 219
5. H. Fritzsch, Phys. Lett. 229 B (1989) 1605
6. G. Veneziano, Mod. Phys. Lett. A 17 (1989) 1605
7. H. Fritzsch, Phys. Lett. 256 B (1991) 75
8. H. Fritzsch, Mod. Phys. Lett. A 5 (1990) 625
9. U. Ellwanger and B. Stech, U. of Heidelberg preprint 1990
10. M. Goldberger and S.B. Treiman, Phys. Rev. 110 (1958) 1178
11. H. Fritzsch, Phys. Lett. 229 B (1989) 122
12. G. Altarelli and G. Roos, Phys. Lett. B 212 (1988) 391
13. R. Carlitz, J. Collins and A. Mueller, Phys. Lett. B 214 (1988) 229
14. S. Brodsky, J. Ellis, and M. Karliner, Phys. Lett. B 206 (1988) 309.

MEASUREMENT OF HOT SPOTS IN THE PROTON AT HERA

J.Bartels

II.Institut für Theoretische Physik, Universität Hamburg

Abstract: Hot spots are regions in the proton with a large density of small-x partons. In this paper we discuss the measurement of these regions by means of an associated jet analysis. After a short outline of the general idea we present an analytical estimate of the associated jet cross section and compare it with a MonteCarlo simulation. We then discuss the jet kinematics in the HERA frame and describe the results of a jet analysis on detector simulated data in order to illustrate the feasibility of this measurement at HERA.

1   Introduction

Perturbative QCD suggests that the gluon density in the proton increases with decreasing Bjorken-x x_B for very small values of x_B. This increase could be as fast as $x^{-\lambda}$ with $\lambda \simeq 0.5$, and at some point will lead to a violation of unitarity. To counter this violation, it is expected that partons will recombine or annihilate when the density gets large, leading to screening corrections which damp the rise. Two types of scenarios are envisaged for the screening corrections: either the growth of the gluon density occurs globally over the full size of the proton, or alternatively the growth will be essentially concentrated in small regions in the proton, e.g. around the valence quarks. In the latter case these high gluon density regions are called "hot spots".

A measurements of structure functions alone cannot distinguish between these two possibilities, since they describe probabilities averaged over the full transverse size of the proton. One therefore has to look for other measurements, designed for investigating parton distributions in limited subregions of the proton. Such a measurement has been suggested by A.Mueller [1,2] and will further be discussed in this talk.

The main motivation for searching for these "hot spots" comes from the interest in establishing the presence of screening corrections in the small-x_B region. As to structure functions, these screening corrections would manifest themselves in the shape of the x_B distribution and its evolution as a function of Q^2. Estimates indicate that they could be at work already in the kinematical domain accessible by the new electron-proton collider HERA [3,4,5,6], but $F_2(x, Q^2)$ measurements alone may not lead to an unambiguous conclusion [7]). When averaged over the full proton, the density of slow gluons may not yet be large enough to make these effects clearly visible. If, on the other hand, subregions with high densities of low-x partons are really present inside the proton, a measurement of the local gluon distribution is expected to give a much clearer signal than the structure function. But even before adressing the question of screening corrections, it would be of interest to measure these "hot spots", since the corresponding cross section tests a piece of QCD dynamics which has not been studied before, namely the BFKL-ladder model [8,9,10] of the Pomeron (Balitsky, Fadin, Kuraev and Lipatov).

In the following section we first describe A.Mueller's idea in more detail. Then we present an analytic estimate of the differential cross section, based upon the leading singularity of the BFKL-Pomeron, and compare with a Monte Carlo simulation. In section 4 we will discuss the kinematics of the jets suitable for the hot spot analysis. In section 5 a jet analysis of detector simulated data is used to evaluate the experimental possibilities of this measurement at HERA.

2   The idea

The idea of the measurement is illustrated in Fig.1a: one searches for a jet in the final state, which is characterized by momentum fraction x and transverse momentum squared k_T^2. Note that the associated jet is not the jet resulting from

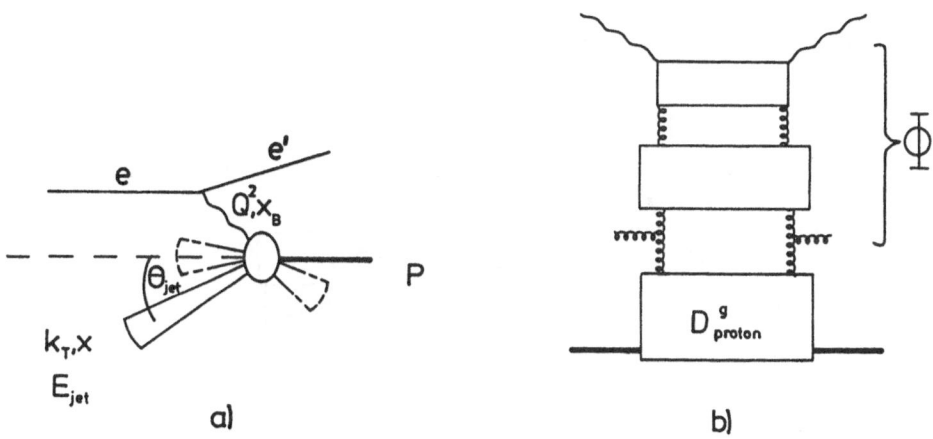

Figure 1: (a) The "hot spot" process ep → e+jet+anything, the dashed jets represent the current jet and the target fragmentation jet: b) diagrammatic illustration of eq.(1)

the quark struck by the virtual photon, which instead has momentum fraction x_B and virtuality Q^2. For the hot spot measurement x_B should be as small as possible, and Q^2 should be large enough to justify the use of perturbative QCD. In order to have a large rapidity difference between the jet and the struck parton, the momentum fraction of the jet should be as large as possible, say of the order unity. k_T^2 is a measure of the square of the average transverse distance between the parent gluon of the jet and the struck parton, hence if $k_T^2 \leq Q^2$, this process explores the parton cloud of diameter $1/k_T$ around the struck parton.

As a result of these requirements, there will be not much evolution in Q^2 between the jet and the photon vertex, whereas the difference in momentum fraction is forced to be large: this is why the ladder diagrams above the jet vertex contain only gluons, and they have to be evaluated in the Regge limit ("Pomeron") and not in the "normal" large-Q^2 limit. The leading approximation within QCD (sum of gluon ladders including the reggeization of the gluon; see Fig.2) has been derived more than 15 years ago [8,9,10], but so far the interest has been somewhat academic. The application to purely hadronic processes has always been hindered by the large QCD coupling which makes it hard to justify the use of perturbation theory. The present situation, on the other hand, is much more favorable: with Q^2 and k_T^2 being large, we are now probing the Regge limit with a small α_s, and hence are allowed to use perturbation theory. Consequently, a measurement of jets with the above kinematics will allow, for the first time, to test this QCD prediction for the Regge limit.

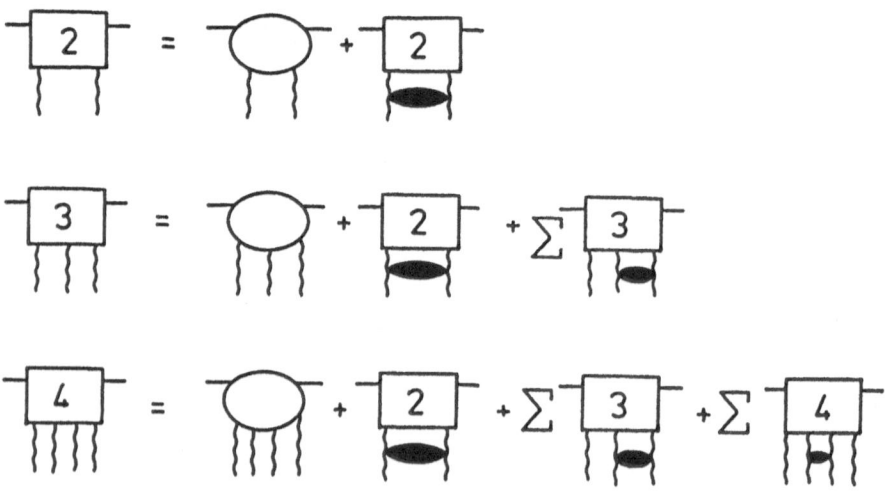

Figure 2: Illustration of the BFKL-ladders (first line only) and the first unitarity corrections to it (full set of equations). The circles denote a closed fermion loop, the wavy lines stand for reggeized gluons.

However, for very large energies (in our case: large $\frac{s}{s_B}$) the BFKL-ladders cannot be correct. With the leading behaviour being proportional to

$$\left(\frac{x}{x_B}\right)^{\frac{4N_c \ln 2}{\pi}\alpha_s}$$

(2.1)

they eventually violate the unitarity bound, in complete analogy with the standard one-ladder approximation which forms the basis of the linear Gribov-Lipatov-Altarelli-Parisi [11] evolution equations. For the latter ones, Gribov, Levin, and Ryskin [12] suggested to restore unitarity by simply adding a nonlinear term in the evolution equations. All existing estimates of screening corrections to deep inelastic structure functions are based upon this equation. However, it has been found recently [13] that this equation does not represent the leading diagrams. In the case of the BFKL-ladders, the first unitarity corrections (Fig.2) have been calculated only very recently citeBa1, and so far no numerical bomputation has been done. It is, however, clear that these corrections will also result in a suppression of the differential jet cross section, compared to the one-ladder approximation of the BFKL Pomeron. How strong this suppression is and at what values of $\frac{x}{x_B}$ they become visible is not known yet.

3 Differential cross section

The differential cross section for the jet production process shown in Fig. 1b is given by:

$$xk_T^2 \frac{d^4\sigma}{dx_B dy dx dk_T^2} = \frac{4\pi\alpha^2}{Q^2} \cdot \frac{1}{2} \cdot \frac{12\alpha_s(k_T^2)}{4\pi}$$

$$\sum_{q_i} [yM\Phi_1(\frac{x_B}{x}, Q^2, k_T^2)$$

$$+\frac{1-y}{y}\frac{\nu}{Mx_B}\Phi_2(\frac{x_B}{x}, Q^2, k_T^2)]$$

$$\cdot\{xD^g(x, k_T^2) + \frac{4}{9}\sum_{\text{flv.}}[xD^q(x, k_T^2) + xD^{\bar{q}}(x, k_T^2)]\}$$

(3.1)

where only the most singular part of the splitting functions was kept. Here $D^A(x, k^2)$ denotes the probability of finding parton A inside the proton, and Φ_i are related to the BFKL QCD gluon ladder graphs with a closed quark loop at its upper end (all evaluated in the Regge limit). The first sum in eq. (3.1) refers to the quarks in the upper loop, the second one belongs to quarks and antiquarks inside the proton. An approximate analytic expression for the functions Φ_i in eq. (3.1), which takes into account only the leading angular momentum plane singularity of the BFKL-Pomeron, has been derived in [14,15,16]. It leads to:

$$xk_T^2 \frac{d^4\sigma}{dx_B dy dx dk_T^2} = \frac{4\pi\alpha^2}{Q^2} \frac{1}{2} \cdot \frac{12\alpha_s(k_T^2)}{4\pi}$$

$$\cdot \sum_{q_i} [y \frac{e_{q_i}^2 \alpha_s}{\sqrt{2}} \frac{9\pi^2}{64} + \frac{1-y}{y} \frac{e_{q_i}^2 \alpha_s}{\sqrt{2}} \frac{11\pi^2}{32}] \sqrt{\frac{Q^2}{k_T^2}} \frac{1}{\sqrt{\ln \frac{x}{x_B}}}$$

$$\cdot \frac{1}{\sqrt{28 N_c \alpha_s (Q^2) \zeta(3)}} (\frac{x}{x_B})^{\tilde{\chi}(0,0)} \exp [-\frac{(\ln k_T^2/Q^2)^2}{2\tilde{\chi}''(0,0) \ln \frac{x}{x_B}}]$$

$$\cdot \{x D^g(x, k_T^2) + \frac{4}{9} \sum_{\text{flv.}} [x D^q(x, k_T^2) + x D^{\bar{q}}(x, k_T^2)]\} \qquad (3.2)$$

with

$$\tilde{\chi}(0,0) = \frac{4N_c}{\pi} \alpha_s(k_T^2) \ln 2$$

and

$$\tilde{\chi}''(0,0) = \frac{28 N_c}{\pi} \alpha_s(k_T^2) \zeta(3)$$

where e_{q_i} is the charge of quark q_i and N_c the number of colours.

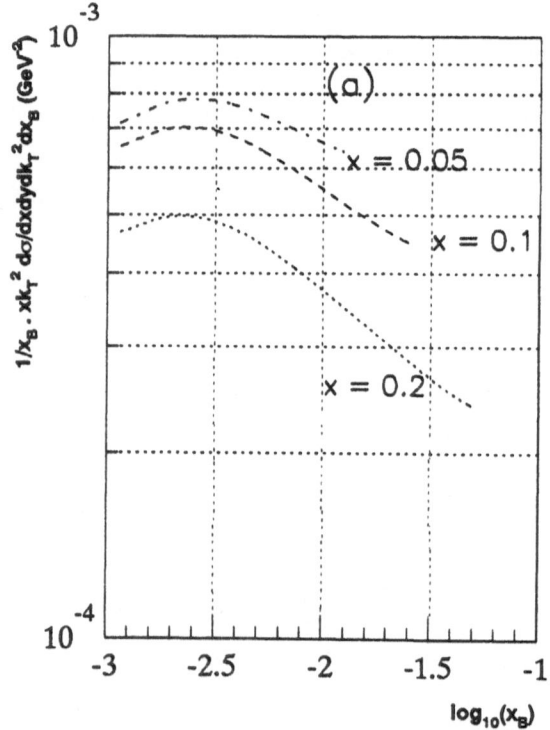

Figure 3: Differential cross section for $Q^2 = 100$ GeV^2 and $k_T^2 = 50$ GeV^2 for different values of x

From eqs.(1) and (2) one can deduce that in a log-log plot the functions $M\Phi_1$ and $\frac{\nu}{Mx_B}\Phi_2$ as function of x_B are approximately straight lines. In order to make this behaviour visible also in the differential cross section of eq.(2) we multiply by $1/x_B$. Fig.3 shows as an example a plot of eq.(2) as function of x_B for various x values and for $Q^2 = 100$ GeV^2 and $k_T^2 = 50$ GeV^2, using the parton distributions EHLQ1 [17] for the proton. The curves are approximately straight lines for $\log(x_B) > -2.5$. For $x \simeq x_B$ eq.(2) diverges due to the $1/\sqrt{\ln(x/x_B)}$ factor. In this region we do not expect that the approximations made are valid any longer, thus the numerical calculation in Fig.3 was limited to values $x_B < x/4$.

Since we do not know a priori, how close the approximation that we used is to the "true" behavior of the BFKL Pomeron, it is important to compare with other numerical calculations. In [14], results from (3.2) have been compared with a Monte Carlo simulation of (3.1). Good agreement is found for not too large x_B, but several caveats are in place. First, the Monte Carlo program used in this calculation is based upon the deep inelastic GLAP-evolution program and not on the BFKL Pomeron. This may be not too serious, since it is believed that in the HERA kinematic region there is not yet much difference between the two schemes. Also, at low x_B different Monte Carlos tend to give quite different answers; in fact, using another Monte Carlo routine, one obtains data points which are slightly higher. Secondly, the program uses a running α_s, whereas the analytic approximation (3.2) to the BFKL Pomeron keeps α_s fixed. When solving the integral equation of the BFKL Pomeron with running α_s, the answer generally lies lower than in case of fixed coupling constant, but it depends rather strongly upon the infrared cutoff which now has to be introduced. This is nicely demonstrated in [16], where the analytic result used in (3.2) is compared with a numerical integration of the BFKL equation with running α_s. Combining all these pieces of information, I presently conclude that the analytic result in (3.2) is rather too low, and the next correction should be calculated. Most important, however, is a numerical solution to the BFKL equation with fixed α_s, which could be compared with (3.2).

4 Kinematics

The translation from the variables k_T and x to HERA variables Θ, E_{jet} for fixed x_B, Q^2 is most easily done with the following two equations:

$$\frac{k_T^2}{x} = 2E_{jet}E_p(1 - \cos\Theta_{jet})$$ (4.1)

$$xsy - \frac{x_B}{x}k_T^2 = 2E_{jet}(E_e - E_e')$$

$$\cdot [1 - \frac{\sqrt{E_e^2 + E_e'^2 - 2E_e E_e' \cos\Theta_e}}{E_e - E_e'}\cos\Theta_{jet\gamma}]$$ (4.2)

Here E_e, E_e', E_p, E_{jet} denote the energies of the incoming electron, outgoing electron, proton and the jet, resp, and $\Theta_{jet}, \Theta_{e'}, \Theta_{jet\gamma}$ stand for the angles between jet and proton, between outgoing electron and incoming electron, and between jet and photon, respectively; s is the center of mass energy squared ($= 98400$ GeV^2 at HERA for an electron beam of 30 GeV and a proton beam of 820 GeV) and y is Bjorken-y. Since the jet direction lies, in general, outside of the scattering plane of the beam axis with the outgoing electron (or the photon), these two equations do not completely specify the jet direction in the HERA system: the azimuthal symmetry of the cross section in the frame where photon and proton were collinear transforms into a slightly deformed cone in the HERA system. Eqs.(4.1) and (4.2) allow, however, to determine the angles under which this cone intersects with the scattering plane. The regions of constant angle in the $x - k_T^2$ plane turn out to be narrow bands, which will be represented by lines in the following, calculated by averaging over the azimuthal degree of freedom.

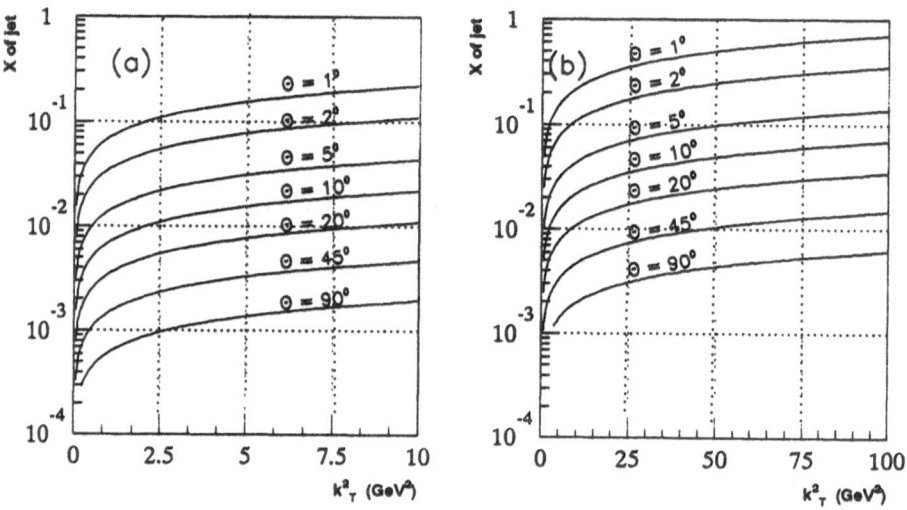

Figure 4: Lines of constant jet angles w.r.t. the proton direction for collisions with (a) $x_B = 2 \cdot 10^{-4}$ and $Q^2 = 10$ GeV^2; (b) $x_B = 10^{-3}$ and $Q^2 = 100$ GeV^2

Fig.4a shows lines of constant angle between the jet direction and the proton beam axis for interactions at HERA with $x_B = 2 \cdot 10^{-4}$ and $Q^2 = 10$ GeV^2. High x values can only be reached for jets with very small angles w.r.t. the proton. Since particles will be lost in the beampipe for experiments at HERA, it will be very difficult to measure jets with angles below 5^0. Practically, this limits the accessable x range at HERA to a maximum value of about 0.04 for interactions at low x_B, Q^2. The lines of constant jet angle are shown for $x_B = 10^{-3}$ and $Q^2 = 100$ GeV^2 in Fig. 4b. Hence somewhat higher x values can be reached for jets with high k_T^2 values: e.g. for jets with an angle of 5^0, one reaches the region $x \geq 0.1$ for k_T^2 values larger than 50 GeV^2.

An important quantity for the analysis of hot spot jets is the accessable range x/x_B where the jet cross section can be determined. The largest range of x/x_B values is obtained with low (Q^2, x_B) events. Fig 4a. shows that maximally $x/x_B \sim 200$ can be reached for $x_B \sim 10^{-4}$.

5   Monte Carlo simulation

Complete *ep* deep inelastic scattering events were generated according to cross section with LEPTO5.2 [18], to study the feasibility of the hot spot jet measurement. In this study we have analysed quark or gluon jets with high *x* values, produced by initial state QCD radiation. This jet production mechanism corresponds to the one shown in Fig. 1b for hot spot jets. Hence, though the event generator does not include the "new physics" of hot spot formation in the proton, events with associated jets with $Q^2 \simeq k_T^2$ will be produced with the correct kinematics. Additionally it turned out that the associated jet production cross section given by the Monte Carlo program is similar to the one presented in Fig. 3 [14]. For the simulation the kinematical variable Q^2 was taken for the QCD scale, and the EHLQ1 distributions were taken for the parton densities in the proton. The jet variables x and k_T^2 were calculated using eqs.(4.1),(4.2).

In order to get a sample of jets candidates for the hot spot analysis the following preselections were made at the parton level: $Q^2 > 10$ GeV^2, $k_T^2/Q^2 >$ 0.1, $E_{parton} > 20$ GeV and $x_B < .01$. Additionally only events which contain one parton surviving these cuts were kept. About 3000 events remain out of a sample corresponding to 10 pb^{-1}. The detector response was calculated with the

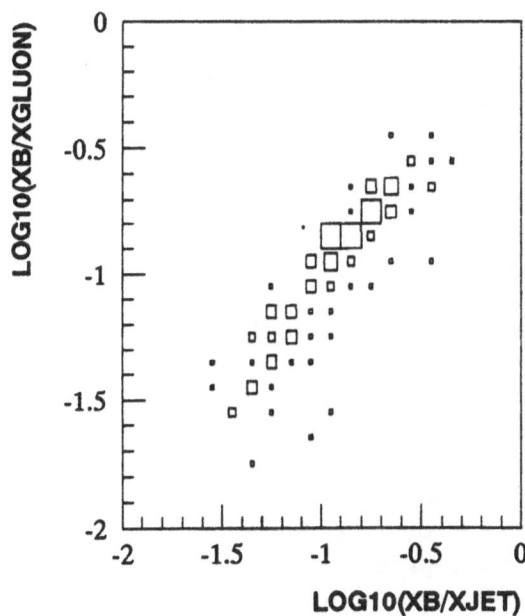

Figure 5: Correlation between x_B/x_{gluon} and x_B/x_{jet} for jets at the detector level

H1 simulation program H1SIM (V2.04), and the calorimeter energy clusters were reconstructed with the reconstrucion program H1REC (V3.01). To reconstruct the parton jet from the hadronic final state the jet algorithm DECO (G. Knies in [7]) was used. A set of cuts was optimized to select a jet sample where DECO reconstructs the jets of interest with high efficiency: $5^0 < \Theta_{jet} < 45^0$, $E_{jet} > 17$ GeV, $p^2_{Tjet}/Q^2 > 0.1$, $p^2_{Tjet} > 2.0$ GeV^2/c^2 and $m_{jet} < 3.0$ GeV. With these selections we effectively enriched the sample of jets where the parton history information is not washed out by the hadronization process [19]

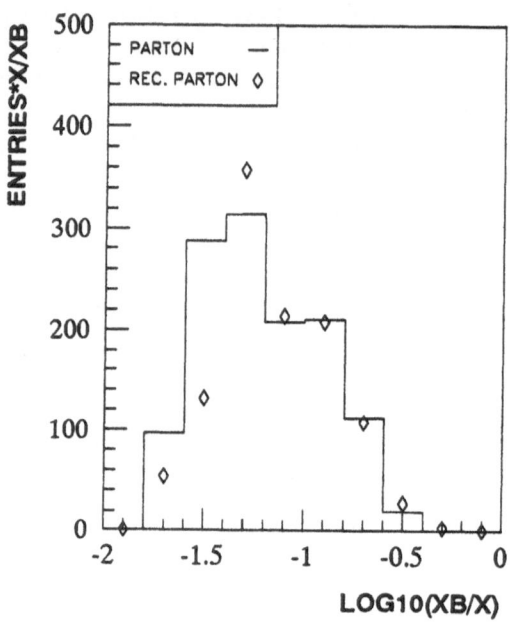

Figure 6: x_B/x distributions for partons (solid line) and reconstructed jets (diamonds) at the detector level

In total 187 events survive the cuts, and the correlation between the original partons and the reconstructed jets is shown in Figs. 5 and 6. It shows that even after detector effects are taken into account a good correlation between x of the parton and x_{jet} persists, except for the region $x_B/x < .025$ where reconstruction losses become large.

Fig. 7 shows the ratio of the $\log(x_B/x)$ distribution for the reconstructed jets to all partons selected at the generation level. It shows that the structure of the input distribution is kept to a large extend, such that one can hope to keep the systematic uncertainties, resulting from jet selection cuts and parton kinematics reconstruction, in control to about 20%-25% in this region. This in turn implies that the deviation of the BFKL Pomeron due to saturation effects should be larger, say at least 20 %, in order to have a chance to observe the effect at HERA.

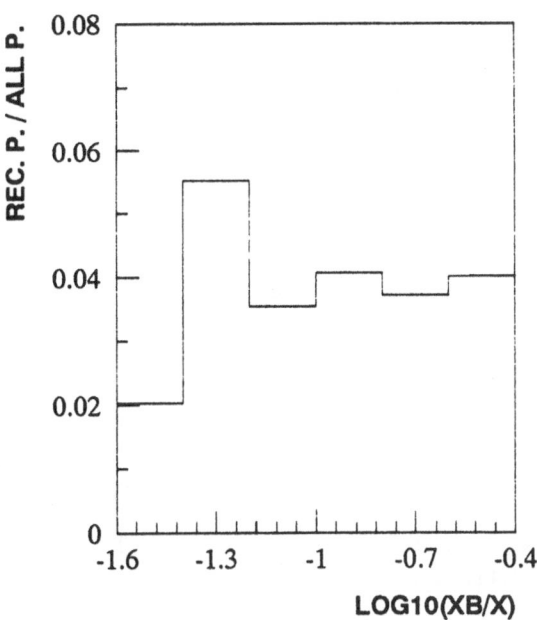

Figure 7: Ratio of the distribution of x_B/x for reconstructed partons to all partons

6 Conclusions

In this paper we followed the ideas of Mueller and studied aspects of the hot spot measurement at HERA. These measurements are of interest both for verifying the existence of screening effects in deep inelastic scattering at small-x values and for testing the gluon-ladder model for the Pomeron in QCD. An analytical expression for the jet cross section was presented which was based upon the leading singularity in the integral equation of Lipatov et al. The signature for hot spots in the proton will be a suppression of this jet cross section, but this suppression still needs to be calculated. The kinematics of the hot spot jet shows the very forward nature of the jets of interest. Practically, with present HERA experiments it will be hard to detect jets with x values larger than 0.1. Monte Carlo event generator estimates shows that the number of potentially useful jets produced at HERA is large (several hundred per pb^{-1}) which makes it tempting to try to measure the cross section. A complete detector simulation and reconstruction analysis was carried out with the encouraging result that the correlation between selected partons and reconstructed jets can be maintained and there is hope to keep the experimental systematics uncertainties to the level of 20% in a region of x_B/x.

Acknowledgements: I gratefully acknowledge the pleasant collaboration with A.DeRoeck and the useful discussions with J.Kwiecinski, A.Mueller, and A.Martin.

References

[1] A.H.Mueller, *Nucl.Phys.B (Proc.Suppl.)* 18C, 125 (1991).

[2] A.H.Mueller and Navelet, *Nucl.Phys.B* 282, 727 (1987).

[3] J.Kwiecinski, A.D.Martin, R.G.Roberts, W.J.Stirling, *Phys.Rev.D* 42, 3645(1991).

[4] J.Bartels, J.Bluemlein, G.Schuler, *Zeitschr.Phys. C* 50, 91(1991).

[5] J.Kwiecinski, A.D.Martin, P.J.Sutton, Durham preprint DTP/91/10.

[6] V.T.Kim, M.G.Ryskin, DESY preprint 91-064.

[7] Proceedings of the Workshop "Physics at HERA", held at DESY, Oct. 1991, ed. by W. Buchmüller and G. Ingelman, Hamburg 1992

[8] E.A.Kuraev, L.N.Lipatov, V.S.Fadin, *Sov. Phys. JETP* 44, 443(1976) ; *Sov. Phys. JETP* 45, 199(1977).

[9] Ya.Ya.Balitsky, L.N.Lipatov *Sov.Jour.Nucl.Phys.* 28, 822(1978) ; *JETP Letters* 30, 355(1979).

[10] L.N.Lipatov, *Sov.Phys.JETP* 63, 904(1986) and references therein.

[11] V.N.Gribov and L.N.Lipatov, *Sov.Journ.Nucl.Phys.* 15, 438 and 675 (1972); G.Altarelli and G.Parisi, *Nucl.Phys.* 126, 297(1977).

[12] L.V.Gribov, E.M.Levin, and M.G.Ryskin, *Phys. Rep.* 100, 1 (1982)

[13] J.Bartels, DESY-preprit DESY 92-114.

[14] J. Bartels, A.De Roeck, M.Loewe, *Z.Phys.* C54, 635(1992).

[15] W.-K.Tang, *Phys.Lett* B278, 363(1991).

[16] J.Kwiecinski, A.D.Martin, P.J.Sutton, *Phys.Lett* B287, 254(1992); *Phys.Rev* D46, 921(1992).

[17] E.Eichten et al., Rev. Mod. Phys. 56 (1984) 579 and erratum 58 (1986)

[18] G. Ingelman, LEPTO5.2 event generator.

[19] J.Bartels, M.Besancon, A.De Roeck, J.Kurzhoefer, in [7], p.203.

Deep Inelastic Scattering
and diquarks

Mauro Anselmino

Department of Theoretical Physics, University of Torino
and INFN, Sezione di Torino
Via P. Giuria 1, 10125 Torino, Italy

The most comprehensive and detailed analyses of the existing data on the structure function $F_2(x, Q^2)$ of free nucleons, from the deep inelastic scattering (DIS) of charged leptons on hydrogen and deuterium targets, have proved beyond any doubt that higher twist, $1/Q^2$ corrections are needed in order to obtain a perfect agreement between perturbative QCD predictions and the data [1].

These higher twist corrections take into account two quark correlations inside the nucleon; it is then natural to try to model them in the quark-diquark model of the proton [2,3]. We know that proton and neutron configurations in which two out of the three valence quarks are closer to each other than to the third quark are the dominant ones [3]; the quark-diquark model of the nucleon treats these two close quarks as an effective constituent which, at least for intermediate values of Q^2, behaves as a single dynamical particle. In so doing all interactions between the two quarks inside the diquark, both perturbative and non perturbative, are supposed to be taken into account. In that the model should not only account for the dominant higher twist corrections but also for higher order and non perturbative ones. At large values of Q^2 only the leading, $1/Q^2$, corrections are significant; the diquarks is then resolved in two quarks and these leading corrections can be somehow related to the $1/Q^2$ corrections one obtains in the usual, pure quark, Operator Product Expansion analysis of deep inelastic scattering. Of course, in the diquark scheme, one might neglect genuine higher twist corrections originating from real three quark configurations of the nucleon, but we suppose the quark-diquark configuration to be the dominant one.

If we picture the nucleon as an effective quark-diquark bound state we must then reformulate the parton model analysis of deep inelastic scattering, allowing for spin 0 (scalar diquarks) and spin 1 (pseudo-vector diquarks) constituents. This has been done in Ref. [4] (and in several previous papers there quoted); the diquark contributions to the unpolarized (F_1 and F_2) and the polarized (g_1 and g_2) nucleon structure functions are given by:

$$F_1^{(S)} = 0$$

$$F_2^{(S)} = e_S^2 S(x) x F_S^2$$

$$F_1^{(V)} = \frac{1}{3} e_V^2 V(x) \left(1 + \frac{\nu}{2m_N x}\right) G_2^2$$

$$F_2^{(V)} = \frac{1}{3} e_V^2 V(x) x \left\{ \left[\left(1 + \frac{\nu}{m_N x}\right) G_1 + \right.\right.$$
$$\left.\left. - \frac{\nu}{m_N x} G_2 + 2 m_N \nu x \left(1 + \frac{\nu}{2m_N x}\right) G_3\right]^2 + 2\left[G_1^2 + \frac{\nu}{2m_N x} G_2^2\right]\right\}$$

$$F_1^{(S-V)} = \frac{1}{2} e_S^2 S(x) x^2 m_N^2 \left(1 + \frac{\nu}{2m_N x}\right) G_T^2$$

$$F_2^{(S-V)} = \frac{1}{2} e_S^2 S(x) x^2 m_N \nu G_T^2 \tag{1}$$

$$F_1^{(V-S)} = \frac{1}{6} e_S^2 V(x) x^2 m_N^2 \left(1 + \frac{\nu}{2m_N x}\right) G_T^2$$

$$F_2^{(V-S)} = \frac{1}{6}e_S^2 V(x)x^2 m_N \nu G_T^2$$

$$g_1^{(V)} = \frac{1}{4}e_V^2 \Delta V \left[\left(2 + \frac{\nu}{m_N x}\right)(G_1 G_2 + x m_N \nu G_2 G_3) - \frac{\nu}{2m_N x}G_2^2 \right]$$

$$g_2^{(V)} = \frac{1}{4}e_V^2 \Delta V \frac{\nu}{2x m_N} \left[\left(2 + \frac{\nu}{m_N x}\right) \times \right.$$

$$\left. (G_1 G_2 + x m_N \nu G_2 G_3) - \left(1 + \frac{\nu}{m_N x}\right)G_2^2 \right]$$

where F_S is the scalar diquark form factor, G_1, G_2 and G_3 are the three form factors appearing in the photon-vector diquark coupling and G_T is the form factor resulting fron the scalar-vector or vector-scalar transition. The apices S, V, $S-V$ and $V-S$ refer, respectively, to scalar diquark, vector diquark, scalar-vector and vector-scalar transitions. The variables in (1) are the usual ones in DIS, $q^2 = -Q^2$, $P^2 = m_N^2$, $x = Q^2/(2P \cdot q)$, $P \cdot q = m_N \nu$ where P and q are respectively the nucleon and virtual photon four-momenta. $S(x)(V(x))$ is the density number of scalar (vector) diquarks with momentum xP inside the nucleon and $\Delta V(x)$ is the difference between the number of vector diquarks with spin parallel to the nucleon spin and those with spin antiparallel.

Let us consider first the results (1) in the limit of pointlike diquarks. In such a case the form factors are given by

$$F_S(0) = 1$$
$$G_1(0) = 1 \quad G_2(0) = 1 + \kappa \quad G_3(0) = 0 \quad (2)$$
$$G_T(0) = 0$$

where κ is the vector diquark anomalous magnetic moment.

Whereas scalar pointlike diquarks do not introduce any Q^2 dependance in the structure functions F_1, F_2, g_1 and g_2, vector pointlike diquarks lead to strong scaling violations:

$$F_1^{(V)} = \frac{1}{3}e_V^2 V(x)\left(1 + \frac{Q^2}{4m_N^2 x^2}\right)(1 + \kappa)^2$$

$$F_2^{(V)} = \frac{1}{3}e_V^2 V(x)x \left[3 + \frac{Q^2}{2m_N^2 x^2}(1 + \kappa^2) + \left(\frac{Q^2}{2m_N^2 x^2}\right)^2 \kappa^2 \right]$$

$$g_1^{(V)} = \frac{1}{4}e_V^2 \Delta V(x,S)(1 + \kappa)\left[2 + \frac{Q^2}{4m_N^2 x^2}(1 - \kappa)\right] \quad (3)$$

$$g_2^{(V)} = \frac{1}{4}e_V^2 \Delta V(x,S)\frac{Q^2}{4m_N^2 x^2}(1 + \kappa)\left[1 - \kappa - \frac{Q^2}{2m_N^2 x^2}\kappa\right]$$

which would be incompatible with experiments. Notice also that the scaling violations in $F_2^{(V)}$ and $g_2^{(V)}$ are of order Q^4, unless the anomalous magnetic moment of the vector diquark, κ, is zero, in which case all scaling violating terms are proportional to Q^2.

Of course, diquarks, bound states of two quarks, are not pointlike objects and any realistic comparison with experimental data should take into account their form factors $F_S, G_{1,2,3}$ and G_T. We know that, apart from the QCD logarithmic ones, power like scaling violations of order $1/Q^2$ are allowed in F_1 and F_2 by the experimental data [1]. We also know the large Q^2 behaviour of the ratio

$$R = \frac{F_2}{2x F_1}\left(1 + \frac{2m_N x}{\nu}\right) \sim \frac{1}{Q^2}, \quad (4)$$

correctly predicted in the quark parton model due to the Callan-Gross relationship $F_2^{(q)} = 2xF_1^{(q)}$. We then demand that these two conditions:

i) scaling violations proportional to $1/Q^2$ or smaller (we do not deal here with the QCD ones)

ii) $R \sim 1/Q^2$

still hold true when introducing diquarks as constituents. We extend point i) to the polarized structure functions g_1 and g_2 as well.

The two above demands can be satisfied if we make the following choices of diquark form factors:

$$F_S \sim \frac{1}{Q^2} \quad G_1 = G_2 \sim \frac{1}{Q^2}$$

$$G_3 \sim \frac{1}{Q^6} \quad D_T \sim \frac{1}{Q^3} \tag{5}$$

or

$$F_S \sim \frac{1}{Q^2} \quad G_1 \sim G_2 \sim \frac{1}{Q^4}$$

$$G_3 \sim \frac{1}{Q^6} \quad D_T \sim \frac{1}{Q^3} \tag{6}$$

The asymptotic behaviours given in Eqs. (6) are indeed those predicted by a perturbative QCD analysis of vector particle form factors [5].

The introduction of diquarks as constituents does not seem to violate any fondamental properties of the nucleon structure functions; the scaling and Callan-Gross relationship violations induced by spin 0 or spin 1 partons are mitigated by the form factors and the final outcome, scaling violations of order $1/Q^2$, can be in agreement with the observed ones [1]. Of course, before drawing any definite conclusion, a detailed analysis of deep inelastic scattering data in the framework of the quark *and diquark* parton model is necessary. Such an analysis has been performed in some particular cases, mainly with scalar diquarks only or simplified vector ones (only one form factor) [6,7]; a complete and general study of DIS data with diquarks, including neutrino data, is still missing.

In Figs. 1-3 we show some recent results obtained by Dugne and Tavernier [8], using scalar diquarks only. In Fig. 1 the proton structure function $F_2^p(x, Q^2)$, obtained assuming the proton to be a quark-scalar diquark state, is shown as a function of Q^2 for different values of x. The Q^2 dependance comes both from the diquark form factor $F_S = M^2/(M^2 + Q^2)$, with $M^2 = 10\,(GeV)^2$, and from the logarithmic QCD evolution of the distribution functions. Excellent agreement with the BCDMS data [9] is obtained. In Fig. 2 the ratio F_2^p/F_2^n is given as a function of x, for two different values, $M^2 = 3$ and $M^2 = 10\,(GeV)^2$, of the parameter appearing in the diquark form factor; the results are compared with the SLAC data [10] and with others obtained in the usual quark model [11]. Fig. 3 plots the ratio R, Eq. (4), versus x, at fixed value of Q^2, and compares with the SLAC data [10] and the Callan-Gross relation, $F_2 = 2xF_1$.

We conclude by mentioning the Gottfried sum rule violation problem and the rôle of diquarks in trying to understand it.

The Gottfried sum rule [12] states that the integral

$$S_G = \int_0^1 dx \, \frac{F_2^p(x) - F_2^n(x)}{x}, \tag{7}$$

is given, in terms of quark distributions, by

$$S_G^q = \frac{1}{3} \int_0^1 dx \, [u(x) + \bar{u}(x) - d(x) - \bar{d}(x)] \tag{8}$$

$$= \frac{1}{3} \int_0^1 dx \, [u_v(x) - d_v(x)] = \frac{1}{3} \tag{8'}$$

As usual, the u and d quark x distributions (number densities of u and d quarks carrying a fraction x of the nucleon momentum) can be written as the sum of the valence and sea contributions, $u = u_v + u_s$ and $d = d_v + d_s$, and the antiquark distributions are given by $\bar{u} = u_s$ and $\bar{d} = d_s$. Eq.(8') holds if one assumes the nucleon sea to contain the same amount of $\bar{u}$ and $\bar{d}$ quarks.

However, the recent NMC measurement of the nucleon structure function ratio F_2^n/F_2^p in the DIS of muons on hydrogen and deuterium has led to the unexpected result, at $Q^2 = 4\,(GeV)^2$ [13]

$$S_G = 0.240 \pm 0.016 \tag{9}$$

This results clearly deviates from the quark parton model result of Gottfried, Eq. (8'). In one attempt of explaining such a discrepancy the prediction of the quark-diquark model of the nucleon for S_G has been computed to be (with some simplifying assumptions concerning the diquark form factors) [14]

$$\begin{aligned}
S_G^{qQ} = &\frac{1}{3} - \frac{4}{9}\sin^2\Gamma + \frac{8}{9}\sin^2\Gamma \int_0^1 dx\, f_{V_{uu}}(x) \left(1 + \frac{Q^2}{6m_N^2 x^2}\right) G^2(Q^2) \\
&+ \frac{4}{9}\left(1 - F^2(Q^2)\right)\sin^2\Gamma
\end{aligned} \tag{10}$$

where $\sin^2\Gamma$ is the probability of having a vector diquark inside the nucleon, $f_{V_{uu}}(x)$ is a (uu) vector diquark distribution function normalized as $\int_0^1 dx\, f_{V_{uu}}(x) = 1$ and G is the diquark form factor. At large Q^2 values this form factor (squared) is expected to vanish making the corresponding term in Eq. (10) negligible. Notice that only (uu) vector diquarks play a rôle in S_G.

The second line of Eq. (10) corresponds to the scattering of the virtual photon off a quark inside the diquark, which breaks up. As usual, we assume the probability of such a break up, $1 - F^2(Q^2)$, to be asymptotically 1, $F(Q^2 \to \infty) = 0$. Eq. (10) then clearly shows that at large values of Q^2 one recovers the Gottfried result, $S_G = 1/3$; at intermediate values of Q^2, where the diquark can still act as a single particle, S_G can differ from $1/3$. However, numerical estimates of (10), taking, as customary, $F(Q^2) \simeq G(Q^2)$, always lead to $S_G > 1/3$. One is forced to conclude that diquarks cannot explain the Gottfried sum rule violation, unless one adopts the rather unconventional and extreme attitude of allowing diquarks to act as single particle, without any break up, even at very large Q^2 values.

References

[1] J.J. Aubert et al., Nucl. Phys. B259, 189 (1985); M. Virchaux and A. Milsztajn, Saclay preprint DPhPE 91-08 (1991); U. Landgraf, talk at this Conference
[2] M.I. Pavkovic, Phys. Rev. D13, 2128 (1976)
[3] See, e.g., M. Anselmino and E. Predazzi, Editors, Proceedings of the Workshop on Diquarks, World Scientific (1989); M. Szczekowski, Int. J. Mod. Phys. A 4, 3985 (1989); M. Anselmino, S. Ekelin, S. Fredriksson, D. Lichtenberg and E. Predazzi, Luleå preprint, TULEA 1992:05
[4] M. Anselmino, F. Caruso, E. Leader and J. Soares, Z. Phys. C48, 689 (1990)
[5] A.I. Vainshtein and V.I. Zakharov, Phys. Lett. 72B, 368 (1978)
[6] S. Fredriksson, M. Jandel and T. Larsson, Z. Phys. C14, 35 (1982); C19, 53 (1983)
[7] E. Leader and M. Anselmino, Z. Phys. C41, 239 (1988)
[8] P. Tavernier and J.J. Dugne, Blaise Pascal University preprint PCCF RI 9205 (1992)
[9] A.C. Benvenuti et al., Phys. Lett. B223, 485 (1989); B237, 592 (1990)
[10] A. Bodek et al., Phys. Rev. D 20, 1471 (1979)
[11] M. Glück, E. Reya and A. Vogt, Z. Phys. C C48, 471 (1990)

155

[12] K. Gottfried, *Phys. Rev. Lett.* **18**, 1174 (1967)
[13] P. Amaudruz *et al.*, *Phys. Rev. Lett.* **66** 2712 (1991)
[14] M. Anselmino, V. Barone, F. Caruso and E. Predazzi, *Z. Phys.* **C55**, 97 (1992)

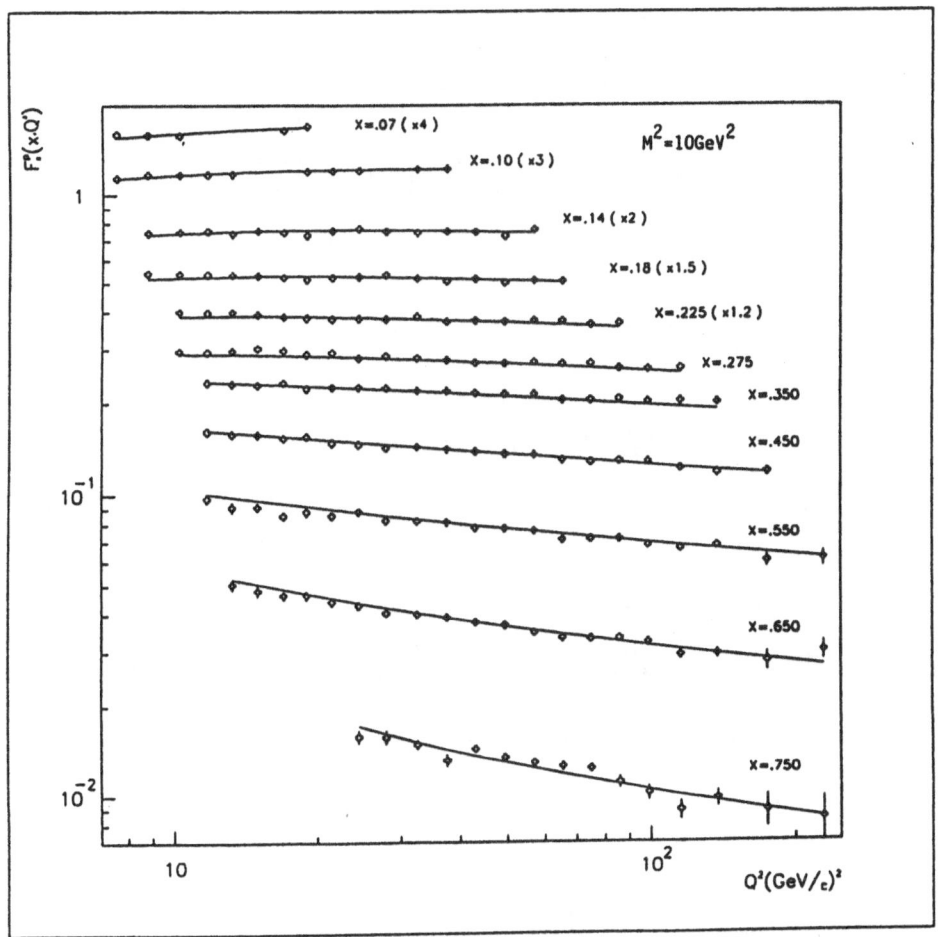

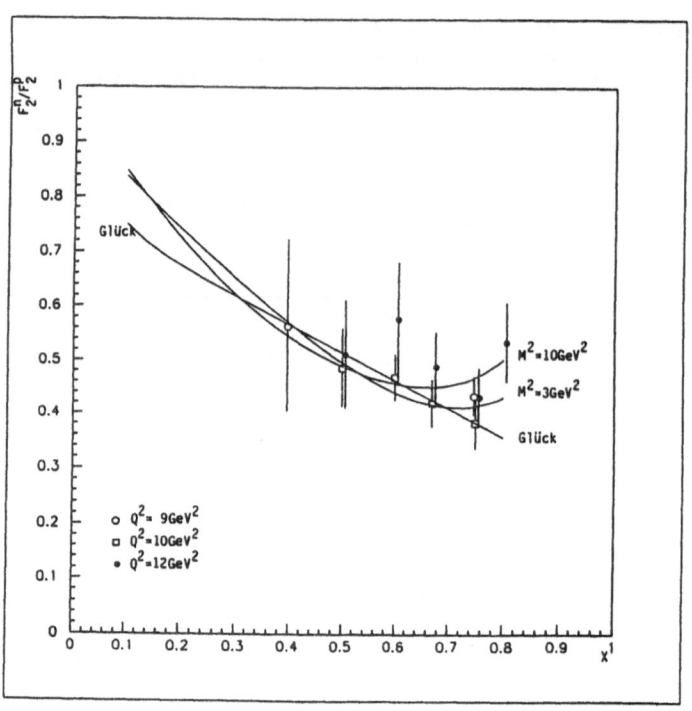

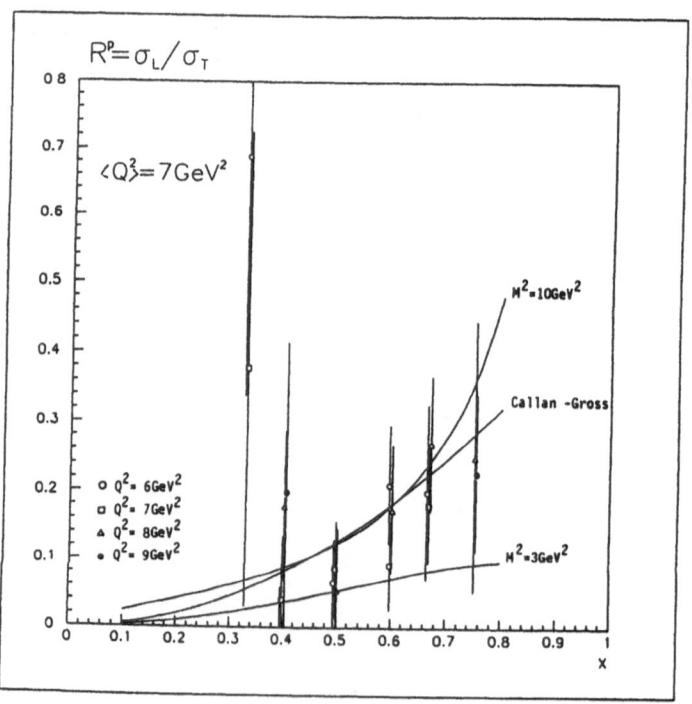

The role of spectator diquarks in deep inelastic structure functions

P.J. Mulders[1] and H. Meyer[2]

[1] National Institute for Nuclear Physics and High Energy Physics (NIKHEF-K), P.O. Box 41882, NL-1009 DB Amsterdam, the Netherlands.
[2] Institut für Theoretische Physik, Universität Regensburg, Universitätsstr. 31, D-8400 Regensburg, Germany.

abstract. The significance of the spectral distribution of diquarks in the nucleon for the $SU(6)$ symmetry breaking in deep inelastic structure functions is investigated in a field theoretical framework, which ensures correct kinematical and symmetry properties for the structure functions.

1. Introduction

Deep inelastic scattering (DIS) off hadronic targets probes the quark distributions in these targets. This is true irrespective of the complexity of the hadronic target as a consequence of the operator product expansion (OPE). For electromagnetic scattering processes, $\gamma^* + H \rightarrow$ anything, where the virtual photon is for instance produced by an electron, the structure functions measured in unpolarized scattering in terms of the quark parton distributions are

$$2\,F_1(x) = \frac{F_2(x)}{x} = \sum_q e_q^2\, q(x).\tag{1}$$

Omitting QCD radiative corrections ($\log Q^2$ dependence) the structure function depends only on the Bjorken scaling variable $x = Q^2/2P\cdot q$ (q is the momentum of the virtual photon, $Q^2 = -q^2$, and P is the momentum of the target). The quark distribution functions $q(x)$ represent specific momentum distributions of the quarks in the nucleon, to be specific $q(p^+/P^+)$ gives the probability to find a quark with +-component p^+. Similarly, for polarized scattering the dominant structure function $g_1(x)$ can be expressed in polarized-quark distributions, $\Delta q(x) = q_\uparrow(x) - q_\downarrow(x)$,

$$2\,g_1(x) = \sum_q e_q^2\, \Delta q(x).\tag{2}$$

Jaffe [1] has discussed in detail the formulation of the quark distributions in terms of the forward antiquark - nucleon amplitude χ,

$$\chi_{ij} = \int d^4x\, e^{ik\cdot x}\, \langle P|T\,\bar{\psi}_i(x)\psi_j(0)|P\rangle,\tag{3}$$

which reads

$$q(x) = \frac{1}{2M} \int \frac{d^4p}{(2\pi)^4}\, Tr\,(\gamma^+\chi(p,P))\, \delta\left(x - \frac{p^+}{M}\right)\tag{4}$$

$$= \frac{1}{4\pi} \int dz \, e^{-izMz} \, \langle P|\bar{\psi}(\xi) \, \gamma^+ \, \psi(0)|P\rangle \Big|_{\xi=(z,0,0,-z)} \tag{5}$$

$$\bar{q}(x) = -\frac{1}{4\pi} \int dz \, e^{-izMz} \, \langle P|\bar{\psi}(0) \, \gamma^+ \, \psi(\xi)|P\rangle \Big|_{\xi=(z,0,0,-z)}, \tag{6}$$

satisfying $\bar{q}(x) = -q(-x)$. Similar expressions for $\Delta q(x)$ with $\gamma^+ \gamma_5$ instead of γ^+, satisfying $\Delta \bar{q}(x) = \Delta q(-x)$, and expressions for gluon distributions exist [2]. The approach can be considered as a resummation of the OPE for free quark fields or it can be derived by substituting free quark currents in the definition of the hadronic tensor $W_{\mu\nu}(P, q)$ used in DIS. It is the intermediate state $|n\rangle\langle n|$ in the forward antiquark - nucleon amplitude that is considered as the definition of 'diquark' spectator in this paper.

It turns out that the spectator system, notably its mass, is a key ingredient for the shape of the quark distributions. This dependence on the spectator system has been discussed in phenomenological parametrizations of parton distributions in ref. [3] and in a bag model calculation in ref. [4].

2. The diquark spectator model

An explicit example of a model calculation of the forward antiquark-nucleon amplitude is the diquark spectator model for nucleon structure functions [5,6]. In this case the amplitude is saturated with a diquark coupling through a nucleon-quark-diquark vertex of the form $ig\bar{\Psi}(x)\psi(x)\phi(x)$, leading for a scalar diquark to the result

$$\frac{1}{2M} Tr(\gamma^+ \chi) = \frac{-g^2(p^2) \left[(2p \cdot P + 2mM)(p^+/M) - (p^2 - m^2)\right]}{(p^2 - m^2 + i\epsilon)^2 \left((P - p)^2 - M_R^2 + i\epsilon\right)}, \tag{7}$$

where m and p indicate the mass and momentum of the quark and M_R indicates the mass of the intermediate diquark state. The quark distribution according to Eq. 4 is then given by

$$q(x) = \int \frac{d^4 p}{(2\pi)^4} \frac{g^2(p^2)}{(p^2 - m^2)^2} \left[(2p \cdot P + 2mM)x - (p^2 - m^2)\right]$$

$$\times (2\pi)\delta \left((P - p)^2 - M_R^2\right) \theta(p^+) \theta(M - p^+) \delta \left(x - \frac{p^+}{M}\right)$$

$$= \frac{\theta(x)\theta(1-x)}{16\pi^2} \int_{\tau_{min}}^{\infty} d(\tau) \, g^2(-\tau)$$

$$\times \left\{ \frac{x \left[(M + m)^2 - M_R^2\right]}{(m^2 + \tau)^2} + \frac{(1-x)}{(m^2 + \tau)} \right\}$$

$$= \frac{\theta(x)\theta(1-x)}{16\pi^2(1-x)} \int_0^{\infty} d(p_\perp^2) \, g^2(-\tau) \frac{(xM + m)^2 + p_\perp^2}{(m^2 + \tau)^2}, \tag{8}$$

where $\tau = -p^2 = \tau_{min} + p_\perp^2/(1 - x)$ and

$$\tau_{min} = \frac{x}{1 - x} \left(M_R^2 - (1 - x) M^2\right). \tag{9}$$

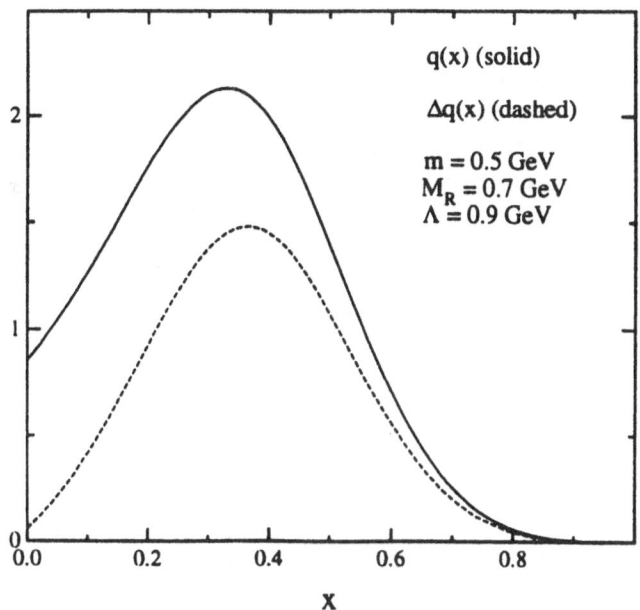

Figure 1: *An example of the distribution functions $q(x)$ and $\Delta q(x)$ in the diquark model using $g(p^2) \propto 1/(p^2 - \Lambda^2)$. For $x < 0$ the functions vanish.*

This exact result is obtained by calculating the integral in the complex (P_R^- plane, P_R being the spectator diquark's momentum, with some care or approximations because of singularities in the vertex function [7]. The θ functions originate from the fact that the singularities coming from the constituent propagator on the one hand and the pole from the spectator propagator on the other only lie only in opposite halves of the complex P_R^- plane if $0 \le x \le 1$.

For the polarized distributions one obtains

$$\Delta q(x) = \frac{\theta(x)\,\theta(1-x)}{16\pi^2(1-x)} \int_0^\infty d(p_\perp^2)\, g^2(-\tau) \frac{(x\,M + m)^2 - p_\perp^2}{(m^2 + \tau)^2} \quad . \tag{10}$$

Note that the axial charge of a quark, $\Delta q \equiv \int_0^1 dx\, \Delta q(x) < \int_0^1 dx\, q(x) = 1$, as the integrand for $\Delta q(x)$ in Eq. 10 is always smaller than the integrand for $q(x)$ because of the perpendicular momentum distribution of the quarks in the nucleon [9].

An explicit example, using a monopole $g(p^2) \propto 1/(p^2 - \Lambda^2)$ for the vertex function, is shown in Fig. (1). The functions $q(x)$ and $\Delta q(x)$ vanish for $x < 0$ and $\bar{q}(x)$ and $\Delta \bar{q}(x)$ vanish for $x > 0$, i.e. there are no antiquarks in the physical region. Furthermore the symmetry relations for structure functions are fulfilled. Note that $q(x)$ is not required to vanish for $x = 0$. Further details regarding this model may be found in refs [5] or [6].

3. Applications

The application that we want to discuss is the spin-isospin structure of the quark distributions due to the fact that the mass spectrum of diquarks depends on the quantum numbers. Assuming that a nucleon-minus-one-quark state, which forms a 'diquark' can be considered as a physical spectator with mass M_R one expects a mass difference of about 200 MeV. This is due to the color-magnetic one-gluon-exchange contribution in the hadron mass [4,5]. Consequently there are for unpolarized scattering two distributions which we refer to as $S(x)$ and $T(x)$ corresponding to the spectator system being a diquark with $I = J = 0$ and $I = J = 1$, respectively. For polarized scattering there are two distributions $\Delta S(x)$ and $\Delta T(x)$. The quark distributions in terms of them read,

$$u(x) = \frac{3}{2} S(x) + \frac{1}{2} T(x), \tag{11}$$

$$d(x) = T(x), \tag{12}$$

$$\Delta u(x) = \frac{3}{2} \Delta S(x) - \frac{1}{6} \Delta T(x), \tag{13}$$

$$\Delta d(x) = -\frac{1}{3} \Delta T(x). \tag{14}$$

The difference between these distributions leads to an x-dependence in quantities like $d(x)/u(x)$, $F_2^n(x)/F_2^p(x)$, $g_1(x)/F_1(x)$, which in the naive parton model are constant. The latter results are found in the limit that $S(x) = T(x) = \Delta S(x) = \Delta T(x)$. Specifically one has (indicating also the $SU(6)$ symmetric results

$$\frac{d(x)}{u(x)} = \frac{2\,T(x)}{3\,S(x) + T(x)} \longrightarrow \frac{1}{2}, \tag{15}$$

$$\frac{F_2^n(x)}{F_2^p(x)} = \frac{S(x) + 3\,T(x)}{4\,S(x) + 2\,T(x)} \longrightarrow \frac{2}{3}, \tag{16}$$

$$\frac{g_1^p(x)}{F_1^p(x)} = \frac{\Delta S(x) - \frac{1}{6}\Delta T(x)}{S(x) + \frac{1}{2}T(x)} \longrightarrow \frac{5}{9}, \tag{17}$$

$$\frac{g_1^n(x)}{F_1^n(x)} = \frac{\Delta S(x) - \Delta T(x)}{S(x) + 3\,T(x)} \longrightarrow 0, \tag{18}$$

$$\frac{g_1^n(x)}{g_1^p(x)} = \frac{\Delta S(x) - \Delta T(x)}{4\,\Delta S(x) - \frac{2}{3}\Delta T(x)} \longrightarrow 0. \tag{19}$$

The results for distributions $S(x)$ and $T(x)$ corresponding to $M_R = 0.6$ and 0.8 GeV and otherwise the same parameters as the distributions shown in Fig. 1 are shown in Fig. 2. The choice of parameters is such that the momentum sum rule yields approximately 1. Note that the result for the integrals $\Delta S = 0.72$ and $\Delta T = 0.52$ corresponds to $g_A = \Delta u - \Delta d = \frac{3}{2}\Delta S + \frac{1}{6}\Delta T = 1.17$ (compared to $g_A^{(exp)} = 1.26$).

We further note that a calculation along these lines applies to an (unknown) low-energy scale. At this scale it is assumed to be the leading twist contribution,

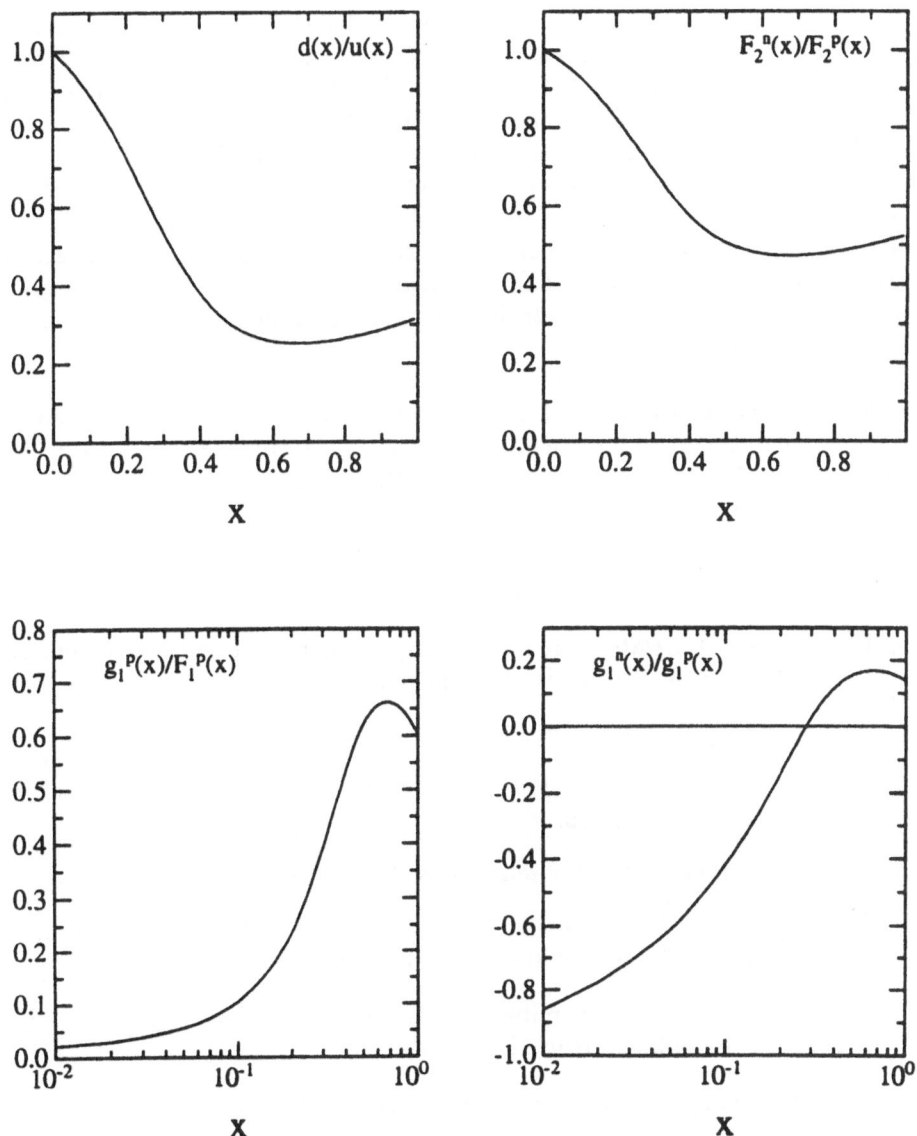

Figure 2: *The x-dependence of ratios of structure functions resulting from the distributions $S(x)$, $\Delta S(x)$ and $T(x)$, $\Delta T(x)$ corresponding to $M_R = 0.6$ and 0.8 GeV with otherwise the same parameters as the distributions shown in Fig. 1*

for which the evolution to the appropriate high energy scale, would allow comparison with experimental distributions. On the other hand the ratios shown in Fig. 2 are not extremely sensitive to evolution. This and some of the problems associated with it have been discussed in ref. [5].

Summarizing we want to emphasize here the important fact that the x-dependence of the deep inelastic structure functions contains important information on the quark structure of the hadrons, specifically the isospin and spin dependences. Furthermore, a field-theoretical approach guarantees correct symmetry and support properties of the structure functions. Of course, given a specific model it remains to calculate the vertices which are the input for such an approach.

Acknowledgement. Part of this work is included in the research program of the Foundation for Fundamental Research on Matter (FOM) and the Dutch Organization for Scientific Research (NWO).

References

1. R.L. Jaffe, in *Relatisvistic Dynamics and Quark-Nuclear Physics*, ed. M.B. Johnson and A. Pickleseimer, Wiley, 1985, p.1.

2. A.V. Manohar, Phys. Rev. Lett. **65** (1990) 2511.

3. R. Carlitz and J. Kaur, Phys. Rev. Lett. **38** (1977) 673; J. Kaur, Nucl. Phys. **B128** (1977) 219.

4. F.E. Close and A.W. Thomas, Phys. Lett. **B212** (1988) 227.

5. H. Meyer and P.J. Mulders, Nucl. Phys. **A528** (1991) 589; see also H. Meyer, Ph.D. thesis, NIKHEF/Free University

6. P.J. Mulders, A.W. Schreiber and H. Meyer, NIKHEF report 92-1 (Feb. 1992), to be publ. in Nucl. Phys. **A**.

7. D. Kusno and M.J. Moravcsik, Phys. Rev. **D20** (1979) 2734.

8. P.M. Fishbane and I.J. Muzinich, Phys. Rev. **D8** (1973) 4015.

9. F.E. Close, *An Introduction to Quarks and Partons*, Academic Press, 1979.

Experimental Indications for Diquarks in Large p_T Physics

Marek Szczekowski

CERN, 1211 Geneva 23 , Switzerland

Abstract. Various phenomena in particle physics suggest that two quarks in the proton can form diquarks – bound two quark objects inside the nucleons. The crucial tests of the diquark models are given by the processes with high momentum transfers. This short review tries to summarize the present status of the diquark hypothesis in large transverse momentum (p_T) collisions at high energies. The data, mainly from the Split Field Magnet detector at the Intersecting Storage Rings at CERN, show abundant proton production at medium values of p_T with properties which cannot be explained by the standard parton models.

1 Introduction

An interesting question concerning the proton structure is whether two quarks in the proton behave collectively, forming bound two-quark objects inside the nucleons. Such diquark concept was introduced in 1966 [1] and from then it was applied in various models of elementary particles phenomena [2]. Although some properties of the diquarks can be tested in "soft", small momentum transfer phenomena, the crucial test of the diquark model is given by the "hard" processes with high momentum transfers. If a diquark is really a tightly bound system of two quarks, it can participate in deep inelastic reactions of leptons and hadrons as a quasi elementary particle. The scattering of diquarks should lead to the substantial production of baryons in certain regions of phase space in lepton-nucleon and in nucleon-nucleon reactions.

This short review tries to summarize the present status of the diquark hypothesis in large transverse momentum (p_T) proton-proton collisions at high energies. The data come mainly from the Split Field Magnet (SFM) detector at the Intersecting Storage Rings (ISR) at CERN. In sec. 2 the properties of inclusive proton production at large p_T in high energy pp collisions are described. Sec. 3 provides more quantitative description of the data with the simple diquark model. In sec. 4 the independent evidence for the diquark scattering is shown in the data on the correlations of particles in events with high p_T proton production. Conclusions are given in sec. 5.

2 Inclusive proton production

Scatterings of protons with large four momentum transfers (Q^2) allow the study of the parton structure of protons [3]. Initial protons can be considered as two beams of quarks and gluons. In a hard collision , two partons interact elastically by the exchange of a gluon. The scattered partons fragment into hadrons and give rise to two jets of particles with high transverse momenta – trigger and away jets. The remaining spectator partons do not interact and they fragment into two other jets with low p_T particles. Such proton-proton collisions show the characteristic four jet structure [4]. The fragmentation of partons into particles can be treated to a fairly good approximation as an independent process described by universal fragmentation functions.

If events with hard collisions of protons are selected in a detector by a requirement of one high p_T particle, the sample of events will be biased. From all the configurations of fragmenting high p_T partons, the trigger will preferentially choose those with one particle much faster than the others in the trigger jet. Such a particle will be called a leading particle. At ISR energies, a leading particle takes on average about 70 % of the parent parton momentum [5]. For events with high p_T particle trigger the distribution of particles in the trigger jet is no longer described by the universal fragmentation function. For such jets however, strong correlations between the quantum numbers of the parent parton and the type of the leading particle have been observed [6]. The identification of the leading particle allows very often the identification of the parent parton.

The reflection of such correlations is the ratio of the inclusive cross-sections for high p_T production of positive and negative pions shown in Fig. 1 [7]. Since there are two u valence quarks and one d valence quark in a proton, we observe $\sigma(\pi^+)/\sigma(\pi^-) \approx 2$ for higher values of p_T, where the scattering of valence quarks is a dominant process. The decrease of $\sigma(\pi^+)/\sigma(\pi^-)$ for smaller values of p_T is explained by the contribution from gluons scattering and their fragmentation with equal probabilities into π^+ and π^-.

Since the fragmentation of a parton is independent of the elementary parton-parton scattering, the production of various leading particles should show a very similar dependence on kinematical variables if the particles come from the same type of parent parton. Considering only the dominant valence quark scattering shown in Fig. 2, we can expect that the cross-sections ratios for high p_T π^+, K^+ and p production will depend only weakly on transverse momenta and scattering angles. Their magnitudes should reflect the relative probabilities for creations of $d\bar{d}$, $s\bar{s}$ and $(ud)(\bar{u}\bar{d})$ pairs during fragmentation of the u quark.

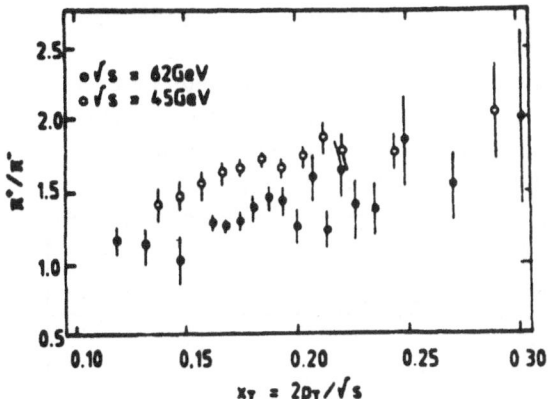

Fig. 1: Ratio of inclusive cross-sections for π^+ and π^- production versus $x_T = 2p_T/\sqrt{s}$ at $\sqrt{s} = 45$ and 62 GeV.

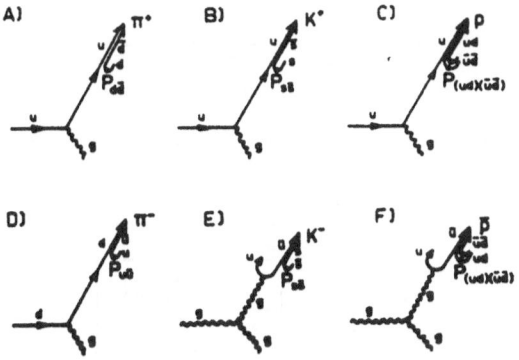

Fig. 2: Dominant fragmentation schemes for π^+, K^+, p, π^-, K^- and $\bar{p}$. $P_{i\bar{i}}$ is a probability to create an $i\bar{i}$ pair of partons in the colour field.

The negative leading particles come from two kinds of partons. Leading π^- mesons are dominantly produced by valence d quarks but leading K^- and $\bar{p}$ arise mainly from gluon fragmentation since the initial protons do not contain valence s, $\bar{u}$ and $\bar{d}$ quarks. Since gluons carry , on the average, a smaller fraction of the proton momentum than do valence quarks, we expect that $\sigma(K^-)/\sigma(\pi^-)$ and $\sigma(\bar{p})/\sigma(\pi^-)$ will decrease with increasing p_T. On the other hand, for gluon scattering the ratio $\sigma(K^-)/\sigma(\bar{p})$ should be fairly independent of θ and p_T and its magnitude should reflect the relative probabilities for the production of $s\bar{s}$ and $(ud)(\bar{u}\bar{d})$ pairs during the gluon fragmentation (Fig. 2). The ratio $\sigma(K^-)/\sigma(\bar{p})$ should be equal to the ratio $\sigma(K^+)/\sigma(p)$.

In Figs.3, 4 and 5, the p_T dependence of the ratios of inclusive cross-sections are shown. The data come from the Split Field Magnet detector [7, 8]. The results for the ratios of pions and kaons agree well with the quark-parton model predictions. The relative cross-section for the production of $\bar{p}$, at the level of a few percent, depends slightly on p_T but shows no dependence on a polar angle θ (Fig. 6). The only exception are the results for the proton production. The relative cross-section for protons is an order of magnitude larger than the corresponding one for antiproton. It depends strongly on θ. For fixed θ, the fraction of protons decreases with increasing p_T.

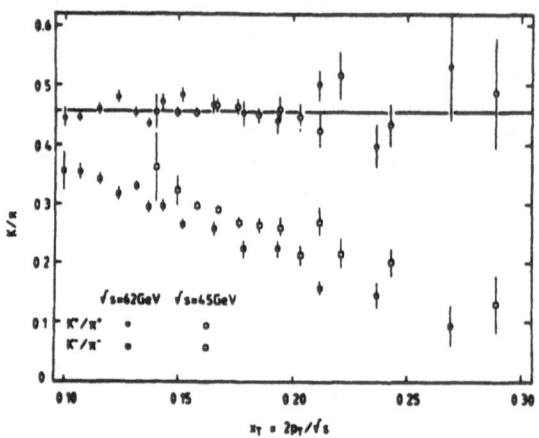

Fig. 3: Production ratios for K^+/π^+ and K^-/π^- in pp collisions at 45 and 62 GeV for $\theta = 50°$, plotted as a function of x_T .

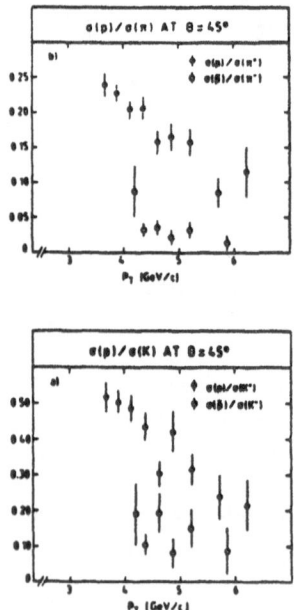

Fig. 4: Particle ratios in pp collisions at $\sqrt{s} = 62$ GeV and $\theta \approx 45°$ plotted as a function of p_T.

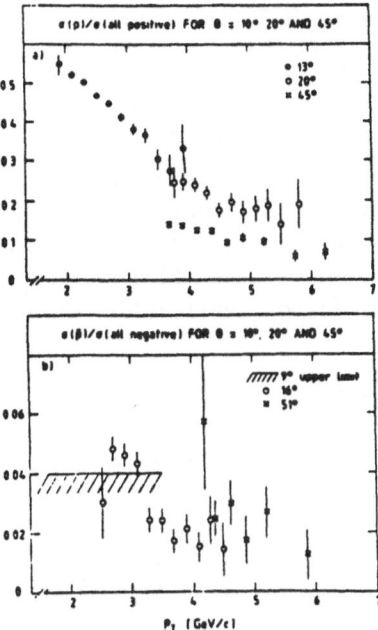

Fig. 5: Particle ratios in pp collisions at $\sqrt{s} = 62$ GeV and three c.m.s. scattering angles $\theta \approx 10°$, $20°$ and $45°$, plotted as a function of p_T.

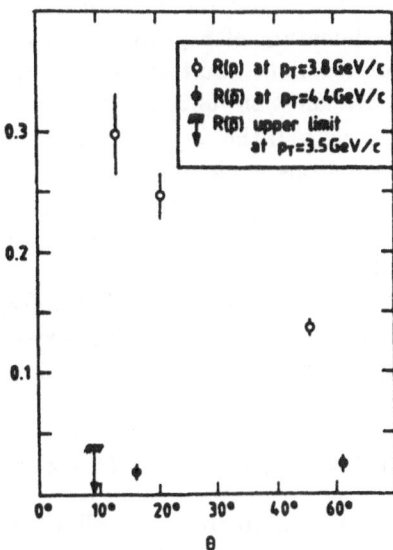

Fig. 6: Angular dependence of $R(p) = \sigma(p)/\sigma(pos.)$ and $R(\bar{p}) = \sigma(\bar{p})/\sigma(neg.)$ in pp collisions at 62 GeV and fixed values of p_T.

The stronger dependence of inclusive cross-section for p on p_T, which leads to $\sigma(p)/\sigma(\pi^+) \sim 1/p_T^4$ [9] could suggest that protons are produced in the fragmentation of gluons. In this case, however, we should observe a similar magnitude of production cross-sections and properties for antiprotons.

If diquark objects can be scattered at large angles, they will provide the additional mechanism for high p_T proton production. With increasing four momentum transfer Q^2, it will be more difficult to scatter diquarks without breaking them. For the description of the extended diquarks, a form factor $F(Q^2)$ should be introduced with $F(Q^2) \sim 1/Q^2$. This additional dependence on Q^2 leads to the decreasing of $R(p) = \sigma(p)/\sigma(positives)$ with increasing θ and p_T.

3 Diquark model

The above qualitative arguments for the presence of diquark scattering at high p_T can be complemented by more quantitative predictions of a simple parton model [10]. The invariant cross-section for inclusive production of a particle in pp scattering can be expressed as the weighted sum of the differential cross-sections $d\sigma/d\hat{t}$ of all possible parton scatterings $(ij \rightarrow kl)$ with a four momentum transfer $\hat{t}$ which can contribute

$$E_h \frac{d^3\sigma}{dp_h^3} = \sum_{ijkl} \int_0^1 dx_i \int_0^1 dx_j f_p^i(x_i) f_p^j(x_j) \frac{d\sigma}{d\hat{t}}(ij \rightarrow kl) \frac{1}{\pi z_k} D_k^h(z_k)$$

where x_i is the fraction of the proton momentum carried by a parton i and z_k is the fraction of the momentum of a parton k carried by hadron h. The cross-section can be calculated if in addition to $d\sigma/d\hat{t}$, the proton structure functions $f_p^i(x_i)$ and the fragmentation functions of a parton k into hadron h — $D_k^h(z_k)$, are known. For diquarks all the required ingredients are unknown. The models make different assumptions about diquark structure functions, diquark scattering amplitudes and form factors and diquark fragmentation functions [10, 11, 12, 13, 14]. To reduce the number of parameters in a model a dominance of scalar ud diquarks is usually assumed.

In the ABCDHW Collaboration model [10, 15] the fragmentation functions for quarks, gluons and diquarks into various hadrons were generated according to the Lund fragmentation scheme. For $0.2 < z < 1.0$ they can be parametrized in the following form

$$z D_k^h(z) = (c_1 + c_2 z + c_3/z)(1 - z)^{c_4}$$

where c_1, c_2, c_3, c_4 are adjustable parameters. For all $z > 0.2$, the fragmentation function of a ud system into a proton is the dominating process.

Since diquarks are not truly elementary objects, the description of the diquark scattering cannot be treated in QCD in a strict way. An approximation can be made by assuming that the nonperturbative part of the diquark scattering amplitude is described simply by the diquark form factor and the perturbative part is dominated by the elementary scattering amplitudes for scalar particles as given by QCD [15].

The proton wave function with a diquark component can be written in the form

$$| p >= (1 - \lambda)^{1/2} | u, u, d, g, sea > + \lambda^{1/2} | \tilde{u}, \widetilde{(ud)}, \tilde{g}, \widetilde{sea} >$$

where λ describes the probability for a diquark state in a proton. This parametrization simplifies a more physical picture of a proton always in a quark-diquark state but with different dynamical configurations of two quarks in the diquark. The description of the first part (qqq) in the formula can be taken from νp deep inelastic scattering data [16]. The second term $(q$-diquark) is unknown. Varying the fraction of the gluon component contained in the diquark, different parametrizations of the diquark structure functions are obtained. For two extreme cases, the distributions are the following:

- Model 1: $\tilde{u}(x) = 0.5u(x)$ $\tilde{d}(x) = 0$
                $\widetilde{ud}(x) = 0.5u(x) + d(x)$ $\tilde{g}(x) = g(x)$
- Model 2: $\tilde{u}(x) = 0.5u(x)$ $\tilde{d}(x) = 0$
                $\widetilde{ud}(x) = 0.5u(x) + d(x) + g(x)$ $\tilde{g}(x) = 0$

For both models $\widetilde{sea}(x) = sea(x)$ is assumed. In Model 1, the diquark does not contain any gluons from the proton. In Model 2, all the gluons from the proton are contained in the diquark. Both models are unphysical but they test the sensitivity of the data to the assumptions about the gluon component.

There are theoretical and experimental arguments that diquarks should appear dominantly at higher Bjorken x values. This can be taken into account if λ increases with x. In Model 3, the structure functions of Model 2 are complemented by a simple parametrization $\lambda = x$.

In all the models, the form factor is parametrized as $F(Q^2) = 1/(1+Q^2/M^2)$. The scale factor M^2 and the probability λ are adjusted to fit the measured ratio $\sigma(p)/\sigma(\pi^+)$ at $\sqrt{s} = 62$ GeV, $\theta = 45°$ and $p_T = 4$ GeV/c :

- Model 1: $\lambda = 0.30$ and $M^2 = 20$ GeV^2

- Model 2: $\lambda = 0.17$ and $M^2 = 10$ GeV^2

- Model 3: $\lambda = x$ and $M^2 = 9$ GeV^2

The obtained values for λ and M^2 are large in all models pointing to a substantial fraction of tightly bound diquarks in a proton.

With fixed λ and M^2 the measured p_T, θ and energy dependence of the proton fractions can be predicted. The comparison of the model predictions with the SFM data are shown in Figs.7, 8 and 9. They describe roughly the data. The θ dependence of $R(p)$ suggests that a more realistic model with only part of proton gluons contained in the diquark can fit the data. Some λ dependence should also be introduced but the simple parametrization in Model 3 gives too strong a dependence on θ and the energy.

170

The predictions of the standard parton model without diquarks are also shown in the figures for comparison. As expected, both models correctly describe the behaviour of antiprotons ratios since there is no contribution of the diquark scattering in these processes. A dramatic difference is however observed for the proton ratios. The standard parton model predicts a similar behaviour of high p_T p and $\bar{p}$ production and fails to describe the proton data.

We can conclude that the unusual features of the inclusive high p_T proton production cannot be explained by the standard quark-parton model. The diquark scattering is at present the only proposed mechanism consistent with the experimental data.

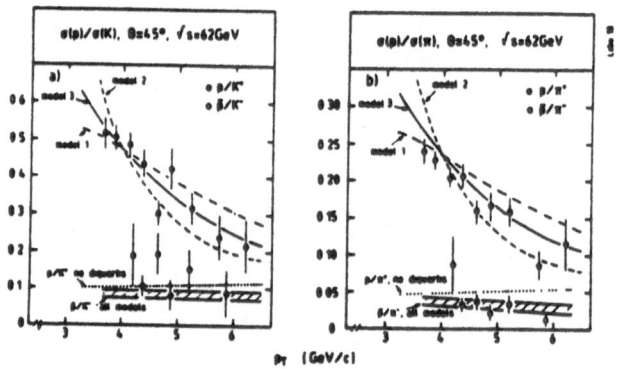

Fig. 7: Particle ratios $\sigma(p)/\sigma(K^+)$, $\sigma(\bar{p})/\sigma(K^-)$ (a), and $\sigma(p)/\sigma(\pi^+)$, $\sigma(\bar{p})/\sigma(\pi^-)$ (b) versus p_T for $\theta \approx 45°$ and $\sqrt{s} = 62$ GeV. The lines represent the predictions of the models described in the text.

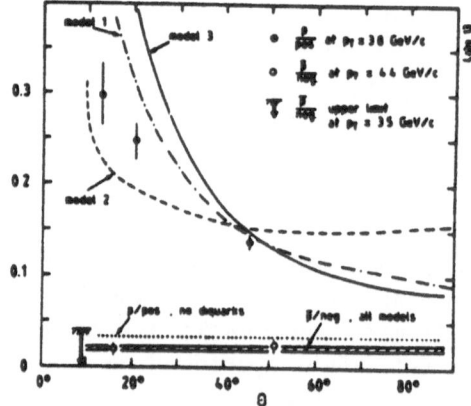

Fig. 8: Angular dependence of the proton $R(p)$ and antiproton $R(\bar{p})$ fractions at fixed p_T. The lines represent the predictions of the models described in the text.

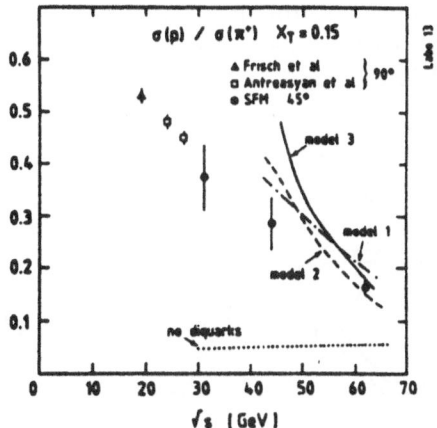

Fig. 9: Energy dependence of $\sigma(p)/\sigma(\pi^+)$ at $x_T = 0.15$ and $\theta \approx 45°$. The Fermilab data at $\theta \approx 90°$ are also shown. The lines represent the predictions of the models described in the text.

4 Correlations of particles in events with large p_T proton production

If the dominant mechanism of high p_T proton production is the diquark scattering, well-defined correlations between particles produced in the trigger jet and the corresponding spectator jet should be observed. Also the study of correlations of particles from the proton trigger jet can reveal the properties of the scattered diquarks. On the other hand , particles from the remaining away and second spectator jets are only weakly correlated with high p_T proton production and their properties are similar to the properties of the particles from the corresponding jets in events triggered by high p_T pions and kaons.

4.1 Correlations between trigger and spectator jets

The correlations between the trigger particle and the quantum numbers of fast particles coming from the fragmentation of noninteracting partons are symbolically shown in Fig.10. For u or d quark scattering which leads to high p_T π^+, π^- or K^+ meson production, there are remaining valence quarks ud or uu which give rise to the spectator jet. These two quark systems easily produce fast protons when they are joined by the third quark during the fragmentation process. Strong correlations between high p_T π^+, π^- or K^+ production and the presence of fast protons in the spectator jet should be observed. The production of high p_T K^- mesons by gluons leads to similar correlations. The same effect is also observed if the high p_T protons are produced in quark scattering with creation of two-quark system in the quark fragmentation. The baryon number is compensated locally in the trigger jet in this case.

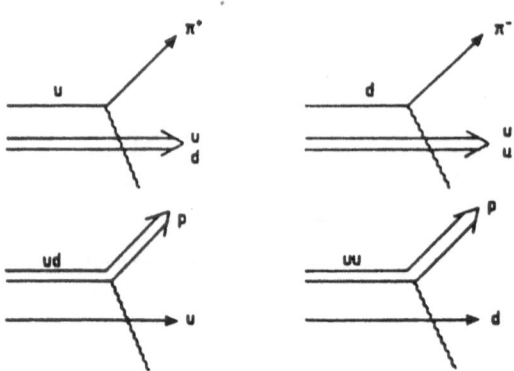

Fig. 10: Expected flavour composition of spectator jets for pion and proton triggers.

For the high p_T protons coming from the fragmentation of hard-scattered diquarks only one valence quark remains in the spectator jet. The production of fast baryons at small angles is then suppressed for high p_T proton triggers.

A fast baryon studied usually in the experiments is a proton. For soft proton-proton collisions final state protons dominate the positive particle distributions for $x_F > 0.5$. In the experiments where the identification of the fast particle is not possible, the assumption is therefore usually made that the majority of fast positive particles with $x_F > 0.5$ are protons.

The densities of positive (ρ^+) and negative (ρ^-) particles associated with proton and pion triggers were measured in The Split Field Magnet experiment [17, 18]. It is convenient to study the ratios of particle densities $R^+(p/\pi) = \rho^+(p)/\rho^+(\pi)$ and $R^-(p/\pi) = \rho^-(p)/\rho^-(\pi)$ where acceptance corrections are cancelled.

The experimental results for R^+ and R^- are shown in Fig. 11. The ratios $R^+(p/\pi^+)$ and $R^+(p/\pi^-)$ decrease substantially with increasing x_F from the values of around 1 for small x_F. It means that the number of fast positive particles (mainly protons) is suppressed for proton triggers.

These negative proton correlations are confirmed in the experiment where both the high p_T trigger particle at $\theta = 90°$ and fast spectator protons could be identified [19]. The depletion of fast protons from the fragmentation of spectator partons for events with a high p_T proton trigger is observed in the data shown in Fig.12.

All the above observations agree qualitatively with the expectations of the hard diquark scattering mechanism for high p_T proton production.

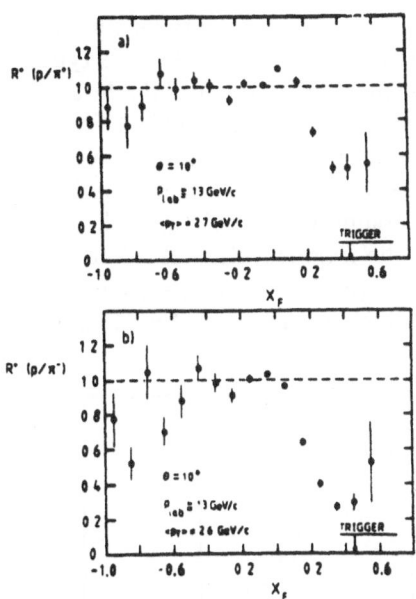

Fig. 11: Ratios of positive spectator secondaries associated with proton and pion triggers at $\theta \approx 10°$, (a) π^+ trigger, (b) π^- trigger.

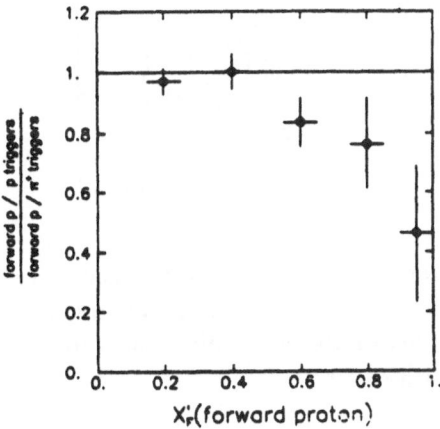

Fig. 12: The ratio of forward protons per proton trigger to forward protons per π^+ trigger as a function of x_F. Both triggers are at $\theta \approx 90°$.

To show more quantitatively the difference between the charged particles densities ρ^+ and ρ^- for fragmentations of various spectator quark combinations the SFM and deep inelastic νp and $\bar{\nu}p$ scattering data can be compared. The neutrino reactions lead to the well-defined two-quark jets in the backward hemisphere of the produced hadrons c.m.s.: uu jets for νp scattering and ud jets for $\bar{\nu}p$ collisions. To compare the results from the two types of experiments, the relation between the fragmentation function variable z used in ν data analysis and x_F measured for spectator jet particles in the SFM detector is needed. For spectator jets in high p_T collisions, it can be obtained using the quark-parton model. Since the relation between x_F and z is mainly kinematical it is not very sensitive to the dynamical assumptions of the model.

The difference between the fragmentation of ud and uu systems is enhanced if the ratio $R_\rho = \rho^-/\rho^+$ is formed. The results for the spectator jets in events with high p_T proton, π^+ and π^- are shown in Fig.13 and compared with νp and $\bar{\nu}p$ scattering data [20].

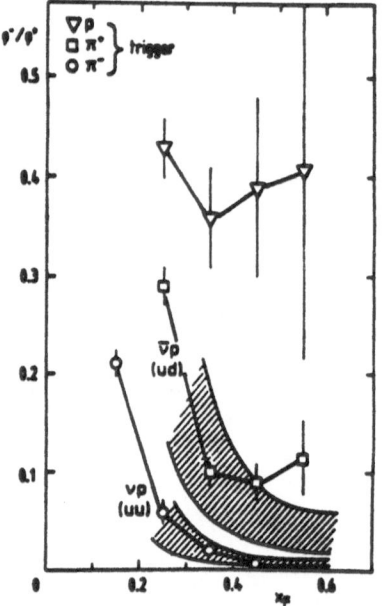

Fig. 13: Charge density ratio $R_\rho = \rho^-/\rho^+$ as a function of x_F for spectator jets in events with high p_T pion and proton triggers in proton-proton collisions. The shaded strips are predictions from deep inelastic neutrino and antineutrino scattering experiments for ud and uu target jet systems.

For the π^- trigger, the properties of the uu spectator system agree very well with R_ρ of about 1 % measured for uu jets in the νp scattering. For the π^+ trigger R_ρ of about 10 % is measured for fast particles from ud systems both in pp and $\bar{\nu}p$ hard scattering. In contrast, the events with the proton trigger have $R_\rho \approx 40$ %, much higher than the events with meson triggers. For high p_T diquark scattering the properties of the spectator quark fragmentation can be compared to the corresponding features of the quark fragmentation in νp and $\bar{\nu}p$ scattering. The comparison of the ratios R_ρ is shown in Fig.14. For the pp data the results depend on the quantum numbers of diquarks in protons. For ud diquarks the spectator quark is always the u quark and R_ρ should correspond to the ratio measured for the u quark fragmentation in νp scattering. The pp data are slightly above the νp data. This result points to a contribution from uu diquarks. If both two quark combinations are equally probable in a proton the mixture of 2/3 of ud diquarks and of 1/3 of uu diquarks should be observed. The ratio R_ρ can be calculated for the correspondingly weighted νp and $\bar\nu p$ data. The results shown in Fig.14 lie far above the pp data.

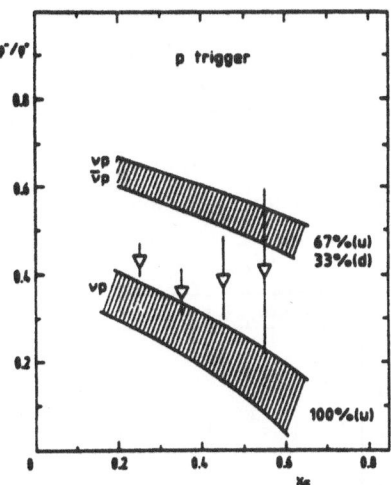

Fig. 14: Charge density ratio $R_\rho = \rho^-/\rho^+$ for spectator jets in events with high p_T proton triggers in pp collisions compared to predictions from deep inelastic ν and $\bar\nu$ scattering for pure u quark fragmentation and for a mixture of 33% d quarks and 67% u quarks.

4.2   Correlations of particles in the proton trigger jet

The particles coming from the fragmentation of the same parton as the trigger particle form a trigger jet. For high p_T pion and kaon triggers, an enhanced density of charged particles in the phase space region close to the trigger particle is observed. The production of particles in the trigger jet is usually analyzed using the variable z_F

$$z_F = \vec{p}_T^a \cdot \vec{p}_T^{Tr} / \mid \vec{p}_T^{Tr} \mid^2$$

where $\vec{p}_T^a$ is the transverse momentum of a particle associated with the trigger particle with transverse momentum $\vec{p}_T^{Tr}$.

Since the properties of particles from the trigger jets are well-measured for meson triggers, it is convenient to compare the observed features of proton trigger jets to those for the meson jets.

If the high p_T protons are produced in the fragmentation of ud diquarks, the second rank π^- mesons in the trigger jet (Fig. 15) should occur as frequently as the second rank π^- mesons in the π^+ trigger jet. In contrast, for uu diquark scattering π^+ mesons are the second rank particles in the high p_T proton production. The ratios of negatively and positively charged particles associated with proton and meson triggers are shown in Fig. 16 as functions of z_F. The errors are large but both ratios are close to 1. This means that for the events with a high p_T proton the same features as in a high p_T meson production are observed — the formation of the trigger jet. The data are consistent with similar $< z_{TR} >$ for both kinds of trigger jets.

Fig. 15: Schemes for u, ud and uu fragmentations into leading and the second rank hadrons.

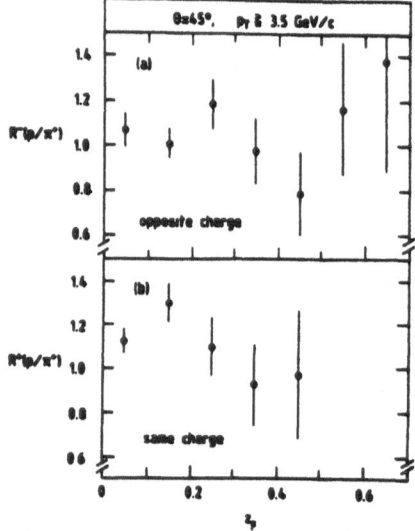

Fig. 16: Ratio of negative (a) and positive (b) particle densities in the trigger jet for proton and π^+ meson triggers at $\theta \approx 45°$ and $p_T > 3.5$ GeV/c as a function of z_F.

The similar properties of second rank particles in the π^+ and p trigger jets point again to the dominance of ud diquark scattering. The small enhancement of positive particles observed for the proton jets for the positive particles could suggest the small admixture of uu diquarks scattering. This conclusion is supported by the observation of high p_T Δ^{++} production in the events with high p_T proton triggers [17].

5 Conclusions

This short review of the experimental results in high p_T proton-proton scattering at ISR energies show that there are strong indications of two quark clustering in protons. The evidence is growing for the dominance of the tightly bound scalar ud diquarks in nucleons. The observation of production of the Δ^{++} resonance at large p_T and the charge density structure in the spectator jets in events with a high p_T proton provides the evidence for a small admixture of uu vector diquarks.

The diquark effects are visible both in low and high momentum transfer processes. In the latter, the diquarks can be observed more directly. The hard processes however can bias the measurements of diquark properties choosing only the special and rare configurations of the wavefunction of quark-diquark baryons. This is suggested by the apparent contradiction between the rather extended, loosely bound two-quark systems obtained in the potential models for static properties of baryons and the tightly bound diquarks required to describe the data presented in this review for the deep inelastic proton-proton collisions.

References

[1] M. Ida and R. Kobayashi, *Prog. Theor. Phys.* **36** (1966) 846;
    D.B. Lichtenberg and L.J. Tassie, *Phys. Rev.* **155** (1967) 1601.

[2] For a review see: M. Szczekowski, *Int. J. of Mod. Phys.* **A4** (1989) 3985.

[3] For a review see: W. Geist *et al.*, *Phys. Rep.* **197** (1990) 263.

[4] A. Breakstone *et al.*, *Z. Phys.* **C23** (1984) 1.

[5] H.G. Fischer, *Proc. of the EPS Conf. on High Energy Physics*, Lisbon, 1981, p.897.

[6] A. Breakstone *et al.*, *Z. Phys.* **C25** (1984) 21.

[7] A. Breakstone *et al.*, *Z. Phys.* **C27** (1985) 205.

[8] A. Breakstone *et al.*, *Phys. Lett.* **147B** (1984) 237.

[9] D. Antreasyan *et al.*, *Phys. Rev.* **D19** (1979) 764.

[10] A. Breakstone *et al.*, *Z. Phys.* **C28** (1985) 335.

[11] S. Ekelin and S. Fredriksson, *Phys. Lett.* **149B** (1984) 509.

[12] T.I. Larsson, *Phys. Rev.* **D29** (1984) 1013.

[13] H. Minakata and T. Shimizu, *Lett. Nuovo Cimento* **27** (1980) 241.

[14] L.V. Laperashvili, *Yad. Fiz.* **35** (1982) 742.

[15] T. Lohse, Ph.D. thesis, Univ. of Dortmund (1985).

[16] H. Abramowicz *et al.*, *Z.Phys.* **C12** (1982) 289.

[17] A. Breakstone *et al.*, *Z.Phys.* **C36** (1987) 567.

[18] A. Breakstone *et al.*, *Nucl. Phys.* **B156** (1979) 309.

[19] A.M. Smith *et al.*, *Phys.Lett.* **184B** (1987) 293.

[20] P. Allen *et al.*, *Nucl.Phys.* **B124** (1983) 369;
      J. Bell *et al.*, *Phys.Rev.* **D19** (1979) 1;
      M. Derrick *et al.*, *Phys.Rev.* **D24** (1981) 1071.

The Rôle of Diquarks in Exclusive Scattering [1]

P.Kroll[a,2], M.Schürmann[a,2], and W.Schweiger[b]

[a] Department of Physics, University of Wuppertal,
W-5600 Wuppertal, P.O.Box 100127, FRG
[b] Institute of Theoretical Physics, University of Graz,
A-8010 Graz, Universitätsplatz 5, Austria

Abstract. The quark-diquark picture of baryons, as emerged from a perturbative description of exclusive hadronic processes at a few GeV of momentum transfer, is reviewed. Various reactions already treated within the diquark model are discussed. Special emphasis is put on Compton scattering off protons and photoproduction of mesons.

Keywords. Diquarks, perturbative QCD, exclusive reactions, baryon form factors, Compton scattering, photoproduction.

1 Introduction

During the Eighties perturbative QCD has developed into the main theoretical tool for studying hard hadronic processes [1]. However, the perturbative description of hadronic reactions is not entirely straightforward. Due to the confinement property of QCD long distance effects are always present, even in the large momentum-transfer (large Q) region. Only the existence of factorization theorems ensures that these long-distance effects can be split off as universal, process-independent factors (structure functions, fragmentation functions, distribution amplitudes). The remaining short-distance subprocesses are then calculable within perturbation theory. This kind of approach proved to be very successful for hard inclusive reactions and has also been elaborated for exclusive scattering [2]. For the latter, however, the situation is very controversial [3]. There is little doubt that perturbative QCD is an adequate tool to study the asymptotic $Q \to \infty$ behaviour of hard exclusive reactions. The main point in dispute rather relates to the question whether the momentum transfers accessible in present days experiments are large enough to justify a perturbative treatment. In order to elucidate this point we briefly recall the so called "hard scattering picture" (HSP) as worked out in Ref.[2].

---

[1] Talk presented by W.Schweiger
[2] Supported in part by the Bundesministerium für Forschung und Technologie, FRG under contract number 06 Wü 765

To leading order in $1/Q$ the perturbative QCD result for hard exclusive scattering is given by the convolution of a hard scattering amplitude $\widehat{T}$ with distribution amplitudes (DA's) ϕ of the incoming and outgoing hadrons. For the magnetic nucleon form factor, for example, it has the form

$$G_M^N(Q^2) = \int_0^1 \left[\prod_{i=1}^3 dx_i \ \delta(1 - \sum_{k=1}^3 x_k) \right] \left[\prod_{j=1}^3 dy_j \ \delta(1 - \sum_{k=1}^3 y_k) \right]$$
$$\phi^{N\dagger}(y_1, y_2, y_3; \tilde{Q}_y) \ \widehat{T}(x_1, \ldots, y_3; Q^2) \ \phi^N(x_1, x_2, x_3; \tilde{Q}_x) \ . \tag{1.1}$$

Its generalization to more complicated processes is straightforward. The hard scattering amplitude $\widehat{T}$ describes the scattering of collinear quarks from a state in which they carry the fractions x_i of the incoming nucleon momentum to a state in which they carry the fractions y_j of the outgoing nucleon momentum. $\widehat{T}$ consists of all possible tree diagrams contributing to $\gamma^* 3q \to 3q$. Basically, $\widehat{T}$ is known, although the calculation of the full set of tree diagrams for more complicated processes may become an enormous undertaking. The distribution amplitude $\phi^N(x_1, x_2, x_3; \tilde{Q}_x)$, which represents the nonperturbative ingredient of the HSP, is the probability amplitude for finding the valence quark Fock state in the nucleon with the quarks collinear up to the scale $\tilde{Q}_x = \min_i(x_i Q)$. It is only well-known in the limits $\tilde{Q}_x \to 0, \infty$ where it becomes [2]

$$\phi^N(x_1, x_2, x_3; \tilde{Q}_x) = \begin{cases} \propto \delta(x_1 - 1/3)\delta(x_2 - 1/3)\delta(x_3 - 1/3) & \text{for} \quad \tilde{Q}_x \to 0 \\ \\ \propto x_1 x_2 x_3 & \text{for} \quad \tilde{Q}_x \to \infty \ . \end{cases} \tag{1.2}$$

It has been recognized very soon that neither of these limiting DA's provides a correct description of the nucleon form factors for experimentally accessible momentum transfers ($Q^2 \lesssim 30 \text{GeV}^2$ for G_M^p) [4]. The asymptotic DA $\propto x_1 x_2 x_3$, e.g., gives a proton magnetic form factor G_M^p that is identically zero and it yields the wrong sign for the neutron magnetic form factor G_M^n. When comparing with experiment it thus occurs to be essential to gain some knowledge about $\phi^N(x_1, x_2, x_3; \tilde{Q}_x)$ for finite values of $\tilde{Q}_x$ (typically about 1 GeV). This partly has been achieved by employing QCD sum rules which provide information on the lowest order moments of various hadron DA's. These moments include the absolute normalization of the DA's, so that the predictions of the HSP are, in principle, free of parameters.

A nucleon DA which reproduces the QCD sum-rule moments and gives a good account of the nucleon magnetic form factors has been derived by Chernyak and Zhitnitsky [5]. As a striking feature this DA exhibits a strong asymmetry which, e.g., gives preference to positive helicity up quarks in positive helicity protons. But just this asymmetry is the reason that the application of the Chernyak-Zhitnitsky DA in a perturbative calculation becomes very problematic, because it strongly enhances soft contributions [3]. At this point it should be mentioned that the hadron DA's derived via QCD sum rules are by no means unique. The reasons are that the lowest order moments can only be obtained up to rather large uncertainties and that higher order moments are also needed to fix the DA's unambiguously. A careful analysis of this problem has been performed recently by Stefanis [6]. Interestingly, he comes to the conclusion that the nucleon DA proposed by Gari and Stefanis [7], which also reproduces the QCD sum-rule moments, indicates **diquark clustering**. A similar observation was made by Dziembowski and Franklin [8]. They relate a nonrelativistic quark-model wave function to the three quark DA on the light cone and find that the introduction of a two-quark correlation in the nonrelativistic wavefunction results in a nucleon DA which matches the QCD sum-rule moments significantly better than a totally symmetric wavefunction. Taking these observations seriously one may imagine that, provided strong two-quark correlations exist in a baryon, a moderately hard photon or gluon cannot resolve the two correlated quarks, but rather sees a diquark.

Another motivation for extending the HSP by the inclusion of diquarks is connected with two characteristic features of the pure quark HSP. These are (fixed angle) **power laws** [9] and the **conservation of hadronic helicity** [10]. Both these properties of the HSP can easily be checked by experiment, which in turn can tell us where the transition from the non-perturbative to the perturbative regime of QCD takes place. Unfortunately, most of the exclusive experiments performed up till now have only explored the region of moderately large momentum transfer ($Q^2 \lesssim 15 GeV^2$). In this region deviations from both, power laws and helicity conservation, can still be observed. For instance, an appreciable spin asymmetry was measured at Brookhaven for Q^2 as large as $\approx 8 GeV^2$ [11].

Thus, very obviously, the HSP needs some modifications at moderately large momentum transfer which account for nonperturbative effects. One possible way for modelling nonperturbative effects is the introduction of diquarks which, in the light of the investigations performed in Refs.[6] and [8], may be regarded as an effective description of two-quark correlations in baryon DA's.

2 The Diquark Model

There are many indications from different fields of hadronic physics, like baryon spectroscopy, deep inelastic scattering, or weak decays, to mention a few, which suggest the presence of diquarks [12]. In order to employ the HSP for the description of intermediate momentum transfer exclusive scattering we simply assume the valence Fock state of a baryon to consist of a quark and a diquark (D) instead of three quarks. Then, at least formally, hadronic quantities look quite the same as in the pure quark HSP. Somewhat loosely written, the magnetic form factor of the nucleon, e.g., reads

$$G_M^N(Q^2) = \sum_i \int_0^1 dx_1 dy_1 \phi_i^{N\dagger}(y_1) \widehat{T}_i(x_1, y_1; Q^2) \phi_i^N(x_1), \qquad (2.1)$$

where the sum extends over the two types of diquarks taken into account, i.e. scalar (S) diquarks and vector (V) diquarks. In Eq.(2.1) the momentum conserving δ-functions have already been exploited. The remaining integration variables

x_1 and y_1 are the fractions of the nucleon momentum carried by the incoming and outgoing quark, respectively.

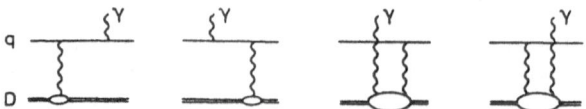

Fig. 1. Diagrams contributing to the electromagnetic form factors of the nucleon

The Feynman diagrams contributing to the hard scattering amplitude $\hat{T}$ of Eq.(2.1) are depicted in Fig.1. The blobs appearing at the gD and γgD vertices symbolize three- and four-point functions which describe the couplings of photons and gluons to diquarks – for Compton scattering and photoproduction of mesons even five-point functions are possible. As illustrated in Fig.2 these n-point functions are first calculated for pointlike diquarks according to standard prescriptions [3] and afterwards multiplied with phenomenological vertex functions (diquark form factors) [13]. The diquark form factors take care of the composite nature of diquarks and are parameterized in such a way that asymptotically the diquark model evolves into the pure quark HSP. Actually, multipole-shaped diquark form factors are employed, i.e.

$$F_S^{(3)}(Q^2) = \delta \frac{Q_S^2}{Q_S^2 + Q^2} \ , \tag{2.2}$$

$$F_V^{(3)}(Q^2) = \delta \left(\frac{Q_V^2}{Q_V^2 + Q^2} \right)^2 \ , \tag{2.3}$$

for the three-point vertex, and

$$F_S^{(n)}(Q^2) = a_S F_S^{(3)}(Q^2) \ , \tag{2.4}$$

$$F_V^{(n)}(Q^2) = a_V \delta \left(\frac{Q_V^2}{Q_V^2 + Q^2} \right)^{n-1} \ , \tag{2.5}$$

for the n-point vertices. In the asymptotic limit $Q^2 \to \infty$ the factor δ defined as

$$\delta = \begin{cases} \alpha_S(Q^2)/\alpha_S(Q_D^2) & Q^2 \geq Q_D^2 \\ 1 & Q^2 < Q_D^2 \end{cases} \tag{2.6}$$

provides for the correct powers of the strong coupling constant $\alpha_S(Q^2)$. $\alpha_S = 12\pi/25 \ln(Q^2/\Lambda_{QCD}^2)$ is used with $\Lambda_{QCD} = 200 \text{MeV}$ and restricted to be smaller than 0.5. a_S and a_V are strength parameters which account for the possibility of diquark excitation and break-up at the n-point vertices.

---

[3] Vector diquarks are allowed to possess an anomalous (chromo)magnetic moment κ_V.

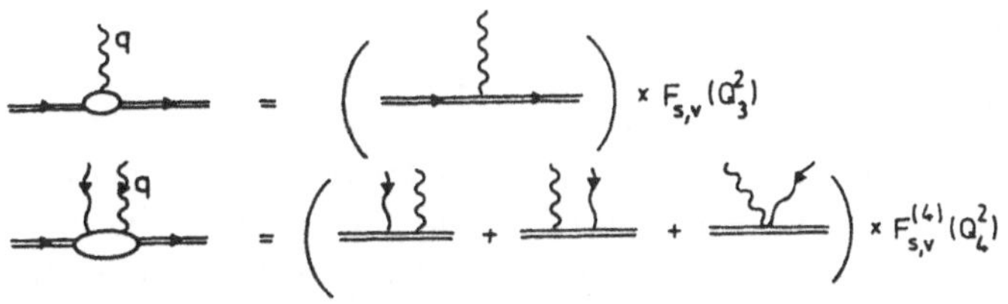

Fig. 2. Born approximation for photon and gluon couplings to diquarks

Having outlined how the perturbative part of the diquark model is treated, still a few words about the nonperturbative ingredients of the model are in order. The DA's $\phi_S^B(x_1)$ and $\phi_V^B(x_1)$ for the scalar and vector diquarks, respectively, are introduced by decomposing the state of a baryon B with helicity λ into its spin-flavour content. Thereby S diquarks are assumed to belong to the $\{\bar{3}\}$ SU(3) spin-flavour multiplet and V diquarks to the $\{6\}$ one. For the proton one obtains

$$|p, \pm\tfrac{1}{2}\rangle = \pm\frac{1}{3} f_V^p \phi_V^p(x_1) \Big[|u_\pm; V_0(ud)\rangle - \sqrt{2}|d_\pm V_0(uu)\rangle - \sqrt{2}|u_\mp; V_{\pm 1}(ud)\rangle$$
$$+ 2|d_\mp, V_{\pm 1}(uu)\rangle \Big] + f_S^p \phi_S^p(x_1)|u_\pm; S\rangle \, .$$

$$(2.7)$$

As already mentioned, the argument x_1 of the DA's denotes the fraction of the baryon momentum carried by the quark (the diquark carries $x_2 = 1 - x_1$). The weak (logarithmic) Q dependence of the DA's caused by the QCD evolution is neglected, since it is of minor importance in the limited Q^2 range we are interested in. For $\phi_S^p = \phi_V^p$ and $f_S^p = f_V^p$ Eq.(2.7) reflects just the SU(6) wavefunction of the proton with quark two and three coupled to a diquark. In accordance with the convention used by Chernyak and Zhitnitsky [5] the DA's $\phi_S^B(x_1)$ and $\phi_V^B(x_1)$ are normalized in such a way that

$$\int_0^1 \phi_{S(V)}^B(x_1) dx_1 = 1 \, .$$

$$(2.8)$$

The constant $f_S^B(f_V^B)$ in principle is determined by the k_T dependence of the $S(V)$ diquark wave function and by the probability of finding the Fock state $|qS\rangle(|qV\rangle)$ in the baryon B. Since for f_S^B and f_V^B only rough estimates can be given they are considered as free parameters to be adjusted. Advice how to choose the functional form of the quark–diquark DA's $\phi_{S(V)}(x_1)$ can be taken from the HSP. Integrating the asymptotic form of the three-quark DA Eq.(1.2) over one of the x_i's gives a quark-diquark DA $\propto x_1(1 - x_1)^3$. Actually, for the lowest lying baryon octet a DA of the form

$$\phi_S^B(x_1) = \phi_V^B(x_1) = \phi_{ho}(x_1) = C x_1(1 - x_1)^3 \exp\left[-b^2 \left(\frac{m_q^2}{x_1} + \frac{m_D^2}{1 - x_1}\right)\right] \quad (2.9)$$

proves to be well suited. This DA is an adaption of a meson DA obtained by transforming a nonrelativistic harmonic-oscillator wave function to the light cone [14]. Hence the masses occurring in Eq.(2.9) are constituent masses ($m_q = 330 MeV$ and $m_D = 580 MeV$ for light quarks and diquarks, respectively). The oscillator parameter b is chosen in such a way that the full wavefunction gives rise to a value of $600 MeV$ for the mean intrinsic transverse momentum $< k_\perp^2 >^{1/2}$ of the proton in accordance with the EMC [15]. The exponential factor guarantees for the suppression of the end-point regions ($x_1 \to 0, 1$) where the application of perturbative QCD becomes problematic. With more and better data in the intermediate and large momentum-transfer region it may turn out that a more refined DA is needed. But at present the simple DA (2.9) does sufficiently well. Be that as it may, it should be emphasized that the extreme sensitivity to the choice of the DA occurring in the pure quark HSP is not observed in the diquark model.

3 Applications of the Diquark Model

Without making explicit calculations a few important consequences of the diquark hypothesis can already be noticed:

i) Since the effective number of hadronic constituents is reduced as compared to the pure quark HSP, power laws are modified in the kinematic range where diquarks appear as nearly elementary particles. In general, the HSP power-law behaviour of an exclusive scattering amplitude at fixed $\theta_{c.m.}$ is $(1/Q)^{(n-4)}$, where n is the number of external particles (quarks, diquarks, leptons, photons, ...) in $\widehat{T}$. This dimensional-counting rule is modified by factors depending on $\ln(Q^2)$ which have their origin in the running strong coupling constant $\alpha_S(Q^2)$ and eventually in the Q^2 evolution of the DA's. In case the momentum transfer Q is large enough, the diquark form factors compensate the missing powers of $1/Q$.

ii) At finite momentum transfer vector diquarks are able to provide for hadronic helicity flips. Only for large values of Q^2 hadronic helicity conservation is recovered via the vector-diquark form factors.

iii) The number of Feynman diagrams to be computed is much smaller than in the pure quark HSP. Therefore problems with delicate cancellations during the numerical summation of a large number of diagrams either do not show up or can be handled much easier by analytical means.

3.1 Electromagnetic Form Factors of the Nucleon [16]

The simplest process to which the diquark model may be applied is elastic electron-nucleon scattering. This reaction, which is usually parameterized in terms of electromagnetic nucleon form factors, has been used to fix the open parameters of the model. Explicit expressions for the electromagnetic nucleon form factors are obtained by convoluting the result for the elementary subprocesses of Fig.1 with the DA (2.9) and comparing with the electromagnetic nucleon current written in the form

$$I^{\mu}_{\lambda_f \lambda_i} = \imath e_0 \bar{u}_N(p_f, \lambda_f) \left[\gamma^{\mu} G^N_M(Q^2) - \frac{\kappa_N}{2m_N}(p_f + p_i)^{\mu} F^N_2(Q^2) \right] u_N(p_i, \lambda_i) . \quad (3.1)$$

The somewhat unorthodox decomposition (3.1) has the advantage that in the Breit frame only helicity non-flip transitions contribute to $G^N_M(Q^2)$, and helicity-flip transitions to $F^N_2(Q^2)$. Whereas due to helicity conservation the applicability of the pure quark HSP is confined to G^N_M, the diquark model also gives results for F^N_2 (and hence G^N_E). In agreement with the pure quark HSP the diquark model predicts G^N_M to behave like Q^{-4} asymptotically. Although F^N_2 decreases like Q^{-6} asymptotically, it contributes to the electric form factor in leading order

$$G^N_E(Q^2) = G^N_M(Q^2) - \kappa_N(1 + \frac{Q^2}{4m^2_N})F^N_2(Q^2) . \quad (3.2)$$

Thus, $G^N_E(Q^2)$ differs from $G^N_M(Q^2)$ even in the limit $Q^2 \to \infty$. A good fit to the elastic electron-nucleon data available for $Q^2 \gtrsim 4 GeV^2$, namely e-p cross sections σ_{ep} measured by Arnold et al.[17] and the ratio σ_{en}/σ_{ep} obtained from the reaction $ed \to epn$ [18], was achieved with the following set of parameters:

$$\begin{array}{ll} Q^2_S = 3.22 \text{ GeV}^2 & Q^2_V = 1.58 \text{ GeV}^2 \\ a_S = a_V = 0.286 & \kappa_V = 1.16 \\ f^p_S = 66.1 \text{ MeV} & f^p_V = 120.2 \text{ MeV} . \end{array} \quad (3.3)$$

The diquark model results for G^p_M, G^p_E, G^n_M, and G^n_E are quite similar to those obtained by Körner and Kuroda [19] with a model based on vector-meson dominance. Unfortunately, data for G^p_E, G^n_M, and G^n_E above, say, $Q^2 \approx 4 GeV^2$ are very sparse and of poor quality so that one cannot discriminate between different models. Another consequence of this fact is that the relative strength of the contributions from vector and scalar diquarks is not well determined at present.

3.2 Compton Scattering [13]

Having fixed the parameters of the diquark model by means of elastic e-N scattering, no freedom is left for other reactions involving only nucleons. For Compton scattering off protons differential cross section data are available up to $|t| \approx 4 GeV^2$ [20]. Thus $\gamma p \to \gamma p$ may be considered as a first test for the diquark model. This reaction is usually described by 6 independent (cms) helicity amplitudes

$$\begin{array}{lll} \phi_1 = M_{1\frac{1}{2}, 1\frac{1}{2}} & \sim s^{-2} & \phi_2 = M_{-1-\frac{1}{2}, 1\frac{1}{2}} \sim s^{-5/2} \\ \phi_3 = M_{-1\frac{1}{2}, 1\frac{1}{2}} & \sim s^{-3} & \phi_4 = M_{1-\frac{1}{2}, 1\frac{1}{2}} \sim s^{-5/2} \\ \phi_5 = M_{1-\frac{1}{2}, 1-\frac{1}{2}} & \sim s^{-2} & \phi_6 = M_{-1\frac{1}{2}, 1-\frac{1}{2}} \sim s^{-5/2} . \end{array} \quad (3.4)$$

In Eq.(3.4) the (fixed angle) large s behaviour of the helicity amplitudes, as resulting from the diquark model, is already indicated. In the pure quark HSP the helicity non-conserving amplitudes ϕ_2, ϕ_4, and ϕ_6 vanish. A set of diagrams contributing to the elementary scattering amplitudes $\hat{T}$ is depicted in Fig.3. The calculation of these diagrams is already a very demanding and complicated job which was actually accomplished by means of symbolic computer programs (SCHOONSHIP and FORM). Nevertheless, the final expressions for the hadronic helicity amplitudes ϕ_i are still simple enough to be printed in a publication (see Ref.[13]).

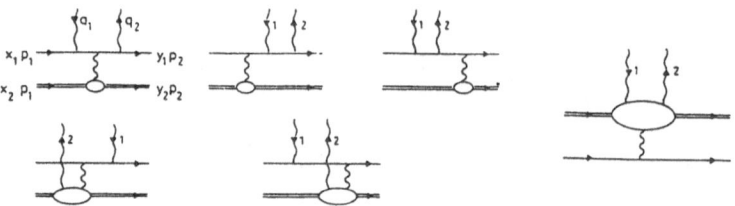

Fig. 3. Diagrams contributing to $\gamma p \rightarrow \gamma p$. The diagrams with the two photons interchanged are not shown

Results of the diquark model are confronted with the data in Fig.4. Although the data are still at the limits of applicability of a model based on perturbative QCD, the diquark model is seen to work surprisingly well. For comparison also the best result obtained within the pure quark HSP by Farrar and Zhang[21] is plotted. It is observed to lie about one order of magnitude below the data. Better agreement has been claimed very recently by Kronfeld and Nižic [22] who essentially redid the calculation of Ref.[21]. They confirm the analytic results for the Feynman diagrams, but disagree with the numerical analysis. They suspect the discrepancy to be caused by the rather crude handling of propagator singularities (see below) in Ref.[21].

A new, interesting feature showing up in Compton scattering is the occurrence of a relative phase between helicity-flip and non-flip amplitudes. The reason is that there are kinematic regions in the space of momentum fractions (x_1, y_1) (cf. Fig.3) where the exchanged gluon can propagate on mass shell. This happens for diagrams where the two photons couple to different constituents (4 point contributions) and holds true for the pure quark HSP as well. The corresponding propagator poles, which are treated in the usual way by means of

$$\frac{1}{z \pm i\epsilon} = \mathrm{P}\left(\frac{1}{z}\right) \mp i\pi\delta(z) , \qquad (3.5)$$

give rise to an imaginary part in ϕ_1, ϕ_3, and ϕ_5. In connection with the non-vanishing helicity flip amplitudes ϕ_2, ϕ_4, and ϕ_6 a nonzero transverse polarization P is generated. As can be seen from Fig.5 P amounts to $\approx 6\%$ for $s = 8.4 GeV^2$. The predictions of the polarization, however, are very sensitive to the choice of the DA. This observation and the fact that due to the absence of Sudakov corrections [23] the imaginary part of the hadronic amplitudes should be well approximated by lowest order perturbation theory are good reasons to believe that valuable information on the proton DA can be gained from polarization measurements.

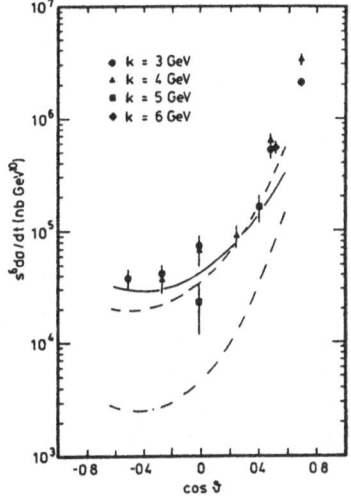

Fig. 4. The Compton cross section vs. $\cos \vartheta$ for two photon energies. The solid (dash–dotted) line represents our result for the DA (2.9) at a photon lab. momentum of $4 GeV$ ($s = 10.2 GeV^2$). The dash–dot dotted line represents a result obtained within the pure quark HSP [21]. Data are taken from Ref.[20]

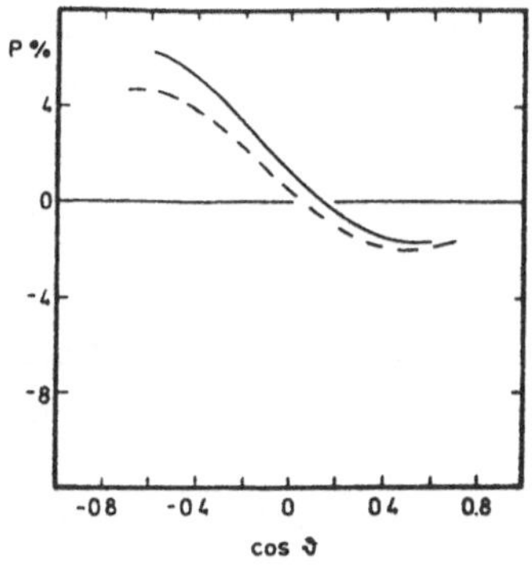

Fig. 5. Predictions for the transverse polarization in Compton scattering. Lines as in Fig.4

Simultaneously with Compton scattering also the crossed process $\gamma\gamma \to p\bar{p}$ has been considered. Since the Feynman diagrams for this reaction were calculated independently it provided a valuable check for the correctness of the $\gamma p \to \gamma p$ amplitudes. For two-photon annihilation there exist, however, no appropriate data to compare with.

3.3 Photoproduction of Mesons

Photoproduction of mesons

$$\gamma p \to M B \qquad (3.6)$$

represents a large class of photon-induced reactions for which data at a few GeV of momentum transfer already exist [20],[24]. Further data at even larger s, t, and u will be provided within the not-too-far future by the DESY e-p collider HERA. From the theoretical point of view, photoproduction means one step further in complexity as compared to Compton scattering. Within the diquark model the number of Feynman diagrams contributing is in general about three times greater than for $\gamma p \to \gamma p$, and triple gluon vertices occur for the first time. However, there are fortunately a few photoproduction reactions for which matters become somewhat simpler - namely those where the baryon B in the final state is a Λ. Since the p and the Λ have in common only the S(ud) diquark such processes proceed solely through the S diquark. For an analogous reason only V diquarks are involved if B becomes a Σ^0. In both cases an additional simplification occurs owing to the fact that those diagrams do not contribute where the quark and the diquark of the incoming proton go into the final baryon B (such contributions, e.g., show up in $\gamma p \rightarrow \pi^0 p$).

Photoproduction of Kaons

$$\gamma p \rightarrow K^+ \Lambda \tag{3.7}$$

has been investigated within the diquark model in Ref.[25]. A few representative examples of Feynman diagrams contributing to this reaction are shown in Fig.6. As in Compton scattering elementary amplitudes $\hat{T}$ become complex due to propagator singularities, which are again treated according to Eq.(3.5). Nevertheless, since $\gamma p \rightarrow K^+ \Lambda$ only proceeds via S diquarks, scattering amplitudes and hence observables which require a flip of the baryonic helicity (like the proton asymmetry or the Lambda polarization) become zero. To be more specific, this means that only two of the four independent amplitudes for photoproduction of pseudoscalar mesons (cf.Ref.[26]) remain nonzero, namely

$$S_1 = M_{-\frac{1}{2},1-\frac{1}{2}} \qquad \text{and} \qquad S_2 = M_{\frac{1}{2},1\frac{1}{2}} . \tag{3.8}$$

At fixed (large) angle both amplitudes scale asymptotically like $s^{-5/2}$. Thus the differential cross section, which is given by

$$\frac{d\sigma}{dt} = \frac{1}{32\pi(s - m_p^2)^2}(|S_1|^2 + |S_2|^2) , \tag{3.9}$$

should decrease asymptotically like s^{-7}.

In order to make quantitative predictions for $\gamma p \rightarrow K^+ \Lambda$ the new pieces entering, namely the DA's of the K^+ and the Λ have still to be specified. Since the DA (2.9) contains already a flavour dependence it is near at hand to choose this form for the scalar and vector DA (actually we need only the scalar one) of the Λ as well, i.e. $\phi_S^\Lambda(x_1) = \phi_{ho}(x_1)$. Following the natural strategy of keeping the model as simple as possible we employ spin-flavour SU(6) to fix f_S^Λ (cf.Eq.(2.7)) to be equal to f_S^p. A form analogous to (2.9) for the K^+ DA reads [14]

$$\phi^{K^+}(x_1) = \tilde{C}x_1(1 - x_1)\exp\left[-\tilde{b}^2\left(\frac{m_u^2}{x_1} + \frac{m_s^2}{1 - x_1}\right)\right] , \tag{3.10}$$

where $\tilde{C}$ is determined by the normalization condition (2.8). As for the proton the parameter $\tilde{b}$ is connected with the mean intrinsic transverse momentum of the K^+. Also the constant f^{K^+} is known to reasonable accuracy from the $K^+ \rightarrow \mu^+ \nu$ decay [27] to be

$$f^{K^+} = \frac{f_{decay}^{K^+}}{2\sqrt{3}} \approx 32 \text{MeV} . \tag{3.11}$$

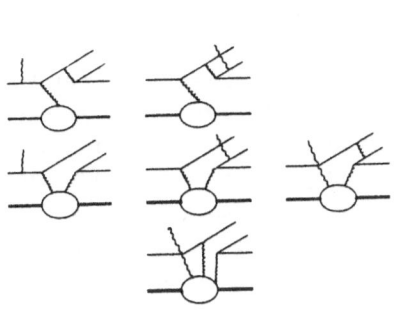

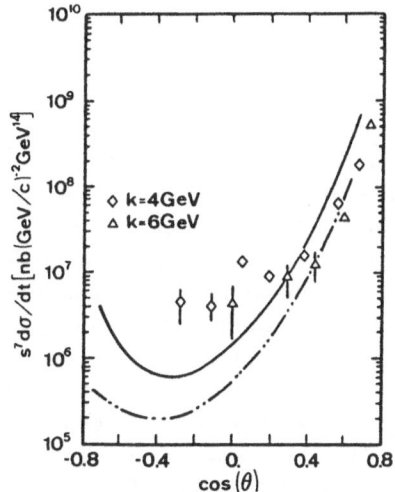

Fig. 6. Some diagrams contributing to $\gamma p \to K^+ \Lambda$

Fig. 7. The photoproduction cross section vs. $\cos\theta$ for $\gamma p \to K^+ \Lambda$. The solid line represents our result for the p and Λ DA of the form (2.9) and the K^+ DA of Eq.(3.10). The dash-dot dotted line represents a result obtained within the pure quark HSP [28]. Data are taken from Ref[24]

With these very plausible choices for the K^+ and Λ ingredients acceptable agreement with the $\gamma p \to K^+ \Lambda$ cross-section data is already achieved without introducing new, open parameters (see Fig.7). With more refined DA's for the hadrons involved the result is even improved [25]. This may be regarded as a big success of the diquark model. In contrast, corresponding calculations within the pure quark HSP fall somewhat short in the kinematic region where they are supposed to be most relevant [28].

In addition to $\gamma p \to K^+ \Lambda$ also $\gamma p \to K^{*+} \Lambda$ and the corresponding electroproduction processes have been considered in Ref.[25]. For these reactions, however, data are confined to small momentum transfers ($t \lesssim 1 GeV^2$) so that a direct comparison with the diquark-model predictions is not meaningful.

4 Concluding Remarks

We have discussed the phenomenological concept of diquarks in connection with exclusive hadronic processes. The diquark model presented here serves the description of the intermediate momentum transfer region - i.e. the kinematic region where, on the one hand, the pure quark HSP is not yet fully operational which, however on the other hand, is just within reach of present days experiments. The main ingredients of the diquark model are baryon DA's in terms of quarks and diquarks, the coupling of gluons and photons to diquarks, and, in order to account for the composite nature of diquarks, phenomenological vertex functions (diquark form factors). The proper choice of the diquark form factors is essential to achieve compatibility of the diquark model with the pure quark HSP at asymptotically large momentum transfers.

We have demonstrated that the diquark model, with a common set of parameters and common DA's, successfully accounts for the cross section data on $\gamma p \rightarrow \gamma p$, $\gamma p \rightarrow K^+ \Lambda$, and for all available information on electromagnetic nucleon form factors. In connection with these reactions also the processes $\gamma\gamma \rightarrow p\bar{p}$, $\gamma p \rightarrow K^{*+}\Lambda$, and $\gamma^* p \rightarrow K^+ \Lambda$ have been considered, although no direct comparison with corresponding data can be performed since these are available only at very low energies. But there are still other reactions, not mentioned already, for which the diquark model has been applied within the same spirit leading to very promising results. These include electroexcitation of nucleon resonances [29], electromagnetic hyperon form factors [30], the axial vector form factors of the nucleon [30], and weak transition form factors between heavy flavour baryons [31]. The same approach (with somewhat different parameters) has been adopted by Anselmino et al.[32] to explain charmonium decays ($\eta_c, \chi_{c0,c1,c2} \rightarrow p\bar{p}$). For the sake of completeness it should still be mentioned that the diquark idea has also been utilized for elastic baryon-baryon and meson-baryon scattering ($pp \rightarrow pp, p\bar{p} \rightarrow p\bar{p}, \pi p \rightarrow \rho p$) [33] and for $p\bar{p} \rightarrow$ heavy flavour baryon-antibaryon pairs [34]. These purely hadronic scattering reactions, however, have not been treated within the full HSP. Apart from the fact that new complications show up due to pinch singularities and the necessary Sudakov resummations, this would be an enormous task even with diquarks. Rather a spectator model (end-point model) has been employed.

Altogether, there exists already an impressive list of hadronic reactions for which the diquark idea works surprisingly well. Diquarks are observed to account effectively for some real nonperturbative physics in exclusive reactions at moderately large momentum transfer (e.g. polarization phenomena). There are, of course, still uncertainties concerning the parameters and DA's of the diquark model. But there is hope that, with the increasing number of reactions considered (e.g. virtual Compton scattering, other photoproduction channels, deuteron form factors, ...) and with more and better data from intermediate energy facilities like CEBAF, one finally will arrive at a clear picture of the rôle which diquarks play in exclusive scattering.

References

1 For a recent survey, see: Perturbative Quantum Chromodynamics, edited by A.H. Mueller (World Scientific, Singapore, 1989).

2 G.P.Lepage and S.J.Brodsky: Phys.Rev.**D22**, 2157 (1980).

3 N.Isgur and C.H.Llewellyn Smith: Nucl.Phys.**B317**, 526 (1989).

4 V.A.Avdeenko, S.E.Korenblit, and V.L.Chernyak: Sov.J.Nucl.Phys.**33**, 252 (1981).

5 V.L.Chernyak and I.R.Zhitnitsky: Nucl.Phys.**B246**, 52 (1984).

6 N.G.Stefanis: Phys.Rev.**D40**, 2305 (1989).

7 M.Gari and N.G.Stefanis: Phys.Lett.**B175**, 462 (1986).

8 Z.Dziembowski and J.Franklin: Phys.Rev.**D42**, 905 (1990);
 see also Z.Dziembowski: these proceedings.

9 S.J.Brodsky and G.R.Farrar: Phys.Rev.Lett.**31**, 1153 (1973);
 V.A.Matveev, R.M.Muradyan, A.V.Tavkhelidze, Lett.Nuovo Cim.**7**, 719 (1973).

10 S.J.Brodsky and G.P.Lepage: Phys.Rev.**D24**, 2848 (1981).

11 W.A.Crabb et al.: Phys.Rev.Lett.**65**, 3241 (1990).

12 See also other contributions in this volume.

13 P.Kroll, M.Schürmann, and W.Schweiger: Int.J.Mod.Phys.**A6**, 4107 (1991).

14 T.Huang: Nucl.Phys.(Proc.Suppl.)**7B**, 320 (1989).

15 J.J.Aubert et al.: Phys.Lett.**B95**, 306 (1980).

16 P.Kroll, M.Schürmann, and W.Schweiger: Z.Phys.**A338**, 339 (1991).

17 R.G.Arnold et al.: Phys.Rev.Lett.**57**, 174 (1986).

18 S.Rock et al.: Phys.Rev.Lett.**49**, 1139 (1982).

19 J.G.Körner and M.Kuroda: Phys.Rev.**D16**, 2165 (1977).

20 M.A.Shupe et al.: Phys.Rev.**D19**, 1921 (1979).

21 G.R.Farrar and H.Zhang: Phys.Rev.**D41**, 3348 (1990).

22 A.S.Kronfeld and B.Nižic: Phys.Rev.**D44**, 3345 (1991).

23 G.R.Farrar, G.Sterman, and H.Zhang: Phys.Rev.Lett.**62**, 2229 (1989).

24 R.L.Anderson et al.: Phys.Rev.**D14**, 679 (1976).

25 M.Schürmann: thesis; paper in preparation.

26 I.S.Barker et al.: Nucl.Phys.**B95**, 347 (1975).

27 X.H.Guo and T.Huang: Phys.Rev.**D43**, 2931 (1991).

28 G.R.Farrar, K.Huleihel, and H.Zhang: Nucl.Phys.**B349**, 655 (1991).

29 P.Kroll, M.Schürmann, and W.Schweiger: Z.Phys.**A342**, 429 (1992).

30 P.Kroll: preprint WU-B 91-17, Wuppertal (1991);
 P.Kroll, M.Schürmann, and W.Schweiger: in preparation.

31 J.G.Körner and P.Kroll: preprint DESY 92-019;
 J.G.Körner: these proceedings.

32 M.Anselmino et al.: Phys.Rev.**D44**, 1438 (1991);
 M.Anselmino: these proceedings.

33 M.Anselmino, P.Kroll, and B.Pire: Z.Phys.**C36**, 89 (1987).

34 P.Kroll and W.Schweiger: Nucl.Phys.**A474**, 608 (1987);
 P.Kroll, B.Quadder, and W.Schweiger: Nucl.Phys.**B316**, 373 (1989).

Segregation of Quarks within the Nucleon

Zbigniew Dziembowski
Department of Physics, Temple University, Philadelphia, PA 19122, USA

Abstract. Measurements of the nucleon form factor and structure function indicate an inhomogeneous distribution of flavor, charge and spin within the nucleon. It is argued that the ordinary constituent quark model can explain the observed inhomogeneity as arising from spin-dependent interaction between quarks in this model. This agreement suggests a specific bound quark picture of the nucleon structure with a moderate diquark correlation.

Keywords. Quark, diquark, nucleon, light-cone, structure functions

1. Introduction

The nonrelativistic CQM dynamics, in its modern QCD-inspired version, was proposed in 1975 by De Rújula, Georgi, and Glashow [1] and subsequently developed by Isgur and Karl [2] into a comprehensive description of the hadron structure. Since then the model has successfully related details of the hadron spectra with basic regularities of electromagnetic, weak and strong couplings and decay rates. The scope of the success gives us some confidence that the nucleon wave function obtained from the CQM may contain reliable information about the *quark correlation*.

Diquark is the generic name given in the development of hadron physics to a correlated pair of quarks or antiquarks [3]. Many contributions to this conference deal with the possible dynamical role of diquarks in processes involving hadrons. There is experimental evidence that in some instances two of the constituent quarks interact collectively as a single entity. But this collective response falls into the category of the *dynamical* quark correlation that can be analyzed only with an explicit reference to the external dynamics of quarks. The aim of this presentation, however, is to study the *static* quark correlations as they may arise from the conventional CQM dynamics. Once the internal quark dynamics is identified, the static quark correlations can be explored from an expectation value of bilocal quark operators in the nucleon ground state

$$< N \mid \overline{\Psi}(\mathbf{r}, t = 0) \, \Gamma \, \Psi(0, t = 0) \mid N >, \tag{1}$$

with the latter being determined by the nucleon rest frame (CM) wave function $\psi_{CM}(\mathbf{r_1}, \mathbf{r_1}, \mathbf{r_3})$. In Sec. 2, I review the CQM picture of the nucleon structure and point to the spin-spin force as the effective quark clustering dynamics. The quark correlations in position space are discussed in Sec. 3.

The richest and most valuable source of information about quark correlations is deep-inelastic inclusive lepton-nucleon scattering (DIS). In the so-called Bjorken limit, all three independent twist-2 structure functions [4] relevant for the DIS are Fourier transforms of static correlation functions in the nucleon ground state. But they are not, however, the equal-time correlation functions of Eq. (1) but instead light-cone correlation functions

$$< N \mid \overline{\Psi}(x^-, x^+ = 0) \, \Gamma \, \Psi(0, x^+ = 0) \mid N > . \tag{2}$$

In the valence approximation, the latter are completely determined by the three-quark momentum (light-cone) wave function $\psi_{LC}(x_1, \mathbf{k}_{\perp 1}, x_2, \mathbf{k}_{\perp 2}, x_3, \mathbf{k}_{\perp 3})$. In Sec. 4, I discuss how, in the nonrelativistic CQM, one can make an approximate connection between ψ_{CM} and ψ_{LC}. Once this is done one can relate any amount of diquark clustering present in the spatial correlation function of Eq. (1) with that observed in the DIS structure function measurements. The quark correlations in the momentum space are discussed in Sec. 5.

2. Quark Model: Nucleon Wave Function

I start with a review of the essential elements of the quark potential model of Ref. [2] and then show how the model's dynamics leads in general to a spatial segregation of quarks within the nucleon.

For a low-momentum-scale description of the nucleon ground state one needs three quarks of two flavors (uud in the proton and ddu in the neutron) with effective masses $m \simeq m_N/3$. Quarks are spin-1/2 particles and carry a color quantum number that influences the permutation symmetry of the nucleon wave function.

The model finds two dynamical elements being dominant for the nucleon ground state properties. First, there is a long-distance, spin-independent confining potential that determines the overall size of the system and gives a symmetric momentum spread to the three quark state of the order of 250 MeV [5]. In addition, there is the contact hyperfine spin-spin interaction motivated by QCD one-gluon exchange and responsible for breaking the nucleon-delta mass degeneracy.

This force is repulsive for quark pairs in a spin-1 state and attractive for those in a spin-0 state. It is known to make the $\Delta(1232)$ (with spin 3/2) more massive than the nucleon (with spin 1/2). But what is more important here is that the force also modifies the nucleon wave function.

Refs. [6, 7] contain perturbation theory estimates showing how the hyperfine interaction modifies the totally symmetric wave function of the harmonic-oscillator confining potential. They use a spin-flavor SU(6) basis and find that the dominant effect of the interaction is the mixing of the $N\,^2S_M$ ($[70, 0^+]$) configuration into the pure $N\,^2S_S$ ([56, 0$^+$]) nucleon. Thus the physical ground state is of the form

$$| N > \simeq cos\varphi \; | N_S > + sin\varphi \; | N_M >,$$
(3)

for which the perturbation theory calculation of Ref. [7] yields $\varphi \simeq -15^0$.

To visualize the quark correlation, I first abandon SU(6) with its merely historical significance and re-write the mixture in the *"uds"* basis [8]. Thus the relevant components are

$$| N_S >= \chi^\lambda \, \Phi^S, \qquad | N_M >= \frac{1}{\sqrt{2}}(\chi^\rho \, \psi^\rho - \chi^\lambda \, \psi^\lambda),$$
(4)

where the χ's are the usual spin-$\frac{1}{2}$ wave functions $\quad \chi^\rho = (s_2 - s_1)/\sqrt{2}$, $\chi^\lambda = (2\, s_3 - s_1 - s_2)/\sqrt{6}$, with $s_1 =\downarrow\uparrow\uparrow$, $s_2 =\uparrow\downarrow\uparrow$, $s_3 =\uparrow\uparrow\downarrow$, and the ψ's and Φ are the harmonic-oscillator wave functions $\Phi^S = exp(-\alpha^2\{\rho^2 + \lambda^2\}/2)$, $\psi^\rho = 2\alpha^2 \; \rho \cdot \lambda \; \Phi^S/\sqrt{3}$, $\psi^\lambda = \alpha^2(\rho^2 - \lambda^2) \, \Phi^S/\sqrt{3}$ with the Jacobi coordinates in the coupling (12,3) given by $\rho = (r_1 - r_2)/\sqrt{2}$ and $\lambda =(r_1 + r_2 - 2r_3)/\sqrt{6}$.

Next, I also abandon the symmetric harmonic-oscillator basis for the distorted one used by Isgur and Karl in their description of the baryon structure in the strange sector. The use of the DHO basis trades off the mixing angle of the perturbative calculation for a more intuitive *deformation parameter* defined below. The latter is unambiguously constrained by the experimental value of the neutron to proton charge radius ratio.

Let me consider the nucleon as being initially in the symmetric Gaussian configuration

$$| N_S >= [(s_3 - s_1) + (s_3 - s_2)]\Phi^S(12,3),$$
(5)

produced by the harmonic oscillator part of the confining potential alone,

$$\Phi^S(12,3) = exp(-\alpha^2\{\rho^2 + \lambda^2\}/2\,).$$
(6)

In the spin state $\chi^\lambda = (s_3 - s_1 + s_3 - s_2)/\sqrt{6}$ the first two quarks with the same flavor are in a spin-1 state whereas each u-d pair is in a spin-0 state. This implies that the u and d quarks are attracted to each other by the spin interaction, but the odd one (i.e., *u* quark in the proton or *d* quark in the neutron) is driven further from the center of mass of the system. In the DHO basis, this effect can be parameterized through the use of a scale parameter $\alpha_\rho^2 = \alpha^2(1 + d)$ (corresponding to a smaller size) for the spin-0 pairs of u-d quarks that is larger than the one $\alpha_\lambda^2 = \alpha^2(1 - d)$ for the coupling of the odd quark to this pair. After symmetrization with respect to the quantum numbers of the first two identical quarks, one gets

$$| N >= (s_3 - s_1) \, \Phi(13,2) + (s_3 - s_2) \, \Phi(23,1),$$
(7)

where the asymmetric spatial wave function for the quark configuration (12,3) is

$$\Phi(12,3) = exp(-\{\alpha_\rho^2\rho^2 + \alpha_\lambda^2\lambda^2\}/2\,).$$
(8)

When writing the asymmetric spatial functions $\Phi(13,2)$ and $\Phi(23,1)$ one should use the relative coordinates appropriate for the given couplings.

Now it is easy to relate the diquark wave function of Eq. (7) to the configuration mixing of Eq. (3). Expanding Eq. (7) up to linear terms in the deformation parameter d (> 0), one gets

$$| N > = \chi^\lambda \, \Phi^S - \frac{\sqrt{3}\,d}{2\sqrt{2}} \frac{1}{\sqrt{2}} (\chi^\rho \, \psi^\rho - \chi^\lambda \, \psi^\lambda). \qquad (9)$$

One can easily recognize in the above expression the mixture of [56,0+] and [70,0+] (written in the "uds" basis) with

$$tan\varphi = - \frac{\sqrt{3}}{2\sqrt{2}} \, d. \qquad (10)$$

To estimate the deformation parameter d and the Gaussian scale α in (8) I calculate the nucleons' ms charge radii. With simple quark additivity one gets

$$< r^2 >_p^{ch} = \frac{1}{\alpha^2}(1 + \frac{1}{2}d)f(d), \quad < r^2 >_n^{ch} = \frac{1}{\alpha^2}(- \frac{1}{2}d)f(d). \qquad (11)$$

where f(d) $= [2(1 - d^2)^{-5/2} + (1 - d^2/4)^{-5/2}]/[2(1 - d^2)^{-3/2} + (1 - d^2/4)^{-3/2}]$. As expected, the spin-spin interaction that drives the odd quark further from the center of mass makes the proton charge radius bigger (compared to the symmetric limit of $1/\alpha^2$) and yields the proper sign for $< r^2 >_n^{ch}$. Using input from the experimental data: $< r^2 >_p^{ch} = 0.743 \pm 0.02$ fm^2 (Ref. [9]) and $< r^2 >_n^{ch} = -0.117 \pm 0.002$ fm^2 (Ref. [10]), one obtains $d = 0.372 \pm 0.020$ and $\alpha = 264 \pm 11$ MeV. With these values the present model produces the correct values for the neutron and proton charge radii and yields a wave function which differs from that of the Isgur-Karl model only by terms of order d^2 provided one neglects (small) contributions from the $N^2 S_{S'}$ and $N^4 D_M$ configurations.

3. Quark Spatial Correlations

Let me mention at this point that it has been recognized for many years that a spatially dependent spin-spin interaction that is more repulsive for quarks with parallel rather than antiparallel spins may induce a segregation of charge within the neutron [11]. This possibility has been used with success in a quark model with a hyperfine interaction to explain quantitatively the charge radius of this particle [6, 7]. Note, however, that the above dynamical argument about segregation is far more general. In fact it implies that the short distance hyperfine interaction together with Fermi-Dirac statistics leads to a spatial segregation of both flavor and spin distributions within the proton and neutron. Thus, one should expect more spectacular manifestations than just the negative charge radius of the neutron.

To see the segregation of quarks in configuration space, I calculate the spin average $(u(r), d(r))$ and spin weighted $(\Delta u(r), \Delta d(r))$ probabilities of finding a u or d quark at a distance r from the center of mass of the nucleon. Fig. 1 gives the neutron charge density $\rho_n(r) = (e_u \, d(r) + e_d \, u(r))$ together with the integral

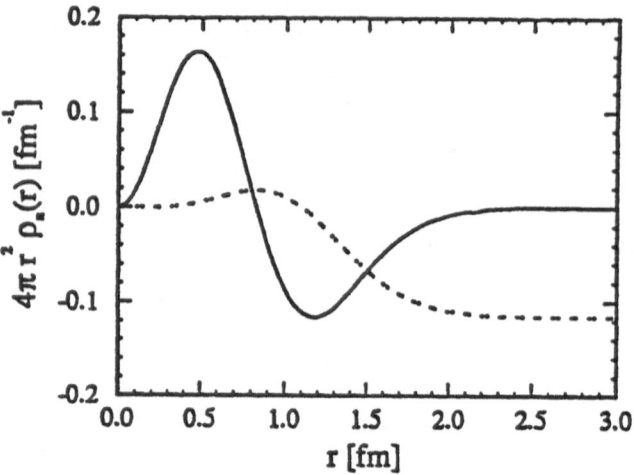

Figure 1: The neutron charge density.

$\int_0^r \rho_n(x)x^2 d^3x$. The latter quantity exhibits the distance over which the total neutron ms charge radius is built up.

Fig. 2 shows the flavor asymmetry $F_2^n(r)/F_2^p(r) = (e_u^2\, d(r)+e_d^2\, u(r))/(e_u^2\, u(r)+ e_d^2\, d(r))$ and the proton spin asymmetry $A_1^p(r) = (e_u^2\, \Delta u(r)+e_d^2\, \Delta d(r))/(e_u^2\, u(r)+ e_d^2\, d(r))$, respectively. These quantities are compared with the predictions of the symmetric model of $|\, N_S >$.

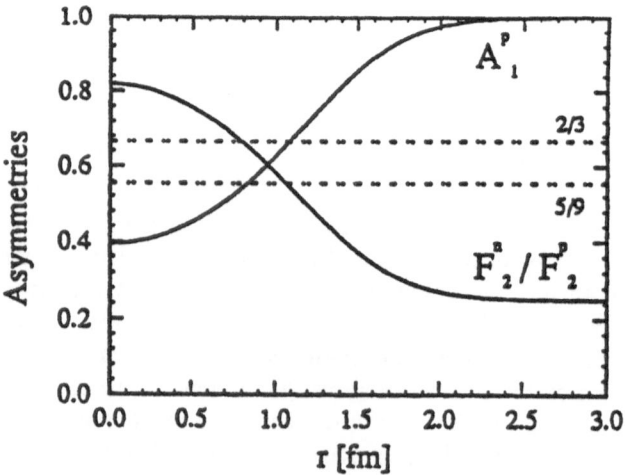

Figure 2: The flavor-spin asymmetries.

All distributions suggest the existence of a soft positive $u - d$ core in the spin-0 state and a broad layer of linear size approximately ~ 1 fm where the polarized u (in the proton) or d (in the neutron) is orbiting. To make this remark more quantitative, I calculate the probability density to find the quarks in the cluster at a distance ρ: $g(\rho) = < N \mid P_{13} \, \delta(\rho - \rho_{13}) + P_{23} \, \delta(\rho - \rho_{23}) \mid N >$ and the probability density to find the odd one at a distance λ from the cluster: $f(\lambda) = < N \mid P_{13} \, \delta(\lambda - \lambda_{13}) + P_{23} \, \delta(\lambda - \lambda_{23}) \mid N >$. Now with these quantities in hand one can follow Fleck *et. al.* [12] in their search for diquark clustering and find the most probable distance between quarks in the cluster together with the most probable distance between the odd quark and the cluster. They are $D_{u-d} = .66$ *fm* and $D_{odd-ud} = .88$ *fm*, respectively. Since D_{u-d} is not much smaller than D_{odd-ud} I think that it is more proper to use the term *quark segregation* rather then *diquark clustering* to describe the inhomogeneous distribution of flavor, charge and spin within the nucleon.

4. Parton Model: Nucleon Wave Function

To understand the nucleon structure as revealed by the DIS experiments at large Q^2, the nucleon is considered to be an infinite collection of several kinds of partons, some of which carry the quark quantum numbers except for having much lighter (*current quark*) masses. It is traditional to formulate the parton model picture in the nucleon's infinite momentum frame or equivalently at fixed light-cone time $x^+ = x^0 + x^z$ [13]. If we let the variable x^+ play the role of time and quantize QCD on the null plane, then a nucleon can be described by

$$\mid N > = \mid qqq > \Psi_{qqq} + \mid qqqg > \Psi_{qqqg} + \mid qqqq\bar{q} > \Psi_{qqqq\bar{q}} + \cdots, \qquad (12)$$

where the first, valence term contains only three quarks but the other terms include gluons and quark-antiquark pairs. The coefficients in this light-cone Fock expansion are the parton wave functions which depend on the light-cone relative momentum variables x_i and $\mathbf{k}_{\perp i}$, as well as on helicity λ_i and other internal variables. The variables x_i are defined as fractions of the plus component of the nucleon momentum carried by the *i-th* parton, so that $p_i^+ = x_i P^+$ ($\sum x_i = 1$), and the variables $\mathbf{k}_{\perp i}$ are related to the parton's transverse momenta by $\mathbf{k}_{\perp i} = \mathbf{p}_{\perp i} - x_i \mathbf{P}_{\perp}$ ($\sum \mathbf{k}_{\perp i} = 0$).

What is the transition from the current to constituent quarks? Some initial suggestions have already been given in the early days of QCD [14, 15] They are based the phenomenon of 'reabsorption' of quark-antiquark pairs and most of the glue into the valence quarks when we go with the perturbative evolution down to the confinement scale. A recent formulation of this idea comes from Jaffe [16] He calls it *minimal identity*: "Quark and gluon distributions evolve perturbatively from some quark model wave function (WF) without surprise."

Acting in this spirit, I shall take the Schrödinger wave function of Eq. (7) from the asymmetric quark model and, after transforming it to light-cone variables, I shall re-interpret the WF of the CQM as the valence part of the nucleon Fock space WF. Next this WF will be used to calculate the light-cone correlation functions at some low scale Q_0^2 which then must be evolved to $Q^2 >> Q_0^2$ in order to be compared with experiment.

As discussed in Ref. [17], any eigenfunction of rest energy and spin obtained

from *non-relativistic* dynamics can be converted (within some reasonable approximations) to a light-cone wave function of the same mass (equal to the rest energy) and spin. Two elements are essential for the conversion:
(1) a Melosh-type transformation to construct the light-cone spin states, and
(2) the Brodsky-Huang-Lepage (BHL) prescription [18] that relates the rest frame momentum WF with the Lorentz-invariant light-cone WF.

The first element of the approach is the Melosh rotation that gives the relation between the equal-time and light-cone spin variables. When used for the spin-0 wave function it gives [17]

$$\chi^{\rho}(12,3) \sim (s_2 - s_1) \quad \rightarrow \quad J(12,3) = \bar{u}_{\lambda_1}(m_N + p_\mu\, \gamma^\mu)\gamma_5 v_{\lambda_2}\bar{u}_{\lambda_3}u_N. \quad (13)$$

Historically the Melosh transformation was constructed as a possible model transformation linking non-interacting current quarks and the constituent quarks of the CQM [19, 20]. It is still unclear what the relation is of such a model to a 'true' light-cone dynamics from QCD, but is, at least, a consistent way to construct a nontrivial unitary representation of the Poincaré group.

The conversion procedure of Ref. [18] starts with the observation that, for a non-relativistic system, equal time and equal x^+ "time" are almost identical. Consequently, the rest frame momentum Schrödinger WF should be almost the same as the light-cone WF. Then, a simple relationship can be found between the equal-time vectors and the light-cone vectors. If a particle has non-relativistic momentum in the nucleon rest frame ($P^+ = P^- = m_N$ and $\mathbf{P}_\perp = 0$), then we can identify

$$p^z \quad \rightarrow \quad x m_N - m. \quad (14)$$

Note, however, that the change of variables in Eq. (14), *that is called sometimes the Licht-Pagnamenta prescription* [21], *should not be used beyond its range of validity, i.e.* $| x m_N - m | \ll m$. Specially sensitive regions are the $x \rightarrow 0$ and $x \rightarrow 1$ endpoints. Note, for example that the limit $x \rightarrow 0$ is an ultra-relativistic limit in a light-cone WF. Recall that in the nucleon rest frame, the light-cone fraction is $x = (p^0 + p^z)/m_N$. Thus the limit $x \rightarrow 0$ generally implies very large constituent momentum $p^z \rightarrow -p^0 \rightarrow -\infty$. At this point I would like to comment that, in a series of influential papers [22] on structure functions and small momentum transfer properties of the nucleon, use is made of the Licht-Pagnamenta prescription in the *whole* range of x ([0,1]) to convert the Schrödinger solution for particles bound in a harmonic oscillator potential into a light-cone WF

$$\Psi_{LC}(x_i, \mathbf{k}_{\perp i}) = exp(-a \sum_i \{\mathbf{k}_{\perp i}^2 + (x_i m_N - m_i)^2\}). \quad (15)$$

This form is sharply peaked at $x_i = m_i/m_N$ and $\mathbf{k}_{\perp i} = 0$ as it should be but does *not* exhibit any strong fall-off at the $x \rightarrow 0$ and $x \rightarrow 1$ endpoints expected for soft non-perturbative solutions in QCD. Because of its unphysical endpoint behavior and the lack of damping of the ultra-relativistic configurations, the function of Eq. (15) leads to the *apparent* conflict between the sign of the neutron charge radius and the structure function ratio behavior in the limit $x \to 1$, noted in Ref. [23].

The BHL prescription avoids the conflict, making a connection that differs from the Licht-Pagnamenta prescription in the endpoint regions, and that exhibits a strong fall-off behavior in these domains. To illustrate this statement, I take the non-relativistic WF of Eq. (8). After Fourier transform, this state is given by the momentum wave function $\Psi_{C.M.}(\mathbf{p}_\rho^2, \mathbf{p}_\lambda^2) = exp(- \{\mathbf{p}_\rho^2/\alpha_\rho^2 + \mathbf{p}_\lambda^2/\alpha_\lambda^2\}/2)$. where $(\mathbf{p}_\rho, \mathbf{p}_\lambda)$ are the normalized non-relativistic Jacobi momenta. Since the WF depends only on the non-relativistic relative kinetic energy, we need to relate the latter to the light-cone, relative kinetic energies [24]

$$(\mathbf{p}_\rho^2, \mathbf{p}_\lambda^2) \rightarrow (\frac{q_\perp^2 + m^2}{x_1} + \frac{q_\perp^2 + m^2}{x_2}, \frac{Q_\perp^2}{x_1 + x_2} + \frac{Q_\perp^2 + m^2}{x_3})$$

where

$$\mathbf{q}_\perp = (x_1 \mathbf{k}_{\perp 2} - x_2 \mathbf{k}_{\perp 1})/(x_1 + x_2), \quad \mathbf{Q}_\perp = (x_3(\mathbf{k}_{\perp 1} + \mathbf{k}_{\perp 2}) - (x_1 + x_2)\mathbf{k}_{\perp 3})/(x_1 + x_2 + x_3)$$

are the light-cone Jacobi momenta (in $\perp$-space).

According to the approach of Ref. [18], one can match physics at the relativistic/non-relativistic interface by equating the off-shell propagator $\mathcal{E} = M^2 - (\sum k_i)^2$ in the two frames:

$$\mathcal{E} = \begin{cases} M^2 - (\sum_{i=1}^{n} p_i^0)^2, & \sum_{i=1}^{n} \mathbf{p}_i = 0 \quad [C.M.], \\ M^2 - \sum_{i=1}^{n} (\frac{\mathbf{k}_\perp^2 + m^2}{x})_i, & \sum \mathbf{k}_{\perp i} = 0, \sum x_i = 1 \quad [L.C.]. \end{cases}$$

In the non-relativistic limit, the off-shell energy is

$$\mathcal{E} = P^+[P^- - \sum_{i=1}^{n} k_i^-] = M^2 - \sum_{i=1}^{n} (\frac{\mathbf{k}_\perp^2 + m^2}{x})_i$$

$$\cong 2M[M - \sum m_i - \sum_{i=1}^{n} (\frac{p_\perp^2 + p_z^2}{2m})_i]$$

For two and three particle systems ($m_1 = m_2 = m_3 = m$), it leads to

$$\mathbf{p}_\rho^2 \approx \frac{1}{3}(\frac{q_\perp^2 + m^2}{x_1} + \frac{q_\perp^2 + m^2}{x_2}) - 2m^2 \tag{16}$$

and

$$\mathbf{p}_\rho^2 + \mathbf{p}_\lambda^2 \approx \frac{1}{3}(\frac{q_\perp^2 + m^2}{x_1} + \frac{q_\perp^2 + m^2}{x_2} + \frac{Q_\perp^2}{x_1 + x_2} + \frac{Q_\perp^2 + m^2}{x_3}) - 3m^2. \tag{17}$$

Now if we use this (approximate) connection in the Gaussian asymmetric wave function, we get

$$\Phi(12, 3) = exp(-\frac{1}{6\alpha_\rho^2}[\frac{m_1^2 + q_\perp^2}{x_1} + \frac{m_2^2 + q_\perp^2}{x_2}] - \frac{1}{6\alpha_\lambda^2}[\frac{m_3^2 + Q_\perp^2}{x_3} + \frac{Q_\perp^2}{x_1 + x_2}]). \tag{18}$$

Finally, combining the Melosh transformation and the BHL prescription one arrives at the following low-scale valence light-cone wave function

$$\Psi(x_i, \mathbf{k}_{\perp i}, \lambda_i) = (J(13, 2)\,\Phi(13, 2) + J(23, 1)\,\Phi(23, 1))/\sqrt{x_1 x_2 x_3}. \tag{19}$$

In this way, the rest frame WF of Eq. (7) which controls binding and hadronic spectroscopy is transformed into an light-cone WF of the three-quark, valence component of the nucleon in the parton model.

5. Quark Momentum Correlations

Given the light-cone WF of Eq. (19), one can compute quark correlations in momentum space. For example, the spin averaged proton structure function is

$$F_2^p(x) = e_u^2\, u(x) + e_d^2\, d(x) \tag{20}$$

where the number density of quarks of type d with momentum fraction x in the proton is

$$d(x) = \sum_{\lambda_i} \int [dx d\mathbf{k}_\perp] \mid \Psi(x_i, \mathbf{k}_{\perp i}, \lambda_i) \mid^2 \delta(x_3 - x) \tag{21}$$

The corresponding u-quark distribution can be obtained from Eq. (21) by replacing $\delta(x_3 - x)$ with $\delta(x_1 - x) + \delta(x_2 - x)$.

Note, when performing the integral in Eq. (21), the presence of the asymmetric exponentials that explain how a flavor-spin asymmetry shows up in the x-dependence of the quark distributions. The outer quark in the wave-function exponentials (particle 1 or 2 in Eq. (19)) has the product $x\,\alpha^2(1\text{-}d)$ in the denominator of the exponential and, with $d > 0$, will have a larger average x than the core u-d quarks. Since the outer quark has the polarization of the nucleon, one arrives at the following picture of the longitudinal momentum-space segregation. As a result of the spin-spin interaction in the quark dynamics, there is a u-quark in the proton that dominates the large-x region with the u-d quarks squeezed into the small-x part of the phase space. This has two immediate consequences: the structure function ratio $F_2^n(x)/F_2^p(x)$ drops to 1/4 as x → 1, and the proton and neutron polarization asymmetry rise to 1 in the large x limit.

Corresponding predictions for the structure functions, both with the QCD evolution (solid) and without (dash-dot), are given in Fig. 3. In order to diminish the influence of the perturbative evolution on my conclusion, I choose to calculate here only ratios of different structure functions in the valence region of $x > 0.2$. At the same time, taking ratios should also reduce the influence of the Gaussian ansatz of Eq. (8).

Now, comparing similarities of the x-space and configuration-space distributions, one concludes that, although the u-quark in the proton is on average at larger distances from the center of mass than is the d-quark, with its Fourier transform being narrower than that of the d-quark, after the change to light-cone variables, the u-quark x-distribution is broader than that of the d-quark.

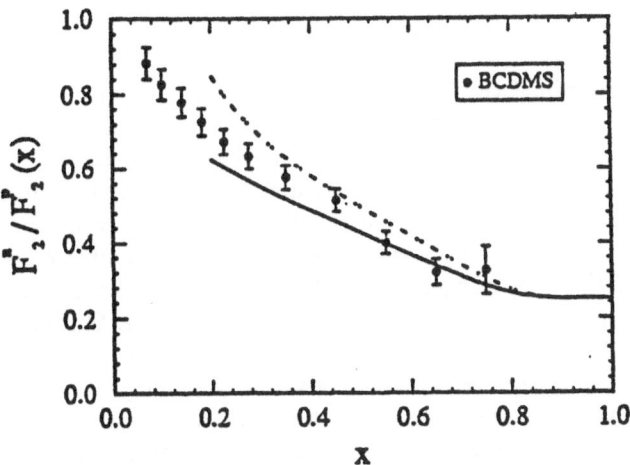

Figure 3: The flavor asymmetry compared with the data of Ref. [25]

I believe that this rather remarkable success of the QCD-inspired quark model suggests that the same spatially dependent spin-spin interaction which breaks the mass degeneracy of the nucleon and delta is also responsible for segregating the quark charge, flavor and spin within the nucleon. Similar conclusions regarding the origin of the structure-function asymmetries have been reached recently by Close and Thomas [27].

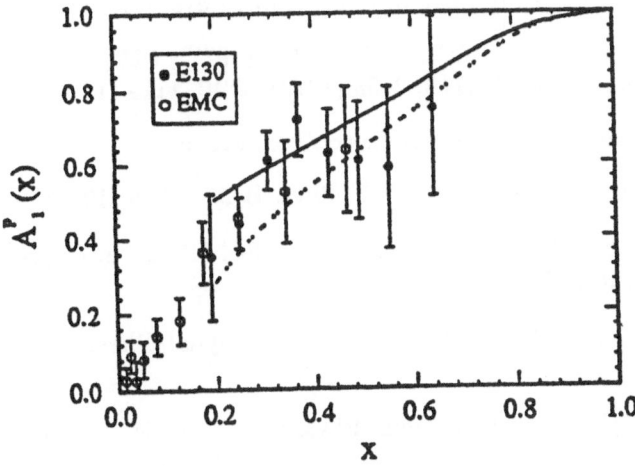

Figure 4: The proton spin asymmetry compared with the data of Ref. [26]

6. Summary

In this talk I re-examined the low-energy nucleon structure which follows from the nonrelativistic constituent quark model. I chose the successful QCD-motivated form of the quark model developed by Isgur and Karl [2], but the conclusions reached here are not restricted to this particular representation. There are good reasons to believe that any quark model that consists of a combination of two dynamical elements: a long-distance confining potential and a short-distance spin-spin force of the type suggested by QCD should predict a spatially inhomogeneous distribution of flavor and spin within the nucleon. Once the emerging quark-diquark picture agrees with such low scale features of the electromagnetic nucleon structure as charge radii, it also gives, after being transformed to the light-cone, the flavor-spin asymmetric quark momentum distributions needed to explain the difference in neutron and proton structure functions or the proton polarization asymmetry measured in the DIS experiments.

Acknowledgements. I wish to thank L. Auerbach, F. Close, J. Franklin, R. Intemann, and D. Lichtenberg for helpful discussions, the Institute for Nuclear Theory at the Univeristy of Washington for its hospitality, and the U.S. Department of Energy for partial support during the completion of this work. I am also grateful to organizers of this conference for their financial support.

References

[1] A. De Rújula, H. Georgi, and S. L. Glashow, Phys. Rev. **D12**, 147 (1975).

[2] N. Isgur and G. Karl, Phys. Lett. **72B**, 109 (1977); **74B**, 353 (1978); Phys. Rev. **D18**, 4187 (1978); **D19**, 2653 (1979); **D20**, 1119 (1979).

[3] D. B. Lichtenberg, contribution to this conference.

[4] R. L. Jaffe and Xiangdong Ji, Nucl. Phys. **B 375**, 375 (1992).

[5] C. Hayne and N. Isgur, Phys. Rev. **D25**, 1944 (1982); S. Capstick and N. Isgur, Phys. Rev. **D34**, 2809 (1986).

[6] A. Le Yaouanc, L. Oliver, O. Pene, and J.C. Raynal, Phys. Rev. **D18**, 1591 (1978).

[7] N. Isgur, G. Karl, and R. Koniuk, Phys. Rev. Lett. **41**, 1269 (1978).

[8] The *"uds"* basis was first introduced by J. Franklin, Phys. Rev. **172**, 1807 (1968) and recently used in the quark model calculations of Refs. [2] and [5]. In this approach, one notes that by isospin symmetry in the u-d sector it is sufficient to consider the flavor states $| uud >$ and $| ddu >$ for the proton and the neutron, respectively. At the same time one carries out the antisymmetrization (or symmetrization if the color is kept implicit) only

with respect to the quantum numbers of the first two identical quarks. Keeping this remark in mind, I will suppress altogether the flavor indices of my wave functions.

[9] G. G. Simon, F. Borkowski, Ch. Schmitt, and V. H. Walther, Z. Naturforsch. **35a**, (1980).

[10] V. E. Krohn, G. R. Ringo, Phys. Rev. **D9**, 1305 (1973) and L. Koester et al., Phys. Rev. Lett. **36**, 1201 (1976).

[11] R. D. Carlitz, S. D. Ellis and R. Savit, Phys. Lett. **68B**, 443 (1977); N. Isgur, Acta Phys. Polon.**B8**, 1081 (1977).

[12] S. Fleck, B. Silvestre-Brac, and J. M. Richard, Phys. Rev. **D38**, 1519 (1988).

[13] S. J. Brodsky and H-Ch. Pauli, in *Lecture Notes in Physics* **396**, Springer-Verlag, (1992).

[14] G. Parisi and R. Petrontzio, Phys. Lett **62B**, 331 (1976).

[15] R. L. Jaffe and G.G. Ross, Phys. Lett. **93B**, 313 (1980).

[16] R. L. Jaffe, SLAC workshop on high energy electroproduction and spin physics, SLAC, 1992, p. 67.

[17] Z. Dziembowski, Phys. Rev. **D37**, 768 (1988).

[18] S. Brodsky, T. Hunag and P. Lepage, in *Particles and Fields 2*, edited by A. Capri and A. Kamal, Plenum Publishing Corporation, 1983.

[19] H. J. Melosh, Phys Rev. **D9**, 1095 (1974).

[20] H. Leutwyler, in *Current Algebra and Phenomenological Lagrange Function*, ed. G. Höhler,(Springer-Verlag, Berlin/New York, 1969); L. A. Kondratyuk and M. V. Terentev, Sov. J. Nucl. Phys. **31**, 561 (1983).

[21] A. L. Licht and A. Pagnamenta, Phys. Rev. **D2**, 1150 (1970).

[22] A. Le Yaouanc, L. Oliver, O. Pene, and J.C. Raynal, Phys. Rev. **D11**, 680 (1975), **D12**, 2137 (1975), **D15**, 844 (1977).

[23] A. Le Yaouanc, L. Oliver, O. Pene, and J.C. Raynal, Phys. Rev. **D18**, 1733 (1978).

[24] J. M. Namyslowski, Prog. Part. Nucl. Phys. **14**, 49 (1984).

[25] A. C. Benvenuti et al., Phys. Lett. **237B**, 599 (1990).

[26] G. Baum et al., Phys. Rev. Lett. **51**, 1135 (1981); J. Ashman et al., Nucl. Phys. **B328**, 1 (1989).

[27] F. E. Close and A. W. Thomas, Phys. Lett. **212B**, 227 (1988).

Asymptotic Dominance of Diquarks in some Exclusive Hard Reactions

Bernard Pire

Centre de Physique Théorique, Ecole Polytechnique, F91128 Palaiseau Cedex.

Abstract. Multiple scattering processes dominate quark counting ones in hadronic exclusive reactions at large transfer. We survey currently available evidence for the importance of these processes at present energies. In the case of fixed angle MB $\rightarrow$ M'B and $B\bar{B} \rightarrow$ MM' reactions, this implies that dynamics selects baryons which exhibit a quark - diquark pattern.

1 Introduction

There are different ways to consider diquarks. One may even say that using the same name for so different physical concepts is quite misleading. I like to oppose fat (dressed) diquarks – those that appear in the study of baryon spectroscopy [1] or in the fragmentation of the left over piece when a quark has been expelled in Deeply Inelastic Scattering [2] – from mini (naked) diquarks – the quasi pointlike clusters that seem to be responsible of proton production at large p_T [3] or are a part of higher twist corrections to deep inelastic structure functions [4]. The relation between fat and thin diquarks is not a simple one and needs a better understanding of soft phenomena than available today.

Exclusive scattering at large momentum transfer has some very unique features which were some years ago amusingly described by G. Farrar as leading to small cross-sections but big rewards. Basically exclusive reactions may be divided into two classes :
– short distance dominated reactions, such as electromagnetic form factors, hard Compton scattering and heavy quarkonium decays. The theory of these processes has been well developed in the last decade [5]. The best known property of these reactions is their energy dependence, which obey the so-called "counting rules" of Brodsky-Farrar [6] slightly modified by logarithms. Another feature is the helicity conservation rule [7] which has very stringent consequences on spin sensitive observables.
– Multiple scattering dominated reactions, which are most hadron-hadron exclusive reactions. At very high energies significant progresses in the understanding of these quite intricated processes have recently been achieved [8]. The energy dependence is slightly

different from the counting rules result, leading to an asymptotic dominance of multiple scattering processes above counting rules processes for all large transfer exclusive reactions involving two mesons and two baryons, or four baryons.

2   Experimental information on hard multiple scattering processes

An asymptotic analysis is of course not sufficient, although many hard strong interactions have a welcome tendency to display early asymptopia. Experimental results for many hard exclusive reactions do indeed exhibit powerlaw energy (or transfer) dependence in good agreement with perturbative Q.C.D. based predictions, and this behaviour begins at transfers $O(10\,\mathrm{GeV}^2)$ sometimes even below. Let us here restrict to meson-baryon and to baryon-baryon elastic scattering (and crossed reactions). Two important pieces of information are available.

2.1   Energy oscillations of fixed angle elastic reaction

The quite precise data on pp (and to some extent πp) elastic scattering at 90° show an oscillatory pattern superposing to the power law energy decrease. A natural explanation [9] is that these wiggles are the signature of interferences between the two competing processes described above. A satisfactory fit of the data is obtained when assuming that the multiple scattering amplitude has a ℓn s-dependent phase coming from its infrared sensitivity (through Sudakov form factors) and that its magnitude is of the same order as the counting-rule amplitude at values of s of the order 10-20 GeV². Fig. 1 shows this fit.

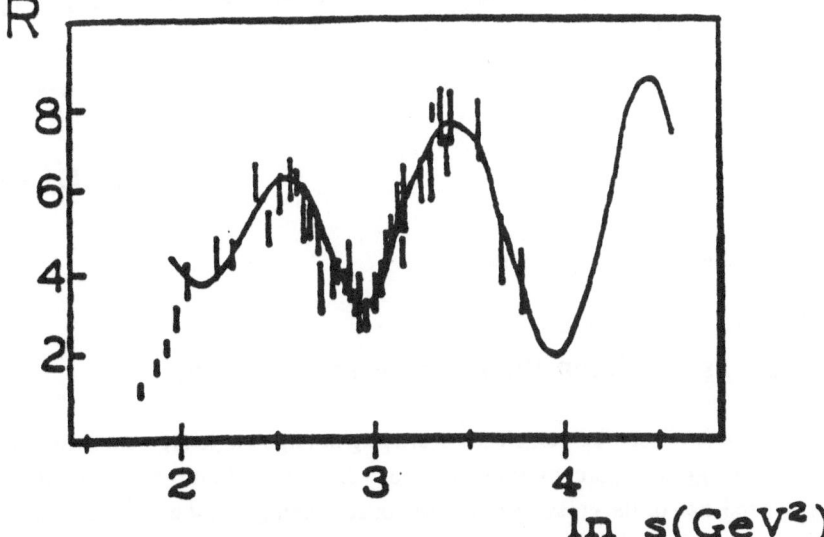

Fig.1. Energy dependence of s^{10} dσ/dt (pp → pp) at 90°. The curve is a fit from ref.[9].

2.2 Color transparency measurements

A very nice property of short distance dominated reactions is that they should exhibit color transparency [10]. This concept means that in a nuclear environment, a hard scattered particle is not sensitive to hadronic absorption, so that the hard cross section is A (or Z) times the free space one. For this picture to be true, it is crucial that the perturbative subprocess is infrared insensitive. Multiple scattering diagrams are plagued with Sudakov form factors which are remnants of the infrared bad behaviour of exclusive on shell colored particle scattering and it is thus no surprise that the multiple scattering contribution to fixed angle exclusive scattering does not exhibit color transparency. In other words, a nuclear environment selectively filters out the multiple scattering component. A simple-minded absorption model [11] allows then to understand the BNL data [12] on pp elastic scattering in Aluminum (Fig. 2). We thus have a second evidence for the importance of both processes in fixed angle pp elastic at present energies.

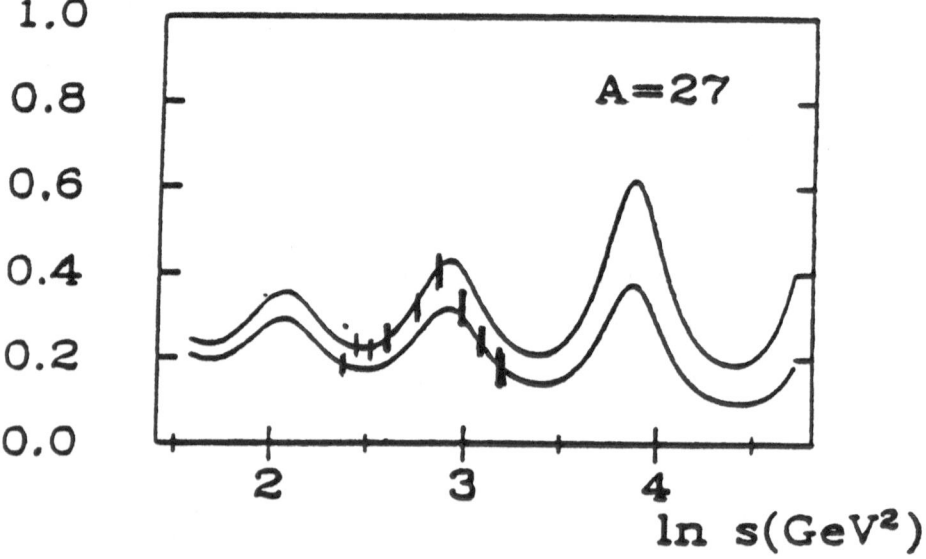

Fig.2. The oscillating color transparency ratio [11].

3 Diquarks in Meson-Baryon Elastic Scattering

Available data on meson-baryon elastic scattering at large angle are compatible with this analysis, although transparency ratios are not yet measured and the oscillations in the energy dependence of the cross sections are not so striking. On the theoretical side, since three independent scatterings are impossible, the dominant contribution comes from diagrams where there are two independent scatterings and one short-distance dominated additional scattering. This is exemplified in Fig.3a. A complete analysis of these diagrams is still missing but one can note two important features :

i. The energy dependence of their contribution to the cross section is, following Sterman and Botts, at asymptotic energies

$$d\sigma/dt \sim s^{-7.83}$$

modulo some logarithmic corrections, to be compared with the s^{-8} behaviour of the counting rule contribution. Needless to say this tiny difference will not permit to test the eventual dominance of this contribution at realistic energies.

ii. The spacial picture developed through the color transparency/nuclear filtering argument shows that these contributions select two quarks which are close together in the initial proton and put them close together in the final proton, while the third quark scatters independently, and is thus likely to be far away, at least in the out of scattering plane direction. The energing picture (see Fig.3b) is thus a quark-diquark system scattering with a meson.

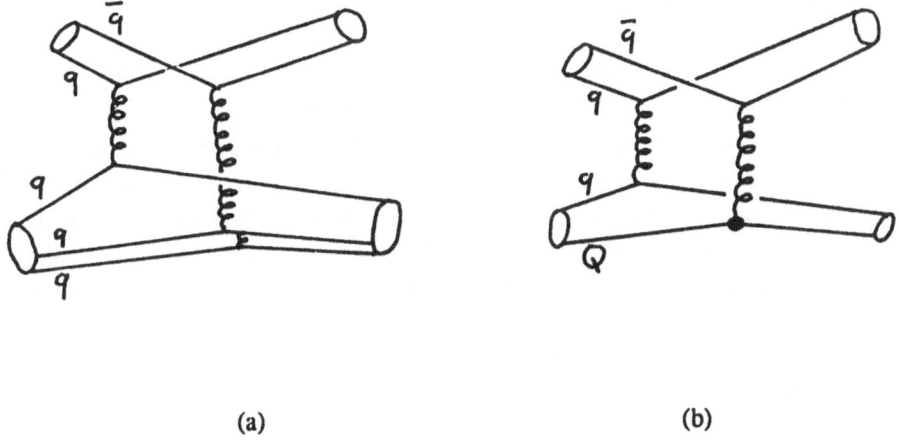

(a)   (b)

Fig.3. A Meson Baryon leading diagram (a) and its spacial representation as a quark-diquark scattering with a meson (b).

There remains to make that picture quantitative. Numerous exclusive scattering processes have been studied in the diquark model [13]. One always needs effective interactions between scalar diquarks, vector diquarks and gluons. There may now be some hope to consistently derive these couplings from the multiple scattering diagrams analyzed à la Botts and Sterman. This requires the understanding of new distribution amplitudes [8], which must contain the effects of the Sudakov from factors and of the chromo-Coulomb phase attached to it.

4 Spin Effects

A nice feature of multiple scattering diagrams, and in particular of those selected in meson baryon exclusive reactions, is their rich potential polarization properties. Contrarily to the quark-counting helicity conservation rule, one may invoke [14] the breaking of rotational symmetry of the hard collision to get hadronic helicity flip from no quark helicity flip, i.e. even in the limit of massless quarks. A chirally-odd exclusive wave function carrying non-zero orbital angular momentum plays a front role in this confinement - controlled transmutation. Added to the rich phase structure of multiple scattering amplitudes, this hadronic helicity violation may explain many unresolved spin puzzles. In particular the non vanishing value of the non diagonal helicity density matrix elements of the ρ meson produced in $\pi p \to \rho N$ scattering at 10 GeV and 90° may be soon understood (see also [13]). A crucial test of this explanation is the disappearance of spin effects in nuclear filtered processes. The energy dependence is an intricate issue, since the size of hadrons diminishes as transfer increases but the oblateness grows.

5 Crossed Diagrams and Superlear Physics

The proton-antiproton initial state is well suited for an extensive study of the diquark asymptotic dominance. It is an easy task to perform crossing on the diagrams of Fig.3. Crossing will allow many different final states : $\pi^+\pi^-$, $\pi^0\pi^0$, $\rho^+\rho^-$, $\rho^0\rho^0$, $\pi\rho$, $K\overline{K}$, $\phi\phi$... These final states select different combinations of multiple scattering amplitudes. Those with a vector meson allow the study of spin effects through the helicity density matrix elements one may deduce from the decay pattern of these resonances. Superlear, with its high luminosity and its convenient energy range should be particularly well suited for these studies.

Acknowledgements

The continuing collaboration with John P. Ralston on many aspects of the work described here is gratefully acknowledged. This research was supported in part under N.S.F. Int. Programs Grant 891626 Centre de Physique Théorique, Ecole Polytechnique, is "Unité Propre du CNRS".

209

References

1. see for instance D. Lichtenberg, these proceedings.

2. M. Fontannaz et al.Phys. Lett. B77, 315 (1978) ; J. Mulders, these proceedings.

3. M. Szczekowski, these proceedings.

4. U. Landgraf, these proceedings.

5. A review by S.J. Brodsky and G.P. Lepage may be found in Perturbative Quantum Chromodynamics, A.H. Mueller ed., World Scientific (1989).

6. S.J. Brodsky and G.R. Farrar, Phys. Rev. Lett. 31, 1153 (1973).

7. S.J. Brodsky and G.P. Lepage, Phys. Rev. D24, 2848 (1981).

8. J. Botts and G. Sterman, Nucl. Phys. B325, 62 (1989) ; J. Botts, Nucl. Phys. B353, 20 (1991) and Phys. Rev. D44, 2768 (1991).

9. B. Pire and J.P. Ralston, Phys. Lett. B117, 233 (1982) ; see also C.E. Carlson et al, College of William and Mary preprint WM92-101.

10. S.J. Brodsky and A.H. Mueller, Phys. Lett. B 206, 685 (1988) and references therein ; B. Pire and J.P. Ralston, Proc. of the 5th Workshop on Perspectives in Nuclear Physics, Trieste (1991), World Scientific.

11. J.P. Ralston and B. Pire, Phys. Rev. Lett. 61, 1823 (1988).

12. A.S. Carroll et al, Phys. Rev. Lett. 61, 1698 (1988).

13. M. Anselmino et al, Z. Phys. C36, 89 (1987) ; W. Schweiger these proceedings.

14. J.P. Ralston and B. Pire, High Energy Helicity Violation in Hard Exclusive Reactions of Hadrons, Ecole Polytechnique preprint A175.0592 (1992).

The η_c, $\chi_{c0,c1,c2} \to p\bar{p}$ decays and a quark-diquark model of the proton

Mauro Anselmino

*Department of Theoretical Physics, University of Torino
and INFN, Sezione di Torino
Via P. Giuria 1, 10125 Torino, Italy*

Exclusive decays of charmonium states into meson or baryon pairs supply a rich and interesting phenomenological ground for testing many of our current ideas about the hadron structure. On one side, we know that the initial ($c\bar{c}$) states are loosely bound so that we can safely treat them in the well known non relativistic approximation; on the other side, we expect the decays to proceed through the $c\bar{c}$ annihilation into hard gluons with the subsequent creation of quark-antiquark pairs, process which can be computed in perturbative QCD. The final quark hadronization into the observed particles is then described by a convolution of the hard part with the soft, non perturbative, hadronic wave functions; the whole process thus involves short and large distance physics. The success of such a scheme relies both on a correct description of the elementary constituent interactions, the short distance process, and an adequate knowledge of the quark distribution amplitudes, the large distance physics; in principle, QCD allows a good description of both of these hadronic processes, at least at very large energies [1-3].

The above statements translate into the following expression for the helicity amplitudes describing the decay of a heavy quark bound state, with quantum numbers J, M, L, S, into a hadronic pair $h\bar{h}$:

$$
\begin{aligned}
A_{JMLS}^{\lambda_h \lambda_{\bar{h}}} = \sum_{\lambda_c \lambda_{\bar{c}}} \left(\frac{2L+1}{4\pi} \right) C_{\lambda_{c_1}, -\lambda_{\bar{c}} \lambda}^{\frac{1}{2} \frac{1}{2} S} C_{0\lambda\lambda}^{LSJ} \\
\times \int d^3k \, M_{\lambda_h \lambda_{\bar{h}}; \lambda_c \lambda_{\bar{c}}} \, D_{M\lambda}^{J*}(\beta, \alpha, 0) \, \psi_c(k) ,
\end{aligned}
\tag{1}
$$

where $\lambda_h, \lambda_{\bar{h}}$ are the final hadron helicities and the C's are Clebsh-Gordan coefficients which appear when expressing the momentum space charmonium wave function in the $c, \bar{c}$ helicity basis; $\vec{k} = k(\sin \alpha \cos \beta, \sin \alpha \sin \beta, \cos \alpha)$ is the $c\bar{c}$ relative momentum and $\lambda = \lambda_c - \lambda_{\bar{c}}$ is the difference between the c and $\bar{c}$ helicities. The amplitude M for the process $c\bar{c} \to h\bar{h}$ is given by

$$
M_{\lambda_h \lambda_{\bar{h}}; \lambda_c \lambda_{\bar{c}}} = \sum_{\{\lambda^{(i)}\}} \int \prod_{i=h,\bar{h}} [dx^{(i)}] \psi_{\lambda_i}^*(x^{(i)}, \{\lambda^{(i)}\}) \, T_{\{\lambda^{(h)}\}, \{\lambda^{(\bar{h})}\}; \lambda_c \lambda_{\bar{c}}}
\tag{2}
$$

where by $\{\lambda^{(i)}\}$ and $x^{(i)}$ we denote respectively the whole sets of helicities and momentum fractions carried by the partons inside the hadron i ($i = h, \bar{h}$). The ψ's are the hadron wave functions and $[dx^{(i)}] = dx_1 dx_2 ... dx_{n_i} \delta(x_1 + x_2 + ... + x_{n_i} - 1)$, with n_i the number of valence constituents inside hadron i. The T's are the helicity amplitudes describing the elementary constituent interactions.

So far only data on charmonium decays are available and the comparison with the theoretical computations, Eqs. (1,2), shows both brilliant successes [4-9] and dramatic failures [10-12]. The reason of all failures can simply be traced down to spin effects; the vector coupling of quarks and gluons conserves, in the massless limit, the quark helicities, leading to the "helicity conservation rule" in exclusive processes [4], which forbids many two-body heavy meson decays [5]. Let us see in some more details how this works: according to the picture outlined above, the initial heavy quarks annihilate into hard gluons, which then create $q\bar{q}$ pairs with the q helicity always *opposite* to the $\bar{q}$ one. The final quarks q_i then hadronize collinearly into a final hadron with helicity $\lambda_h = \sum_i \lambda_{q_i}$; similarly do the antiquarks, $\lambda_{\bar{h}} = \sum_i \lambda_{\bar{q}_i}$. We are thus led to final states with only opposite helicity particles, $\lambda_h = -\lambda_{\bar{h}}$, which immediately forbids many charmonium decays. For example, the decays η_c, $\chi_0 \rightarrow p\bar{p}$ should not occur, because a spin zero particle cannot decay into two opposite helicity fermions; similarly forbidden decays are $\eta_c \rightarrow VV$ ($V =$ vector meson) and $J/\psi \rightarrow \pi\rho$, $K^*\bar{K}$ (in general $J/\psi \rightarrow$ any pseudoscalar-vector meson pair).

With the only possible exception of $\chi_0 \rightarrow p\bar{p}$ all of the above decays have been observed to occur with relative large branching ratios [13]

$$B(\eta_c \rightarrow \rho\rho) = (2.6 \pm 0.9) \times 10^{-2} \tag{3a}$$

$$B(\eta_c \rightarrow K^*\bar{K}^*) = (9 \pm 5) \times 10^{-3} \tag{3b}$$

$$B(\eta_c \rightarrow \phi\phi) = (3.4 \pm 1.2) \times 10^{-3} \tag{3c}$$

$$B(\eta_c \rightarrow p\bar{p}) = (1.04 \pm 0.19) \times 10^{-3} \tag{3d}$$

$$B(J/\psi \rightarrow \rho\pi) = (1.28 \pm 0.10) \times 10^{-2} \tag{3e}$$

$$B(J/\psi \rightarrow K^*\bar{K}) = (3.8 \pm 0.7) \times 10^{-3} \tag{3f}$$

$$B(\chi_0 \rightarrow p\bar{p}) < 9.0 \times 10^{-4} \tag{3g}$$

The experimental data (3) show beyond any doubt the limits of the perturbative QCD approach in describing exclusive hadronic charmonium decays. The most obvious consideration at this stage is that of admitting that, in this energy region, non perturbative, higher order effects or even new decay mechanisms can still be at work and cannot be neglected. Of course, the same consideration should not hold for, e.g., $(b\bar{b})$ decays, which would involve much higher values of Q^2; the same discrepancies between theory and experiment in case of bottomonium decays would be a much more serious challenge to perturbative QCD.

It is then natural to consider some of these "forbidden" decays, which have nevertheless been observed, trying to understand which corrections to the perturbative QCD scheme or alternative decay mechanisms might be able to give an explanation. We stress again that we do expect significant modifications to the theoretical framework outlined in Eqs. (1) and (2); the modest values of the Q^2 involved in charmonium decays (Q^2 up to few (GeV)^2) make non perturbative or higher order corrections non negligible.

In one of the attempts to overcome the difficulties encountered by the massless perturbative QCD scheme in describing charmonium decays, quark-diquark models of the nucleon have been proposed and applied to many physical processes [14-17]. Two quark correlations, induced by colour forces, are indeed present inside baryons [18]; in intermediate energy regions these correlations might behave as actual particles, scalar or vector *diquarks*, participating as single entities to the underlying dynamics. Vector diquarks, in particular, might help with the spin problems: the coupling of gluons to spin 1 diquarks may change their helicity, thus avoiding the troublesome helicity conservation rule.

One can then introduce diquarks in the scheme of Eqs. (1,2): it simply amounts to consider the final baryons as made of a quark and a diquark (rather than 3 quarks), with the appropriate quark-diquark wave function. The elementary interactions are then $c\bar{c}$ annihilations into $q\bar{q}$ and diquark-antidiquark pairs. The quark-diquark model of the nucleon has been consistently applied to the description of several charmonium decays into $p\bar{p}$, in order to fix the parameters of the model and the properties of diquarks [14,15]:

$$\Gamma(\eta_c \rightarrow p\bar{p}) = (12.1 \pm 7.9)\, KeV \tag{4a}$$

$$\Gamma(\chi_0 \rightarrow p\bar{p}) < 12\, KeV \tag{4b}$$

$$\Gamma(\chi_{c1} \rightarrow p\bar{p}) = (69 \pm 13)\, eV \tag{4c}$$

$$\Gamma(\chi_{c2} \rightarrow p\bar{p}) = (180 \pm 31)\, eV \tag{4d}$$

Of the above decays the first two are "forbidden" ones, whereas the last two are not and are indeed well described in perturbative QCD [9].

Let us now recall the main features of the quark-diquark model starting from the proton wave function [19-21]

$$\psi_{p,\pm}(x) = \frac{\pm F_N}{3}\left\{\left[\sqrt{2}V_{\pm 1}(ud)u_{\mp} - 2V_{\pm 1}(uu)d_{\mp}\right.\right.$$
$$\left.+ \sqrt{2}V_0(uu)d_{\pm} - V_0(ud)u_{\pm}\right]\phi_V(x)\sin\Gamma \tag{5}$$
$$\left.\mp 3S(ud)u_{\pm}\phi_S(x)\cos\Gamma\right\}$$

The $\phi_{S,V}(x)$ are the scalar, vector diquark distribution amplitudes; $V_{+1}(ud)$ stands for a vector (ud) diquark with helicity $+1/2$ and so on. F_N is the hadronization constant, with the dimension of [mass], somewhat analogous to the pion decay constant F_π. We have also introduced a certain amount of $SU(6)$ violation: $\sin^2\Gamma(\cos^2\Gamma)$ is the probability of finding a vector (scalar) diquark in a proton; when $\Gamma = \pi/2$ one recovers the usual $SU(6)$ wave function.

The distribution amplitudes ϕ_S and ϕ_V have been taken to be of the general form

$$\phi_S = N_1 x^{\alpha_1}(1-x)^{\beta_1}$$
$$\phi_V = N_2 x^{\alpha_2}(1-x)^{\beta_2} \tag{6}$$

where $N_{1,2}$ are the normalization constants such that $\int_0^1 dx\,\phi_{V,S}(x) = 1$. By varying α and β we get wave functions with different "average" values of x, the fraction of the mass and the momentum of the proton carried by the diquark.

We expect the average mass of scalar diquarks to be smaller than the average mass of vector diquarks: this is supported by the analogy with the $q\bar{q}$ bound states (the π mass versus the ρ mass) and by explicit calculations [22] which indicate $m_S < m_V \lesssim 2m_S$. A similar conclusion has been reached by studying the contribution of diquarks to deep inelastic scattering [23]. We have used in our computations [14,15] different sets of wave functions; the most significant ones are:

$$\phi_S = N_S\,x(1-x)^3 \qquad \phi_V = N_V x^3(1-x) \tag{7a}$$
$$\phi_S = N_S\,x(1-x)^{2.5} \qquad \phi_V = N_V x^{2.5}(1-x) \tag{7b}$$

which are representative of the dependence of the numerical results on the distribution amplitudes. Such dependence turns out to be very weak. We have also checked that more elaborate kinds of wave functions [19,20] do not improve the numerical results.

The elementary interactions involve the couplings of gluons to scalar and vector diquarks; we can parametrize them in the most general form allowed by Lorentz, gauge and parity invariance as:

$$S^\mu \equiv -ig_s T_{ij}^a(Q-\bar{Q})^\mu F_s$$
$$V^\mu \equiv ig_s T_{ij}^a\{(\epsilon_Q^* \cdot \epsilon_{\bar{Q}}^*)(Q-\bar{Q})^\mu G_1$$
$$-[(Q \cdot \epsilon_{\bar{Q}}^*)(\epsilon_Q^*)^\mu - (\bar{Q} \cdot \epsilon_Q^*)(\epsilon_{\bar{Q}}^*)^\mu]G_2$$
$$-(\epsilon_Q^* \cdot \bar{Q})(\epsilon_{\bar{Q}}^* \cdot Q)(Q-\bar{Q})^\mu G_3\} \tag{8}$$

where the T^a are Gell-Mann colour matrices; Q e $\bar{Q}$ are respectively the diquark and antidiquark four momenta; F_S, G_1, G_2 and G_3 are the diquark form factors, and ϵ_Q, $\epsilon_{\bar{Q}}$ are the diquark polarization vectors.

The diquark form factors are parametrized in the most simple way which reproduces the correct pointlike ($Q^2 = 0$) and asymptotic ($Q^2 \to \infty$) [24] behaviours:

$$F_S = \frac{Q_S^2}{Q_S^2 + Q^2}$$

$$G_1 = \left(\frac{Q_V^2}{Q_V^2 + Q^2} \right)^2 \tag{9}$$

$$G_2 = (1 + \kappa)G_1$$

$$G_3 = 0$$

where κ is the vector diquark anomalous magnetic moment. The values of $Q_{S,V}^2$ set the scale for the transition from the small Q^2 region, where diquarks act as elementary objects, to the large Q^2 one, where they start being resolved in two quarks. It is generally agreed that scalar diquarks are more pointlike than vector diquarks; accordingly we take $Q_S^2 = 10\,(GeV)^2$ and $Q_V^2 = 2\,(GeV)^2$. Small variations of these values do not lead to relevant changes in the numerical results.

The last quantities to be discussed before presenting the outcome of our computations are the parameters appearing in the charmonium non relativistic wave functions

$$(L = 0) \qquad \psi_{\eta_c}(k) = \sqrt{\frac{\pi}{2}} R(0) \frac{1}{k^2} \delta(k)$$

$$(L = 1) \qquad \psi_{\chi}(k) = -3i\sqrt{2\pi} R'(0) \frac{1}{k^2} \frac{d}{dk} \delta(k) \tag{10}$$

$R(0)$ and $R'(0)$ are the values of the radial wave function and its first derivative at the origin. These values can be fixed from a computation of the decay rates

$$\Gamma(\eta_c \to \gamma\gamma) = \frac{16}{27} \frac{\alpha^2}{m_c^2} |R(0)|^2$$

$$\Gamma(\chi_0 \to gg) = 6 \frac{\alpha_s^2}{m_c^4} |R'(0)|^2 \tag{11}$$

$$\Gamma(\chi_2 \to gg) = \frac{8}{5} \frac{\alpha_s^2}{m_c^4} |R'(0)|^2$$

and a comparison with the experimental data [13,25] (assuming $\Gamma(\chi \to gg) \simeq \Gamma(\chi \to \text{hadrons})$). This yields

$$|R(0)| = 0.63 \pm 0.25\,(GeV)^{3/2}$$

$$|R'_{\chi_0}(0)| = 0.46 \pm 0.10\,(GeV)^{5/2} \tag{12}$$

We have now remained with essentially two parameters, the "proton decay constant" F_N and the mixing angle Γ, which fixes the relative abundance of scalar and vector diquarks. The best fits to the experimental data (4) which we can obtain are typically of the kind (all results are in MeV, unless explicitly indicated)

$\Gamma(\chi_0 \to p\bar{p})$	$\Gamma(\chi_1 \to p\bar{p})$	$\Gamma(\chi_2 \to p\bar{p})$	$\Gamma(\eta_c \to p\bar{p})$	
200 ± 130	70 ± 30	184 ± 100	$\simeq 1$	(13a)
460 ± 200	71 ± 30	185 ± 110	$\simeq 1$	(13b)
< 12000	69 ± 13	180 ± 31	(12.1 ± 7.9) KeV	exp.

Eqs. (13a,b) refer, respectively, to the wave functions (7a,b); the last line of the above Table gives the experimental results. The values of F_N and Γ found to fit the data are $F_N \simeq 63$

MeV, $\Gamma \simeq 16°$ with the distribution amplitudes (7.a), and $F_N \simeq 72$ MeV, $\Gamma \simeq 26°$ with the distribution amplitudes (7.b).

The clear conclusion of our application of the quark-diquark model to the $p\bar{p}$ decays of charmonium states is that, like in perturbative QCD with massless quarks [9], one obtains very good results for $\Gamma(\chi_{1,2} \to p\bar{p})$; $\Gamma(\eta_c \to p\bar{p})$, however, although different from zero, turns out to be much smaller (a factor $\sim 10^{-4}$) than data, thus not solving one of the problems of pQCD. The real difference between quark-diquark and massless quark models lies in the prediction for $\Gamma(\chi_0 \to p\bar{p})$; our model gives sizeable results, comparable to the measured values of $\Gamma(\chi_{1,2} \to p\bar{p})$, contrary to pQCD which predicts zero. The only existing experimental information on $\Gamma(\chi_0 \to p\bar{p})$ is the large upper bound (4b); a precise measurement of $\Gamma(\chi_0 \to p\bar{p})$ would then greatly help in understanding the rôle of diquarks in protons.

Lacking experimental data on $\chi_0 \to p\bar{p}$ decay, another way of discriminating between the two models is that, recently suggested [26], of measuring the spin density matrix elements of χ_2 states created in $p\bar{p}$ annihilations; in one case (pQCD) only states with $S_z = \pm 1$ can be formed, whereas in case of diquark active presence also states with $S_z = 0$ are allowed.

We conclude by considering again the elusive η_c decays, not only $\eta_c \to p\bar{p}$ but also the decays listed in Eqs. (3), which up to now have defeated any attempt of explanation in terms of constituents and by mentioning a possible alternative decay mechanism.

In order to do that let us recall first that according to the helicity conservation rule both the $J/\psi(\psi') \to \rho\pi$, $K^*\bar{K}$ decays should be forbidden and their experimental decay widths should be either zero or very small. However, this holds true only for the ψ', but not for the J/ψ, the so called $J/\psi (\psi') \to \pi\rho$, $K^*\bar{K}$ puzzle [27]. This might be explained by assuming that the J/ψ couples directly to a glueball with its same quantum numbers and a similar mass value, so that the decay proceeds via the sequence $J/\psi \to$ glueball $\to \pi\rho$, $K^*\bar{K}$; for the ψ' this contribution would be much smaller, due to a larger mass difference with the glueball (mass difference which appears in the denominator of the glueball propagator).

A similar explanation might hold for the η_c, which has so far caused so many troubles; a 0^{-+} trigluonium state, with a mass close to that of the η_c, could explain the otherwise misterious η_c decays [28]. The expectation of such a glueball in the η_c mass region is indeed plausible, if one assumes the existence a 1^{--} trigluonium state in the J/ψ mass region. For this explanation to hold, the analogous decays of the η_c', similarly to what happens for the ψ', should be strongly suppressed, according to perturbative QCD. Data on η_c' decays should be available in the near future.

References

[1] G.P. Lepage and S.J. Brodsky, *Phys. Rev.* D **22**, 2157 (1980); S.J. Brodsky and G.P. Lepage, *Perturbative Quantum Chromodynamics*, A.H. Mueller Editor, World Scientific (1989)
[2] A.H. Mueller, *Phys. Rep.* **73**, 237 (1981)
[3] V.L. Chernyak and A.R. Zhitnitsky, *Phys. Rep.* **112**, 173 (1984)
[4] S.J. Brodsky and G.P. Lepage, *Phys. Rev.* D **24**, 2848 (1981)
[5] V.L. Chernyak and A.R. Zhitnitsky, *Nucl. Phys.* B**201**, 492 (1982)
[6] A. Andrikopolou, *Z. Phys.* C **22**, 63 (1984)
[7] V.L. Chernyak and I.R. Zhitnitsky, *Nucl. Phys.* B**246**, 52 (1984)
[8] P.H. Damgaard, K. Tsokos and E. Berger, *Nucl. Phys.* B**259**, 285 (1985)
[9] V.L. Chernyak, A.A. Ogloblin and I.R. Zhitnitsky, *Z. Phys.* C **42**, 569 (1989); **42**, 583 (1989)
[10] S.J. Brodsky, G.P. Lepage and S.F. Tuan, *Phys. Rev. Lett.* **59**, 621 (1987)
[11] M. Anselmino, F. Caruso and F. Murgia, *Phys. Rev.* D **42**, 3218 (1990)

[12] M. Anselmino, R. Cancelliere and F. Murgia, University of Torino *preprint*, DFTT 4/92 (1992), to appear in *Phys. Rev. D*

[13] Particle Data Group, *Phys. Lett.* **B239**, 1 (1990)

[14] M. Anselmino, F. Caruso and S. Forte, *Phys. Rev. D* **44**, 1438 (1991)

[15] M. Anselmino and F. Murgia, University of Torino *preprint* DFTT 5/92 (1992)

[16] C. Carimalo and S. Ong, *Z. Phys. C* **52**, 487 (1991)

[17] E.-H. Kada and J. Parisi, Collège de France *preprint* LPC 9140 (1991)

[18] See, e.g., M. Anselmino and E. Predazzi, Editors, *Proceedings of the Workshop on Diquarks*, World Scientific (1989); M. Szczekowski, *Int. J. Mod. Phys. A* **4**, 3985 (1989); M. Anselmino, S. Ekelin, S. Fredriksson, D. Lichtenberg and E. Predazzi, *Luleå preprint*, TULEA 1992:05

[19] M. Anselmino, F. Caruso, S. Forte and B. Pire, *Phys. Rev. D* **38**, 3516 (1988)

[20] M. Anselmino, F. Caruso, P. Kroll and W. Schweiger, *Int. J. Mod. Phys. A* **4**, 5213 (1989)

[21] M.I. Pavkovic, *Phys. Rev. D* **13**, 2128 (1976); R.T. Van der Walle, in *Erice Lectures 1979*, edited by A. Zichichi (Plenum, New York, 1979), p. 477

[22] J. Praschifka, R.T. Cahill and C.D. Roberts, *Int. J. Mod. Phys. A* **4**, 4929 (1989)

[23] S. Fredriksson, M. Jandel and T.I. Larsson, *Z. Phys. C* **14**, 35 (1982); S. Fredriksson and M. Jandel, *ibid.* **14**, 41 (1982)

[24] A.I. Vainshtein and V.I. Zakharov, *Phys. Lett.* **72B**, 368 (1978)

[25] T.A. Armstrong *et al.*, *Nucl. Phys.* **B373**, 35 (1992)

[26] M. Anselmino, F. Caruso and R. Mussa, University of Torino *preprint* DFTT 53/91 (1991), to appear in *Phys. Rev. D*

[27] S.J. Brodsky, G.P. Lepage and S.F. Tuan, *Phys. Rev. Lett.* **59**, 621 (1987)

[28] M. Anselmino, M. Genovese and E. Predazzi, *Phys. Rev. D* **44**, 1597 (1991)

Meson–Diquark Bosonization in the Nambu–Jona–Lasinio Model and in QCD

D. Ebert[1] and L. Kaschluhn[2]

[1] Fachbereich Physik, Humboldt–Universität, Berlin, Germany
[2] Deutsches Elektronen–Synchrotron – Institut für Hochenergiephysik,
Berlin–Zeuthen, Germany

Abstract. Meson and diquark properties are studied at zero and finite temperature within an extended QCD-motivated Nambu–Jona–Lasinio model using path integral techniques. Furthermore, the extension of these methods to the bilocal case of QCD is demonstated by considering QCD in two dimensions.

1 Introduction

In this talk we want to demonstrate the powerfulness of the path–integral bosonization technique [1, 2] for describing low–energy bound states of colour singlet mesons and intermediate colour antitriplet diquarks. At first we investigate the meson–diquark system arising in an effective lagrangian obtained from an extended Nambu–Jona–Lasinio (NJL) model [3] at zero temperature. This model successfully reproduces the low–energy meson physics [4] and allows one to calculate analogously diquark properties [5 − 7].

To make contact with the problem of chiral phase transitions from hadronic matter to a quark plasma which is expected to appear in ultra–relativistic heavy ion collisions we introduce a nonvanishing temperature (T) and a baryon number density (chemical potential μ) into the underlying NJL model [8,9]. At the end we will consider the exact bilocal meson–diquark bosonization of QCD_2 for $T = 0$ within the light–cone gauge [10].

2 The effective meson–diquark lagrangian at zero temperature

Let us consider the $SU(2)_L \otimes SU(2)_R \otimes SU(3)_c$ NJL model defined by the lagrangian [7]

$$\mathcal{L}_{NJL} = \bar{q}(i\not\partial - m_0)q + \mathcal{L}_{int}^{(4)} , \tag{1}$$

$$\mathcal{L}_{int}^{(4)} = \sum_{a=1}^{4} \{G_a(\bar{q}\mathcal{M}_M^{ac}q)(\bar{q}\mathcal{M}_M^{ac}q) + \tilde{G}_a(\bar{q}\mathcal{M}_D^{a\rho g}q^c)(\bar{q}^c\mathcal{M}_D^{a\rho g}q)\} \tag{2}$$

(In the following we will for simplicity omit the symbol for summation over a.).
Here $q(x)$ are quark fields with three colours and two flavours, $q^c \equiv C\bar{q}^T$ with C being
the charge conjugation matrix (T means transposition). Furthermore, m_0 is the mass
of the light current quarks. The term $\mathcal{L}_{int}^{(4)}$ describes the four–quark interaction of $q\bar{q}$
and qq pairs leading to the formation of meson and diquark bound states. $\mathcal{M}_M^\tau$ and
$\mathcal{M}_D^\theta$ are meson and diquark projection matrices, respectively,

$$\mathcal{M}_M^\tau \equiv \mathcal{M}_M^{ae} = \frac{1}{\sqrt{3}}\mathcal{K}^a 1_c \mathcal{F}^e \,, \quad \mathcal{M}_D^\theta \equiv \mathcal{M}_D^{a\rho g} = \frac{i}{\sqrt{6}}\mathcal{K}^a \epsilon^\rho \mathcal{H}^g \,. \tag{3}$$

The quantities $\mathcal{K}^a$ are Dirac matrices

$$\{\mathcal{K}^a, a = 1, 2, 3, 4\} = \{1, i\gamma^5, \frac{1}{\sqrt{2}}\gamma^\mu, \frac{1}{\sqrt{2}}\gamma^\mu\gamma^5\} \,, \tag{4}$$

$(\epsilon_\rho)_{\alpha\gamma}$ is the Levi-Civita tensor in colour $SU(3)_c$ space and $\mathcal{F}^e$ and $\mathcal{H}^g$ are generators
in flavour $SU(2)_f$ space

$$\begin{aligned}
\{\mathcal{F}^e, e = 0, 1, 2, 3\} &= \{\frac{1}{\sqrt{2}}, \frac{\sigma^1}{\sqrt{2}}, \frac{\sigma^2}{\sqrt{2}}, \frac{\sigma^3}{\sqrt{2}}\}, \\
\{\mathcal{H}^g, g = 1, 2, 3, 4\} &= \{\mathcal{F}^e, e = 2, 0, 1, 3\}
\end{aligned}$$

with σ^n being the Pauli matrices. Finally, the constants G_a and $\tilde{G}_a$ are chosen to be
equal in pairs

$$\begin{aligned}
G_1 = G_2 \equiv G \,, \quad G_3 = G_4 \equiv -G' \,, \\
\tilde{G}_1 = \tilde{G}_2 \equiv \tilde{G} \,, \quad \tilde{G}_3 = \tilde{G}_4 \equiv -\tilde{G}' \,.
\end{aligned}$$

Next, it is convenient to introduce the generating functional of the NJL model
(1),(2)

$$Z = \int \mathcal{D}q\mathcal{D}\bar{q} \exp\left\{i \int d^4x [\mathcal{L}_{NJL} + \mu_0(\bar{q}\gamma_0 q)]\right\} \,, \tag{5}$$

where μ_0 has the meaning of a (bare) chemical potential. Using a standard procedure for introducing collective meson (ϕ) and diquark ($\omega^\dagger, \omega$) fields one obtains the
expression [7]

$$Z = \int \mathcal{D}\phi\mathcal{D}\omega^\dagger\mathcal{D}\omega \exp iW_{eff}[\phi, \omega^\dagger, \omega] \tag{6}$$

with the effective action

$$W_{eff}[\phi, \omega^\dagger, \omega] \equiv \int d^4x \mathcal{L}_{eff}[\phi, \omega^\dagger, \omega] = -i\mathrm{Tr}\ln S_\phi^{-1} - \frac{i}{2}\mathrm{Tr}\ln(1 + 4S_\phi^T \Omega^\dagger S_\phi \Omega)$$
$$- \int d^4x \left[\frac{1}{4G_a}(\phi^{ae})^2 + \frac{1}{\tilde{G}_a}\omega^{\dagger a\rho g}\omega^{a\rho g}\right]. \tag{7}$$

Here the trace Tr runs over internal and spinor indices and includes an integration over
space–time variables. S_ϕ^T means the transpose of S_ϕ, whereby

$$\begin{aligned}
S_\phi^{-1} &= S_0^{-1} + \mathcal{M}_M^\tau \phi^\tau \,, \tag{8} \\
S_0^{-1} &= i\partial\!\!\!/ - m_0 + \mu_0\gamma_0 \,,
\end{aligned}$$

and

$$\Omega^\dagger \equiv \omega^{\dagger\theta} C^T \mathcal{M}_D^\theta \ , \quad \Omega \equiv \mathcal{M}_D^\theta C^T \omega^\theta \ .$$

Let us remark that the expression for the terms Tr ln ... in (6) contains diverging integrals which should be regularized. Since this model aims to describe spin–1 mesons a gauge–invariant regularization is chosen [4, 7].

For the description of the physical quantities we determine the quark condensate $< \bar{q}q >_0 = \sqrt{6}/(2G) < \sigma >_0$ and the vacuum excitation value of the quark number $< \bar{q}\gamma_0 q >_0 = -\sqrt{6}/(2G') < \hat{\omega}_0 >_0$. The minimum conditions for $\mathcal{L}_{eff}$ lead to equations for the scalar field component $\phi^{10} = \sigma$ and the time component $\hat{\omega}_0$ of the $\hat{\omega}$–meson field

$$\left. \frac{\delta \mathcal{L}_{eff}}{\delta \sigma} \right|_{\sigma = <\sigma>} = -\frac{1}{2G} < \sigma >_0 -i \mathrm{Tr}[(\mathcal{M}_M^{10}) S_{\phi_0}] = 0 \ , \tag{9}$$

$$\left. \frac{\delta \mathcal{L}_{eff}}{\delta \hat{\omega}_0} \right|_{\hat{\omega}_0 = <\hat{\omega}_0>} = \frac{1}{2G'} < \hat{\omega}_0 >_0 -i \mathrm{Tr}[(\mathcal{M}_M^{30})_0 S_{\phi_0}] = 0 \ , \tag{10}$$

where the inverse of the modified "free" quark propagator S_{ϕ_0} is now given by

$$S_{\phi_0}^{-1} = i\not{\partial} - m + \mu \gamma_0 \ ,$$

with the constituent quark mass defined by the gap equation

$$m = m_0 - \frac{1}{\sqrt{6}} < \sigma >_0 = m_0 - \frac{G}{3} < \bar{q}q >_0 \ , \tag{11}$$

and the dressed chemical potential fulfilling the gap equation

$$\mu = \mu_0 + \frac{1}{\sqrt{6}} < \hat{\omega}_0 >_0 = \mu_0 - \frac{G'}{3} < \bar{q}\gamma_0 q >_0 \ . \tag{12}$$

All other fields have $< \phi^\tau >_0 = 0$.

In order to introduce physical fields σ and $\hat{\omega}$ with vanishing vacuum expectation values we have to perform the change of variables

$$\sigma = < \sigma >_0 + \sigma' \ , \quad \hat{\omega} = < \hat{\omega}_0 >_0 + \hat{\omega}_0' \ , \quad \hat{\omega}_i = \hat{\omega}_i' \ , \quad i = 1, 2, 3 \ . \tag{13}$$

Then, the inverse quark propagator takes the form

$$S_\phi^{-1} = \left[\gamma_0 (i\frac{\partial}{\partial t} + \mu) + (\gamma \Delta) - m \right] + \mathcal{M}_M^\tau \phi'^\tau \equiv S_{\phi_0}^{-1} + \mathcal{M}_M^\tau \phi'^\tau \ .$$

In the following we will use for shortness the notation S_0 for S_{ϕ_0}.

Let us now derive the expressions for the meson and diquark masses as well as for the coupling constants. To do this we expand the lagrangian $\mathcal{L}_{eff}$ to second order in the meson and diquark fields. Then, the masses of all mesons and diquarks as well as the corresponding kinetic terms are expressed in the one–loop approximation by divergent integrals of the type

$$L^{aa'}(q) = \alpha \, \mathrm{tr}_\gamma \int \frac{d^4 p}{(2\pi)^4} \frac{1}{\not{p} - m} \mathcal{K}^a \frac{1}{\not{p} - \not{q} - m} \mathcal{K}^{a'} \ , \tag{14}$$

where the parameter α equals to 1 in the case of mesons and to 1/3 for diquarks; $\mathcal{K}^a$ is defined by (4). In a low-momentum expansion the integrals of the type (14) may be expressed by logarithmically and quadratically divergent parts

$$
\begin{aligned}
I_1 &= i \int^\Lambda \frac{d^4 p}{(2\pi)^4} \frac{1}{p^2 - m^2} \\
&= \frac{1}{8\pi^2}\left[\Lambda_3 \sqrt{\Lambda_3^2 + m^2} - m^2 \ln\left(\frac{\Lambda_3}{m} + \sqrt{1 + \frac{\Lambda_3^2}{m^2}} \right) \right],
\end{aligned}
\tag{15}
$$

$$
\begin{aligned}
I_2 &= -i \int^\Lambda \frac{d^4 p}{(2\pi)^4} \frac{1}{(p^2 - m^2)^2} \\
&= \frac{1}{8\pi^2}\left[\ln\left(\frac{\Lambda_3}{m} + \sqrt{1 + \frac{\Lambda_3^2}{m^2}} \right) - \left(1 + \frac{m^2}{\Lambda_3^2} \right)^{-1/2} \right].
\end{aligned}
\tag{16}
$$

Here the quantity Λ_3 denotes a non-covariant cut-off in the three-momentum space introduced after integration over p_0. Then, after redefining the fields to provide the right coefficients of the kinetic terms and eliminating the pseudoscalar/axial-vector meson (scalar/vector diquark) mixing terms by a transformation of the axial-vector meson fields ϕ^{4c} (vector diquark fields $\omega^{4\rho g}$) one gets the following result for the free lagrangian [7]:

$$
\begin{aligned}
\mathcal{L}^{(2)}_{free} = &- \sum_{a=1,2} \left[\frac{1}{2}(\phi^a \partial^2 \phi^a + M_{Ma}^2 \phi^{a2}) + \omega^{\dagger a}\partial^2 \omega^a + M_{Da}^2 \omega^{\dagger a}\omega^a \right] \\
&+ \sum_{a=3,4} \left[-\frac{1}{4}(\partial_\mu \phi_\nu^a - \partial_\nu \phi_\mu^a)(\partial^\mu \phi^{a\nu} - \partial^\nu \phi^{a\mu}) + \frac{1}{2}M_{Ma}^2 \phi_\mu^a \phi^{a\mu} \right. \\
&\left. - \frac{1}{2}(\partial_\mu \omega_\nu^{\dagger a} - \partial_\nu \omega_\mu^{\dagger a})(\partial^\mu \omega^{a\nu} - \partial^\nu \omega^{a\mu}) + M_{Da}^2 \omega_\mu^{\dagger a}\omega^{a\mu} \right].
\end{aligned}
\tag{17}
$$

Here the squared particle masses of $SU(2)_f$ quartets are given by the relations

$$ M_{M1}^2 \equiv M_\sigma^2 = 4m^2 + M_\pi^2 Z \qquad\qquad M_{D1}^2 \equiv M_{D_P}^2 = 4m^2 + M_{D_S}^2 \tilde{Z} \tag{18} $$

$$ M_{M2}^2 \equiv M_\pi^2 = (-2\frac{I_1}{I_2} + \frac{1}{4GI_2})Z^{-1} \qquad M_{D2}^2 \equiv M_{D_S}^2 = (-2\frac{I_1}{I_2} + \frac{3}{4\tilde{G}I_2})\tilde{Z}^{-1} \tag{19} $$

$$ M_{M3}^2 \equiv M_\rho^2 = \frac{3}{4G'I_2} \qquad\qquad M_{D3}^2 \equiv M_{D_A}^2 = \frac{9}{4\tilde{G}'I_2} \tag{20} $$

$$ M_{M4}^2 \equiv M_{a_1}^2 = 6m^2 + M_\rho^2 \qquad\qquad M_{D4}^2 \equiv M_{Dv}^2 = 6m^2 + M_{D_A}^2. \tag{21} $$

The constants

$$ Z = 1 - \frac{6m^2}{M_{a_1}^2}, \tag{22} $$

$$ \tilde{Z} = 1 - \frac{6m^2}{M_{Dv}^2} \tag{23} $$

arise from the redefinition of the pion field and scalar diquark field due to mixing. From the field renormalizations one gets for the coupling constants the relations

$$Z_S \equiv 3g_{S\bar{q}q}^2 = \frac{1}{4I_2} \,, \tag{24}$$

$$Z_V \equiv 3\left(\frac{g_\rho}{2}\right)^2 = \frac{3}{8I_2} \,, \tag{25}$$

$$g_{\pi\bar{q}q} = g_{S\bar{q}q} Z^{-1/2} \,. \tag{26}$$

We should mention that the values of the index a in (17) correspond for $a = 1$ to scalar mesons and pseudoscalar diquarks, for $a = 2$ to pseudoscalar mesons and scalar diquarks, for $a = 3$ to vector mesons and axial–vector diquarks and for $a = 4$ to axial–vector mesons and vector diquarks. Note that there exists an important difference between M_π^2 and $M_{D_S}^2$ despite their formal similarity. In fact, because $\tilde{G} \neq 3G$ the diquark mass $M_{D_S}^2$ cannot be rewritten in the form of the pion mass,

$$M_\pi^2 = \frac{1}{4I_2}\frac{m_0}{mG}Z^{-1} \,,$$

obtained by using the gap equation (11) and being proportional to m_0. Thus, for $m_0 \to 0$ one has $M_{D_S}^2 \neq 0$ in contrast with $M_\pi^2 \to 0$. This simply reflects the fact that diquarks cannot be Goldstone bosons.

The meson sector of the model (17) has been extensively investigated earlier in [4]. Therefore we will restrict ourselves to the calculation of the diquark masses. The parameters of the NJL model may be fixed from the input values $F_\pi = 93\text{MeV}$, $g_\rho^2 \approx 12\pi$, $M_\pi = 140\text{MeV}$ and $M_\rho = 770\text{MeV}$. Then, one obtains estimates for the constituent quark mass $m = 280\text{MeV}$, for the cut–off parameter $\Lambda_3 = 1.03\text{GeV}$ and for the meson couplings $G = 10.45(\text{GeV})^{-2}$ and $G' = 48(\text{GeV})^{-2}$. Using these relations one gets for the current quark mass $m_0 \approx 2\text{MeV}$. The diquark couplings will be represented in the form

$$\tilde{G} = \kappa G \,, \quad \tilde{G}' = \kappa' G'$$

with κ and κ' as proportionality factors. In [7] several possibilities have been discussed. For the case $\kappa = \kappa' = 1$ the scalar diquark mass comes out to be too large, $M_{D_S} \approx 1500\text{MeV}$. Therefore, it seems to be reasonable to consider the case $\kappa \neq 1$. Then, κ is determined by fitting the scalar diquark mass to the value of $\approx 600\text{MeV}$ to obtain a realistic baryon mass in the order of 900MeV. Now, the other diquark masses can be calculated in dependence on κ'. For $\kappa' = 1$ their numerical values are given in Table 1.

Table 1.: Masses of P, A and V diquarks for fitted scalar diquark mass $M_{D_S} = 600\text{MeV}$ with $\kappa = 2.55$ and $\tilde{Z} = 0.540$.

κ'	M_{D_P}/MeV	M_{D_A}/MeV	m_{D_V}/MeV
1	769	1334	1541

3 NJL model at finite temperature and baryon number density

Now, let us describe the properties of the constituent quarks, mesons and diquarks within the NJL model at finite temperature and baryon number density closely following ref. [9]. The corresponding grand partition function Z can be represented in terms of the path integral

$$Z = \mathcal{N} \int \mathcal{D}q \mathcal{D}\bar{q} \exp\left\{i \int_0^\beta d\tau \int d^4x [\mathcal{L}_{NJL}(\tau, x) + \mu_0 N]\right\} , \tag{27}$$

which is very similar to the zero temperature expression (5). Here $\beta = 1/T$ and $\tau = ix_0$. $N = \int d^3x \bar{q}\gamma_0 q$ is the number operator for the u and d valence quarks. The integration in (27) has to be performed over antiperiodic Grassman fields, $q(0) = -q(\beta)$.

In the "real time" formalism [11] the expression for the quark propagator in momentum space reads

$$\begin{aligned}
S(p, T, \mu_0) &= (\not{p} + m_0)\Big\{ \frac{1}{p^2 - m_0{}^2 + i\epsilon} \\
&\quad + i2\pi\delta(p^2 - m_0^2)[\theta(p_0)n(\mathbf{p}, \mu_0) + \theta(-p_0)\bar{n}(\mathbf{p}, \mu_0)]\Big\} ,
\end{aligned} \tag{28}$$

where

$$\begin{aligned}
n(\mathbf{p}, \mu_0) &= [1 + \exp(E - \mu_0)\beta]^{-1} , \\
\bar{n}(\mathbf{p}, \mu_0) &= [1 + \exp(E + \mu_0)\beta]^{-1} ,
\end{aligned}$$

are the Fermi–Dirac functions for quarks and antiquarks, respectively, and E is defined as $E = \sqrt{\mathbf{p}^2 + m_0^2}$.

After introducing collective fields as in the zero–temperature case, the path integral can be rewritten in a form analogous to (6),(7), where in S_ϕ^{-1}, eq. (8), the inverse propagator S_0^{-1} has to be replaced by the inverse of $S(p, T, \mu_0)$. Finally, due to the nonvanishing thermal expectation values $< \sigma >$, $< \hat{\omega}_0 >$ the bare parameters in (28) have to be shifted according to $m_0 \to m$, $\mu_0 \to \mu$ (see eqs. (11)–(13)).

Let us first look at the analoguous formulae of eqs. (9), (10) for the quark condensate and the averaged quark number. Using the Green function $S(p, T, \mu)$ and performing a contour integration in the complex p_0–plane we obtain

$$\begin{aligned}
< \bar{q}q > &\equiv -6i \int \frac{d^4p}{(2\pi)^4} \mathrm{tr}_\gamma S(p, T, \mu) \\
&= -\frac{6m}{\pi^2} \int_0^{\Lambda_3} dp \frac{p^2}{E}\left[1 - n(\mathbf{p}, \mu) - \bar{n}(\mathbf{p}, \mu)\right] , \tag{29}
\end{aligned}$$

$$\begin{aligned}
< \bar{q}\gamma_0 q > &\equiv -6i \int \frac{d^4p}{(2\pi)^4} \mathrm{tr}_\gamma \left[\gamma_0 S(p, T, \mu)\right] \\
&= \frac{6}{\pi^2} \int_0^{\Lambda_3} dp p^2 \left[n(\mathbf{p}, \mu) - \bar{n}(\mathbf{p}, \mu)\right] ,
\end{aligned}$$

where Λ_3 is a non–covariant three–momentum cut–off which is adjusted to reproduce various meson properties at $T = \mu = 0$.

Using (29) we may represent the thermal gap equation for the constituent quark mass in the form

$$m(T,\mu) = m_0 + 8G\, m(T,\mu)\, I_1(m,T,\mu)\,, \qquad (30)$$

where

$$I_1(m,T,\mu) = \frac{1}{(2\pi)^2} \int_0^{\Lambda_3} dp \frac{p^2}{E}\left[1 - n(\mathbf{p},\mu) - \bar{n}(\mathbf{p},\mu)\right] \qquad (31)$$

denotes the T– and μ–dependent generalization of the integral I_1 introduced in (15). Analogously, the expression of I_2, eq. (16), now reads

$$I_2(m,T,\mu) = \frac{1}{2(2\pi)^2} \int_0^{\Lambda_3} dp \frac{p^2}{E^3}\left[1 - n(\mathbf{p},\mu) - \bar{n}(\mathbf{p},\mu)\right]\,. \qquad (32)$$

To calculate the meson masses we have to evaluate the loop integrals (14) substituting $S(p) \to S(p,T,\mu)$, i.e.

$$L^{aa'}(q) = \alpha \mathrm{tr}_\gamma \int \frac{d^4p}{(2\pi)^4} S_\beta(p,T,\mu) \mathcal{K}^a S_\beta(p-q,T,\mu) \mathcal{K}^{a'}\,. \qquad (33)$$

In the treatment of the NJL model in a hot and dense medium [9] the meson and diquark masses get contributions from the constant parts of the integral (33) in the limit q $\to$ 0. They are now functions of T and μ having, however, the same form as in the zero–temperature case, eqs. (18)–(21). Thereby, m, I_1, I_2 have to be replaced by the quantities $m(T,\mu), I_1(m,T,\mu), I_2(m,T,\mu)$ defined by (30)–(32), respectively. The same concerns the renormalization constants (22), (23) as well as the coupling constants (24)–(26).

Let us now present some graphics indicating the behaviour of the physical quantities on T. The behaviour of the constituent quark mass $m(T,\mu)$ as a function of T and μ is found by a self–consistent solution of the thermal gap equations (29),(30) with the inclusion of (12). The dependence of the function $m(T,\mu)$ on T for fixed values of μ is shown in Fig.1.a. For $m_0 = 0$ the restoration of chiral symmetry is indicated by the vanishing of the order parameter $< \bar{q}q > \to 0$ at critical values of the temperature and chemical potential, and consequently by $m(T,\mu) \to 0$. When $m_0 \neq 0$ the sharp phase boundary disappears. For the critical temperature one gets $T_c \approx 200\text{MeV}$, the critical density is $\mu_c \approx 300\text{MeV}$. These results are in agreement with other works [8, 12]. Next, the pion decay constant is an important physical quantity which determines the scale of low–energy chiral meson theories. Its T–dependence is depicted in Fig.1.b. Clearly, the sharp decrease in T near $T \approx T_c$ is associated to the constituent quark mass, as follows from the Goldberger–Treiman relation. Recall that in the above model F_π includes the effects of $\pi - A$ mixing.

Figure 2 exhibits the behaviour of the meson (a) and diquark (b) masses as functions of T for $\mu = 0$. Let us comment in some detail the meson mass graphics. As T increases the mass of the σ meson decreases sharply as a result of the decrease of the constituent

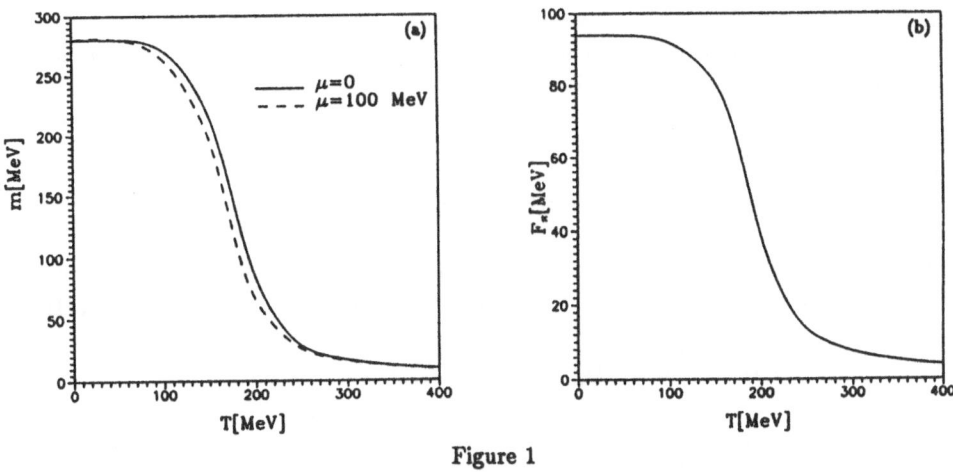

Figure 1

Fig.1.a: The T–dependence of the quark mass m for a constant chemical potential μ

Fig.1.b: The T–dependence of the pion decay constant F_π

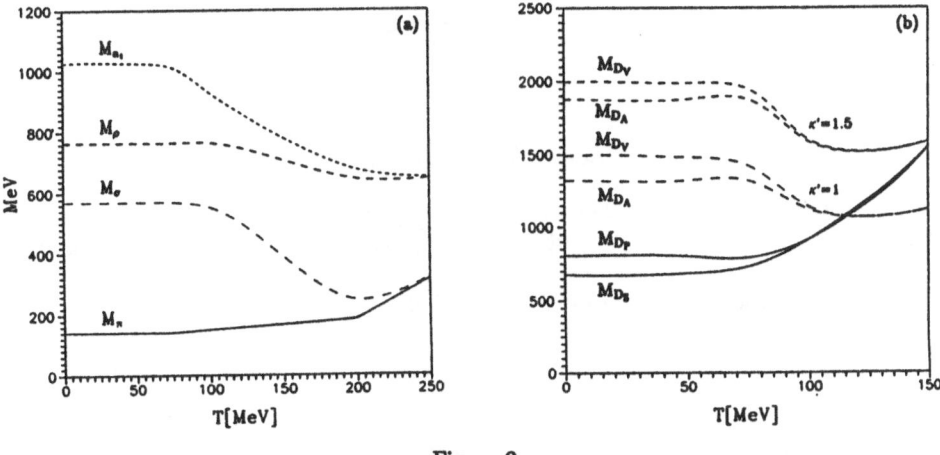

Figure 2

Fig.2.a: The behaviour of the meson masses as functions of the temperature T for vanishing chemical potential μ

Fig.2.b: The behaviour of the diquark masses as functions of the temperature T for vanishing chemical potential μ

quark mass. On the other hand, the mass of the (would be) Goldstone pion will persistently stay constant until the critical conditions for chiral restoration are reached, beyond which it ceases to exist. Furthermore, M_ρ is merily independent of T (a weak T–dependence for $T \geq 0\text{MeV}$ arises from $g_\rho(T,\mu)$, whereas M_{a_1} shows a sharp decrease similar to that of M_σ. Finally, above the critical temperature one obtains $M_{a_1} = M_\rho$, as is expected for a chiral symmetric phase. Note, that the diquark masses approach each other already at a lower temperature ($T \approx 100\text{MeV}$) and chemical potential. For the behaviour of the meson coupling constants and the renormalization factors we again refer to [9].

4 Meson–diquark bosonization of QCD$_2$

We start with the lagrangian of two–dimensional QCD with exact local colour symmetry $SU(3)_c$ and, for symplicity, also with exact global flavour symmetry $SU(3)_f$

$$\mathcal{L} = -\frac{1}{2} G_{\mu\nu,\alpha\beta} G^{\mu\nu}_{\beta\alpha} + \bar{q}_{a\alpha}(i\gamma^\mu D_{\mu,\alpha\beta} - m\delta_{\alpha\beta})q_{\alpha\beta} .$$

Here

$$G_{\mu\nu,\alpha\beta} = \partial_\mu A_{\nu,\alpha\beta} - \partial_\nu A_{\mu,\alpha\beta} + ig[A_\mu, A_\nu]_{\alpha\beta} ,$$

is the field strength tensor,

$$A_{\mu,\alpha\beta} = \sum_{n=1}^{8} \frac{\lambda^n_{\alpha\beta}}{2} A^n_\mu$$

is the gauge field and

$$D_{\mu,\alpha\beta} = \partial_\mu \delta_{\alpha\beta} + ig A_{\mu,\alpha\beta}$$

is the covariant derivative. The indices $\alpha, \beta = 1, 2, 3$ denote colour and $a = 1, 2, 3$ flavour. $\lambda^n/2$ are the generators of the colour group $SU(3)_c$ in the Gell–Mann representation.

It is convenient to consider QCD_2 in the light–cone gauge

$$A_- = A^+ = \frac{1}{2}(A_0 - A_1) = 0 , \quad (a \cdot b = a_+ b^+ + a_- b^- = a_+ b_- + a_- b_+) .$$

In this case there exists only one independent dynamical quark variable $\hat{q} \equiv \begin{pmatrix} 0 \\ q_2 \end{pmatrix}$ and gluonic self–interactions as well as Faddeev–Popov ghosts are absent [1, 10]. The generating functional of Green functions of quarks reads

$$Z[\eta, \bar{\eta}] = C \int \mathcal{D}A_+ \mathcal{D}q \mathcal{D}\bar{q} \exp\left\{ i \int d^2x [\mathcal{L} + \mathcal{L}_s] \right\}$$

with the source term

$$\mathcal{L}_s = \bar{q}\eta + \bar{\eta}q .$$

After integrating over the gluon fields and performing a Fierz transform of the interaction terms into $1_c(q^*q)$ and $3_c^*(qq)$ channels one obtains [10] [1]

$$
\begin{aligned}
Z[\eta, \eta^*] = C_1 \int \mathcal{D}q \mathcal{D}q^* \exp i \int d^2x d^2y \Big\{ & q^*(x) i S^{-1}(x,y) q(y) \\
& + \frac{i}{2}(2g)^2 \Big[[q^*(x)\mathcal{M}_M^e q(y)] D(x-y) [q^*(y)\mathcal{M}_M^e q(x)] \\
& + [q^*(x)\mathcal{M}_D^\theta q(y)] D(x-y) [q^*(y)\mathcal{M}_D^\theta q(x)] \Big] \\
& + [q^*(x)\eta(y) + \eta^*(x)q(y)]\delta^{(2)}(x-y) \Big\} .
\end{aligned}
$$
(34)

Here the projection matrices $\mathcal{M}_M^\tau$ and $\mathcal{M}_D^\theta$ are defined as generalizations of eqs. (3) to the case of $SU(3)_f$. In (34) S is the quark propagator,

$$
S(x-y) = \frac{-\partial_-^x}{-2\partial_+^x \partial_-^x - m^2 + i\epsilon} \delta^{(2)}(x-y) ,
$$

and D the gluon propagator

$$
D(x-y) = \frac{i}{(\partial_-^x)^2} \delta^{(2)}(x-y) .
$$

Introducing collective fields one finally gets

$$
Z[\eta, \eta^*] = C_2 \int \mathcal{D}\phi \mathcal{D}\omega \mathcal{D}\omega^+ \exp\{i W_{eff}[\phi, \omega, \omega^+]\} \, Z[\eta, \eta^* | \phi, \omega, \omega^+]
$$
(35)

(here "+" denotes the hermitean conjugate and "*" the Grassmann conjugate) with

$$
\begin{aligned}
W_{eff}[\phi, \omega, \omega^+] = & -i \operatorname{Tr} \ln S_\phi^{-1} - \frac{i}{2} \operatorname{Tr} \ln(1 - 4 S_\phi^T \Omega^+ S_\phi \Omega) \\
& + \int \int \left[\frac{i}{2(2g)^2} \frac{\phi^e \phi^e}{D} + \frac{2i}{(2g)^2} \frac{\omega^{+a}\omega^a}{D} \right] ,
\end{aligned}
$$
(36)

$$
Z[\eta, \eta^* | \phi, \omega, \omega^+] = \exp i \int \int \left(-\eta^* S_n \eta + \frac{1}{2}\eta^* S_a \eta^* + \frac{1}{2}\eta S_a^+ \eta \right) .
$$
(37)

In (37) the normal and anomalous Green functions for the quarks moving in external fields ϕ, ω, ω^+ are defined as

$$
S_n = -i H^{-1} S_\phi , \quad S_a = -i H^{-1} S_\phi 2\Omega S_\phi^T
$$

with

$$
H = 1 - 4 S_\phi \Omega S_\phi^T \Omega^+
$$

---

[1] Here and in the following q means the only nontrivial component q_2 of the spinor q. q^* is the Grassmann conjugate of q.

and

$$iS_\phi^{-1} = iS^{-1} - \mathcal{M}_M^e \phi^e .$$

The stationarity condition on the effective bilocal action (36) leads to dynamical equations for the quark spectrum of QCD_2,

$$\frac{W_{eff}}{\delta \phi^a} = 0 , \quad \frac{W_{eff}}{\delta \omega^\theta} = 0 , \quad \frac{W_{eff}}{\delta \omega^{+\theta}} = 0 ,$$

from which one gets the following system of equations:

$$\Phi \equiv \mathcal{M}_M^e \phi^e = -4g^2 D\{\mathrm{tr}[\mathcal{M}_M^e S_n(\Phi, \Omega, \Omega^+)]\}\mathcal{M}_M^e ,$$

$$\Omega \equiv \mathcal{M}_D^\theta \omega^\theta = 2g^2 D\{\mathrm{tr}[\mathcal{M}_D^\theta S_a(\Phi, \Omega, \Omega^+)]\}\mathcal{M}_D^\theta ,$$

$$\Omega^+ \equiv \mathcal{M}_D^\theta \omega^{+\theta} = -2g^2 D\{\mathrm{tr}[\mathcal{M}_D^\theta S_a^+(\Phi, \Omega, \Omega^+)]\}\mathcal{M}_D^\theta .$$

Expanding the integrand of (35) around the stationary solutions $\Phi_0 \neq 0, \omega = \omega^+ = 0$ and shifting the integration variable $\phi^e \to \phi^e - \Phi_0 \delta^{e0}$ one can derive the homogeneous Bethe–Salpeter equations for the vertex functions of the corresponding bound states in the ladder approximation. From these equations one can determine the discrete meson and diquark [1, 10] spectrum. Qualitatively, one obtains that the squared meson masses $(m_M^k)^2$ are finite for $\lambda \to 0$, where λ is an infrared cut–off parameter. However, the squared diquark masses look like

$$(m_D^k)^2 = \frac{const.}{\lambda} ,$$

so that they tend to infinity for $\lambda \to 0$. This behaviour is analogous to that of the constituent quark masses. It reflects the fact that coloured states are not observable – they are confined. Let us add that diquarks have also been investigated in an approximate treatment of QCD in four dimensions [13].

5 Summary

By considering the local current×current NJL interaction we have demonstrated a simultaneous bosonization of the model in meson and diquark sectors for $T = 0$ and $T \neq 0, \mu \neq 0$. Thereby, vector and axial–vector mesons have been included explicitly (they have not been Fierz transformed away). Furthermore, the pseudoscalar/axial-vector meson and scalar/vector diquark mixings have been taken into account. In this type of models the restoration of chiral symmetry takes place at $T_c \approx 200\mathrm{MeV}, \mu_c \approx 300\mathrm{MeV}$.

Concerning nonlocal current×current interactions we have further considered the meson–diquark bosonization of QCD_2. Thereby bilocal collective fields have been used. For the diquark masses confinement properties like those of quarks have been established due to infrared divergencies.

Finally, the results obtained here may be the starting point for further investigations including the gluon condensate [14] or devoted to the derivation of a realistic effective meson–baryon theory. First results in the latter direction have been published in refs. [15, 5, 16].

References

[1] D. Ebert and V.N. Pervushin, *Theor. Math. Phys.* **36** (1978) 759.

[2] D. Ebert, V.N. Pervushin and H. Reinhardt, *Sov. J. Part. Nucl.* **10** (1979) 1114;
see also: H. Kleinert, in: *Understanding the Fundamental Constituents of Matter* (Erice Lectures 1976), ed. A. Zichichi (Plenum, New York, 1978).

[3] Y. Nambu and G. Jona-Lasinio, *Phys. Rev.* **122** (1961) 345, **124** (1961) 246;
T. Eguchi, *Phys. Rev.* **D14** (1976) 2755;
H. Kikkawa, *Progr. Theor. Phys.* **56** (1976) 974.

[4] D. Ebert and M.K. Volkov, *Yad. Phys.* **36** (1982) 1265, *Z. Phys.* **C16** (1983) 205;
M.K. Volkov, *Ann. Phys.* **157** (1984) 282, *Sov. J. Part. Nucl.* **17** (1986) 433;
A. Dhar, R. Shankar and S.R. Wadia, *Phys. Rev.* **D31** (1985) 3256;
D. Ebert and H. Reinhardt, *Nucl. Phys.* **B271** (1986) 188.

[5] R.D. Ball, *Int. J. Mod. Phys.* **A5** (1990) 4391;
H. Reinhardt, *Phys. Lett.* **B244** (1990) 316, **B257** (1991) 375.

[6] D. Kahana and U. Vogl, *Phys. Lett.* **B244** (1990) 10;
U. Vogl, *Z. Phys. A - Atomic Nuclei* **337** (1990) 191.

[7] D. Ebert, L. Kaschluhn and G. Kastelewicz, *Phys. Lett.* **264** (1991) 420.

[8] V. Bernard, U.-G. Meissner and I. Zahed, *Phys. Rev. Lett.* **59** (1987) 966;
V. Bernard and U.-G. Meissner, *Nucl. Phys.* **A489** (1988) 647;
T. Hatsuda and T. Kuhiniro, *Phys. Lett.* **B145** (1984) 7, **B185** (1987) 309, **B198** (1987) 126;
M. Asakawa and K. Yazaki, *Nucl. Phys.* **A504** (1989) 668;
S. Klimt, M. Lutz and W. Weise, *Phys. Lett.* **B249** (1990) 386;
M. Lutz, S. Klimt and W. Weise, Preprint *TPR 91-12*, Regensburg 1991.

[9] D. Ebert, Yu.L. Kalinovsky, L. Münchow and M.K. Volkov, Preprint *JINR* **E2-92-134**, Dubna 1992.

[10] D. Ebert and L. Kaschluhn, *Nucl. Phys.* **B355** (1991) 123.

[11] L. Dolan and R. Jackiw, *Phys. Rev.* **D9** (1974) 3320.

[12] J. Gasser and H. Leutwyler, *Phys. Lett.* **B184** (1987) 83;
P. Gerber and H. Leutwyler, *Nucl. Phys.* **B321** (1989) 387.

[13] R.T. Cahill, J. Praschifka and C.J. Burden, *Aust. J. Phys.* **42** (1989) 161;
J. Praschifka, R.T. Cahill and C.D. Roberts, *Int. J. Mod. Phys.* **A4** (1989) 4929.

[14] D. Ebert and M.K. Volkov, *Phys. Lett.* **B272** (1991) 86.

[15] R.T. Cahill, *Aust. J. Phys.* **42** (1989) 171.

[16] D. Ebert and L. Kaschluhn, Preprint *DESY* **92-110**, Zeuthen (1992).

Hadronization of QCD [†]

H. Reinhardt

Institut für Theoretische Physik, Universität Tübingen, D-7400 Tübingen, FRG

Abstract. By means of functional integral techniques, QCD is converted into a quantum hadrodynamics, an effective theory of interacting mesons and baryons. While mesons appear as composite quark antiquark fields baryons are built up from diquark fields in the $\bar{3}_c$ representation plus a quark. In the large N_c limit the diquark correlations disappear and the effective hadron theory becomes an effective meson theory where baryons appear as chiral solitons. Numerical results are presented for the $s = \frac{1}{2}$ baryon octet.

1. Introduction

According to our present knowledge QCD is the theory of strong interactions and should therefore explain all hadronic systems in particular the structure of the individual hadrons. Due to the complexity of non–Abelian gauge theories the description of hadrons is so far mainly based on phenomenological models. We can distinguish two principally different classes of hadron models. One class of models is based on the valence quark picture of hadrons as for example the MIT bag model. The other class is based on large–N_c QCD considerations of 'tHooft [1]) and Witten [2]), namely that for a large number of colours QCD reduces to the effective theory of weakly interacting mesons and glueballs. Baryons appear here as solitons of the meson field. The prototype of this class of baryon models is the Skyrme model [5]). Furthermore a phenomenological unification of these two types of models is provided by the chiral bag model.

It is a big challenge to understand how these two types of models are related to each other and how they emerge from QCD.

In my talk I will demonstrate, how QCD can be converted into an effective theory of interacting mesons and baryons. For this purpose QCD is first reduced to an effective quark theory of NJL type. This is accomplished by means of the socalled field strength approach to Yang-Mills theories [13]). In leading order semiclassical approximation this approach converts QCD into an effective quark theory which looks at low energies very similar to the Nambu-Jona-Lasinio model. A local four-fermion interaction is also obtained in leading order of the strong coupling approximation to lattice QCD [14]). We have therefore good evidence to assume that NJL type of models mimic the low energy quark flavour dynamics of QCD. The NJL model is then hadronized [11]), i.e. converted into and effective theory

[†]supported by Deutsche Forschungsgemeinschaft under contract Re 856/2-1 and by COSY under contract 41170833

of interacting mesons and baryons which are described by composite $q\bar{q}$ and qqq states. If one expands this effective hadron theory in leading order in the quantum fluctuations of the fields around their vacuum values one arrives at a quantum hadrodynamics where mesons couple to the baryons via fully fledged form factors. Only when these form-factors are approximated at zero energy by contact couplings the obtained quantum hadrodynamics reduces to a Walecka type of model field theory [10].

In the limit where the number of colours goes to infinity the effective hadron theory reduces to an effective meson theory which contains no longer baryon fields as composite three-quark fields but baryons emerge as solitons of the meson fields.

I will present numerical results for the masses of the $s = \frac{1}{2}^+$ baryon flavour octet obtained in both the valence quark and the soliton pictures.

Finally let me mention that an alternative hadronization approach to an effective four-quark interaction was advocated in ref. [15] and subsequently considered in ref. [16]. I believe, however, that the approach of ref. [11] is more general and also more rigorous.

2. Fierzing the quark interaction

For simplicity I will assume the following effective quark lagrangian

$$\mathcal{L} = \mathcal{L}_0 + \mathcal{L}_{int} \tag{1a}$$

$$\mathcal{L}_0 = \bar{q}(x)(i\,\rlap{/}D - m_0)q(x) \tag{1b}$$

$$\mathcal{L}_{int} = \frac{g}{2}j_\mu^a(x)j^{a\mu}(x) \tag{1c}$$

Here q denotes the quark field, $m_0 = diag(m_u^0, m_d^0, m_s^0)$ is the current quark mass and $j_\mu^a(x) = \bar{q}(x)\gamma_\mu\frac{\lambda^a}{2}q(x)$ is the colour (octet) current of the quarks. I will confine myself to the light flavours u, d, s.

Like QCD the model (1) has global chiral invariance (for $m_0 = 0$) but lacks confinement and is not renormalizable.

In spite of the principally different nature of baryons and mesons, from the point of view of hadron spectroscopy within the valence quark picture baryons can be treated analogously to the mesons by noticing that in a colourless baryon any two of the three valence quarks must be in a $\bar{3}_c$ representation of the colour group and therefore transform under colour as an antiquark. This symmetry between baryons and mesons based on the equivalence of (bosonic) diquarks and (fermionic) quarks with respect to colour is referred to as 'supersymmetry' by Lichtenstein [21]. This symmetry is also exploited in the old string picture of hadrons. Furthermore the gluon exchange between the two quarks in the $\bar{3}_c$ representation is known to be attractive leading to the formation of bound states referred to as diquarks. In fact, there is mounting evidence for the importance of diquark correlations in low-energy hadron spectroscopy, as this meeting has shown. This suggests that baryons might be understood as bound diquark-quark states.

The interaction (1c) has the same colour and Lorentz structure as the one gluon exchange. The latter is attractive in the $(q\bar{q})_{1_c}$ and $(qq)_{3_c}$ channels. Since we are interested in bound hadron states we rewrite this interaction exclusively in the attractiv channels by making extensive use of Fierz transformations for colour, flavour and Dirac matrices. The unique result is [15,11]

$$\mathcal{L}_{int} = \frac{g_1}{3}(\bar{q}\Lambda_\alpha q)(\bar{q}\Lambda^\alpha q) + \frac{g_2}{3}(\bar{q}\Gamma_\alpha q^c)(\bar{q}^c\Gamma^\alpha q) = \mathcal{L}_{int}^{q\bar{q}} + \mathcal{L}_{int}^{qq} \quad (2)$$

where

$$g_1 = g_2 = g$$

and

$$q^c = C\bar{q}^T = C\gamma^{0\ T}q^* = i\gamma^2 q^*$$
$$\bar{q}^c = (q^c)^\dagger\gamma^0 = q^T C$$

are the charge conjugated spinors. Furthermore we have introduced the following vertex operators for the quark bilinears

$$\Lambda^\alpha = 1_c t_F^a O^\alpha \quad , t_F^a \in \{t_F^0 = \sqrt{\frac{2}{n}}, t_F^{a\neq 0}\}$$

$$\Gamma_\alpha = (i\epsilon/\sqrt{2})_c t_F^a O^\alpha \quad , t_F^a \in \{A^{a=2,5,7}, S^{A=0,1,3,4,6,8}\}$$

and $\alpha = (A, a, \alpha)$ denotes collectively the indices: A for colour vertex, a for the flavour vertex and α for the Dirac vertex. Furthermore A^a and S^a are the symmetric and antisymmetric generators ($t^a = A^a$, $a = 2, 5, 7$ and $t^a = S^a$, $a = 0, 1, 3, 4, 6, 8$) of SU(n=3), which correspond to the $\bar{3}$ and the 6 representation of SU(3).

The first term on r.h.s. of (2) gives the interaction of colour singlet (1_c) quark-antiquark pairs, $\bar{q}\Lambda^\alpha q$, while the second one describes the interaction of diquarks in the $\bar{3}_c$ representation. Let me stress that both interactions $\mathcal{L}_{int}^{q\bar{q}}$ and $\mathcal{L}_{int}^{qq}$ are attractive and are both separately invariant under global chiral (flavour) transformations $U(n)_L \times U(n)_R$.

In passing we note a crucial difference between the quark-antiquark and the diquark vertices due to the Pauli principle, which acts restrictive in the diquark channel but is inactive in the quark-antiquark channel. While in the latter the Dirac vertex O^α can occur with any flavour vertex t^a in the former, the Pauli principle allows for the Dirac vertices $O^\alpha = \{1, i\gamma_5, i\gamma^\mu\gamma_5/\sqrt{2}\}$ for diquark states in the $\bar{3}_F$ representation while diquarks in the 6_F representation can occur only for the Dirac vertex $O^\alpha = \{i\gamma^\mu/\sqrt{2}\}$. Finally we also note that under Lorentz transformations $\bar{q}^c i\gamma_5 q$ behaves as a scalar.

3. Hadronization of quark flavour dynamics

The quantum theory of the NJL model (1) is defined by the path integral representation

$$Z = \int \mathcal{D}q \mathcal{D}\bar{q} e^{i\int d^4x\ \mathcal{L}} \quad (4)$$

of the quantum transition amplitude. Our aim is to convert this theory of the unobservable quarks into an effective theory of the physical mesons and baryons.

For this purpose we introduce composite meson and baryon fields built up from the quark degrees of freedom in accord with their valence quark content. Once these fields are introduced the unobservable quark fields can be integrated out exactly.

The meson fields are introduced as colourless quark-antiquark fields via the constraint

$$1 = \int D\xi \delta(\xi - \bar{q}q) = \int D\xi D\phi e^{i \int d^4 x \phi(\xi - \bar{q}q)}. \tag{5}$$

Note that ξ, ϕ are flavour and Dirac matrices with flavour singlet and octet, as well as Lorentz scalar, pseudoscalar vector and axial-vector components ($\xi = \xi_\alpha \Lambda_\alpha$ etc.). Analogously the baryon field ψ is introduced as composite colourless three (constituent) quark field via the identity

$$
\begin{aligned}
1 &= \int \prod_{\alpha,\mu} D\Psi_\mu^\alpha D\Psi_\mu^\alpha \delta(\bar{\psi}_\mu^\alpha - \bar{q}_\mu(\bar{q}\Gamma^\alpha q^c))\delta(\psi_\mu^\alpha - (\bar{q}^c\Gamma^\alpha q)q_\mu) \\
&= \int \prod_{\alpha,\mu} D\psi_\mu^\alpha D\bar{\psi}_\mu^\alpha D\chi_\mu^\alpha \mathcal{D}\bar{\chi}_\mu^\alpha exp \left\{ i \int d^4x \, d^4X \right. \\
&\quad \left[(\bar{\psi}_\mu^\alpha(x,X) - \bar{q}_\mu(x)[\bar{q}(X)\Gamma^\alpha q^c(X)])\chi_\alpha^\mu(x,X) \right. \\
&\quad \left. \left. + \bar{\chi}_\alpha^\mu(x,X)(\psi_\alpha^\mu(x,X) - [\bar{q}^c(X)\Gamma^\alpha q(X)]q_\mu(x)) \right] \right\} \tag{6}
\end{aligned}
$$

where the colourless (three valence quark) configuration has been expressed by a diquark in the $\bar{3}_c$ representation $\bar{q}^c\Gamma_\alpha q$, and a quark q_μ. In view of the importance of diquarks as building blocks of baryons and also because of the presence of the diquark interaction $\mathcal{L}_{int}^{qq}$ it is convenient to introduce explicitly diquark fields

$$
\begin{aligned}
1 &= \int \prod_\alpha D\kappa_\alpha D\kappa_\alpha^* \delta(\kappa_\alpha^* - \bar{q}\Gamma_\alpha q^c)\delta(\kappa_\alpha - \bar{q}^c\Gamma_\alpha q) \\
&= \int \prod_\alpha D\kappa_\alpha D\kappa_\alpha^* D\Delta_\alpha D\Delta_\alpha^* exp \left\{ i \int d^4x \right. \\
&\quad \left. \left[(\kappa_\alpha^*(x) - \bar{q}(x)\Gamma_\alpha q^c(x))\Delta^\alpha(x) + \Delta_\alpha^*(x)(\kappa^\alpha(x) - \bar{q}^c(x)\Gamma^\alpha q(x)) \right] \right\} \tag{7}
\end{aligned}
$$

By constructions κ and Δ are complex bose fields while ψ and χ are fermion fields. From the symmetry relations

$$
\begin{aligned}
(\bar{q}\Lambda_\alpha q)^\dagger &= \pm(\bar{q}\Lambda^\alpha q) \\
(\bar{q}\Gamma_\alpha q^c)^\dagger &= \pm(\bar{q}^c\Gamma_\alpha q)
\end{aligned}
$$

follows that ϕ_α and ξ_α are either hermitian or antihermitian while $\kappa_\alpha, \Delta_\alpha$ and $\kappa_\alpha^*, \Delta_\alpha^*$ are either complex conjugate or anti-complex conjugate to each other.

Inserting the identities (5), (6) and (7) into the generating functional (4) and exploiting the constraints defined by the δ-functions one can replace $\bar{q}\Lambda_\alpha q, \bar{q}^c\Gamma_\alpha q$ and $\bar{q}\Gamma_\alpha q^c$ by $\xi_\alpha, \kappa_\alpha$ and κ_α^* , respectively. As a result the quark fields are removed from the two-body interaction and can be integrated out exactly. Furthermore the fields ξ and κ, κ^* can also be integrated out exactly. The quantum transition amplitude (4) can then be cast into the form

$$Z = \int D\phi D\psi D\bar{\psi} D\chi D\bar{\chi} exp\{iW[\chi, \phi] + i \int (\bar{\psi}\chi + \bar{\chi}\psi)\} \tag{8}$$

where $W[\chi, \phi]$ is the generating functional of the baryon (three-(valence) quark) Green functions in the background of the fluctuating meson field ϕ with χ figuring as colourless three-quark-source field, which eventually has to be integrated out.

The generating functional $W[\chi, \psi]$ is still exact but unfortunately only implicitly defined as functional integral over the diquark field Δ (for the explicit representation see ref. [11]).

Since both Δ and χ are not observable fields one would eventually like to integrate them out, leaving a theory defined entirely in terms of the physically observable meson ϕ and baryon ψ fields. This can be accomplished approximately [11]). In terms of the dynamical source field χ the generating functional has the expansion

$$W[\chi, \phi] = S[\phi] + \int d^4x\, d^4y\, d^4X\, d^4Y\ \bar{\chi}(x, X)G_B[\phi](x, X, y, Y)\chi(y, Y) + \cdots \tag{9}$$

Here

$$S[\phi] = W[\chi = 0, \phi] \tag{10}$$

is the effective meson action and

$$G_B[\phi](x, X, y, Y) = \frac{\delta^2 W[\chi, \phi]}{\delta\bar{\chi}(x, X)\delta\chi(y, Y)}|_{\chi=0} \tag{11}$$

is (after projection onto physical flavour channels) the connected baryon Green function. The omitted higher order terms represent baryon correlations that cannot be generated by meson exchange, which, however, can be modified or dressed by mesonic exchanges. Such correlations are also obtained in traditional quark models. Since we have at present little empirical evidence for these correlations we will ignore them in the following.

Truncating the expansion (9) at second order in the source field χ, this field can be integrated out yielding

$$Z = \int D\phi e^{iS[\phi]} Det G_B[\chi] \int D\psi D\bar{\psi} e^{i \int \bar{\psi} G_B^{-1}[\phi]\psi} \tag{12}$$

What we have obtained here is a (non-local) coupled meson-baryon quantum field theory (quantum hadrodynamics) where the baryons described by the propagator $G_B[\phi]$ (11) move in the background of the fluctuating meson field ϕ. The baryon determinant $DetG_B[\phi]$ in front of the baryon integral in (12) may, at first sight, appear strange, since phenomenological quantum hadron field theories (e.g. the Walecka model [10])) do not have it. However, this term, which natural arises in the hadronization procedure of ref. [11]), is quite important, at least from the conceptual point of view, in the sense that it makes the effective hadron theory (12) anomaly free. It ensures that no anomaly, in particular the chiral anomaly, arises from the integration measure of the baryon fields. The anomalies arise from the fundamental fermions, the quarks, and are entirely already included in the effective mesonic action $S[\phi]$ (10) through the quark determinant (see below and ref. [11])) which arises from integrating out the quark fields. Were $DetG_B[\phi]$ not present in (12) one would have a double counting of anomalies.

If one is not explicitly interested in baryons (implying no external baryon or quark sources are present) the ψ-field can be integrated out in (12) leaving in the absence of valence baryons the effective meson theory

$$Z = \int D\phi e^{S[\phi]}, \qquad (13)$$

which was previously obtained in the bosonization of the Nambu-Jona-Lasinio model. This shows that the hadronization of ref. [11]) is consistent with previous bosonizations [3]).

The effective meson action (10) is minimized by the vacuum configuration ϕ_0

$$\frac{\delta S[\phi]}{\delta\phi}|_{\phi=\phi_0} = 0 \qquad (14)$$

Expanding the meson action up to second order in the fluctuations $\varphi = \phi - \phi_0$ and taking into account (19) we find

$$S[\phi] = S[\phi_0] + \frac{1}{2}\int d^4x\, d^4y\, \varphi(x)\mathcal{D}_m^{-1}[\phi_0](x,y)\varphi(y) + \cdots \qquad (15)$$

where

$$\mathcal{D}_m^{-1}[\phi_0](x,y) = \frac{\delta^2 S[\phi]}{\delta\phi(x)\delta\phi(y)}|_{\phi=\phi_0} \qquad (16)$$

is the free meson propagator. Similarly expanding the inverse baryon propagator

$$G_B^{-1}[\phi](x,y) = G_B^{-1}[\phi_0](x,y) + \int dz F[\phi_0](x,y,z)\varphi(z) + \cdots \qquad (17)$$

we obtain in zeroth order the baryon propagator of the baryons in the vacuum ϕ_0 and in leading order in the fluctuations the meson baryon form factor

$$F[\phi_0](x,y,z) = \frac{\delta G_B^{-1}[\phi](x,y)}{\delta\phi(z)}|_{\phi=\phi_0} \qquad (18)$$

Note that this form factor is in general non-local. Inserting the expansions (15) and (17) into (12) one obtains (except for the presence of $DetG_B[\phi]$) a standard type quantum hadrodynamics.

$$Z = const \int D\varphi e^{i\frac{1}{2}\varphi D_m^{-1}[\phi_0]\varphi} Det G_B[\phi_0 + \varphi] \int D\psi D\bar{\psi} e^{i\int \bar{\psi} G_B^{-1}[\phi_0]\psi + i\int \bar{\psi} F[\phi_0]\varphi\psi}$$

(19)

Consistent with the expansion of the effective hadron action in powers of the fluctuations of the meson field φ one should also expand $Det G_B[\phi_0 + \varphi]$. Using

$$Det G_B = exp Tr log G = exp[-Tr log G_B^{-1}]$$

and eq. (17) this yields in leading order in φ

$$Det G_B[\phi_0 + \varphi] = Det G_B[\phi_0] exp(-Tr G_B[\phi_0] F[\phi_0]\varphi)$$

(20)

If one discards the non-local structure of the form factor $F[\phi_0]$ (13) and also $Det G_B$ eq. (12) reduces precisely to a Walecka type of theory [10]), frequently used to describe nuclear matter and finite nuclei in a relativistic fashion.

4. Explicit evaluation of mesons and baryons

From the generating functional $W[\chi, \phi]$ (see eq. (8)) we can in principle evaluate all meson and baryon properties. Unfortunately the generating functional $W[\chi, \phi]$ is still given by an functional integral over the diquark field Δ (see ref. [11])). This integral can be worked out by using the method of stationary phases and resorting to the loop expansion. In leading order the diquark states are then defined by the Bethe–Salpeter equation in ladder approximation shown in fig. 1. Similary the meson propagator becomes an analogous partial sum of bubble diagrams illustrated in fig. 2. Finally in zeroth order the resulting baryon propagator describes the independent propagation of a diquark and a quark both in the fluctuating meson field ϕ (see fig. 3 (a)). It is, however, not consistent to stick to this zero order approximation for the following reason: The diquark field (propagator) has been constructed by summing partially all diquark bubbles (see fig. 1). Thus the leading order baryon diagram (fig. 3 (a)) contains in fact already all the diagrams arising from fig. 1 (b) by adding a quark line. The leading order diagram of this class is shown in fig. 3 (b). This diagram contains three intermediate quarks which should be in a completely antisymmetric state. But since the diquark was constructed irrespectively of the presence of the third quark this intermediate three-quark state will in general not fulfill the Pauli principle. To obtain a consistent description of the baryons which takes fully care of the Pauli principle, in addition to the leading order diagram, fig. 3 (a), one has to include at least the corresponding exchange diagrams shown in fig. 4. Due to the colour of the quarks these exchange correlations are attractive and, as we will see later, bind diquark and quark into baryons. The leading order exchange diagram corresponding to the leading order diquark bubble (fig. 3 (b)) is shown in fig. 3 (c).

Similary the higher order exchange diagrams shown in fig. 4. have to be included to correct for the Pauli principle in the higher order loop diagrams. All these exchange diagrams naturally emerge in the hadronization approch of ref. [11]).

Figure 1: The diquark propagator (b) and its inverse (a) in the ladder approximation.

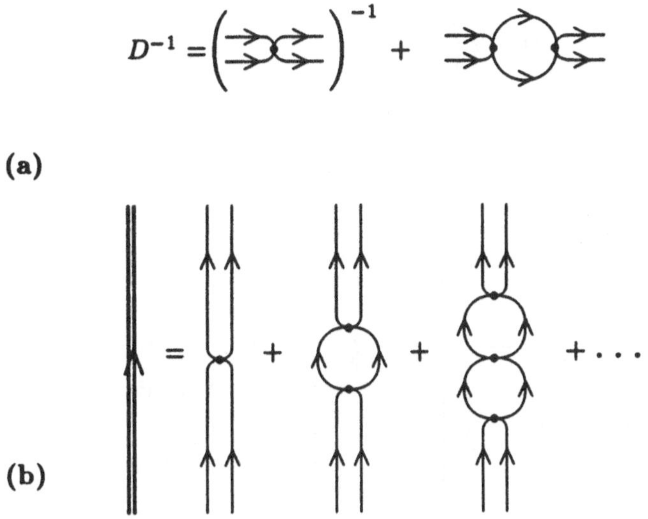

(a)

(b)

The baryon states are defined by the poles of the baryon Green function $G_B[\phi]$. The pole conditon results in the Faddeev type of equation shown in fig. 5 (a). The physical baryon states in the vacuum are obtained by projecting this equation onto colour singlet states with appropriate flavour content and solving the projected equation for the vacuum configuration $\phi = \phi_0$ of the meson field.

At this point let me emphasize that in principle the baryon propagator $G_B[\phi](40)$ depends via the quark $G[\phi]$ and diquark $\mathcal{D}[\phi]$ propagator on the fluctuating meson field ϕ, and the free baryon Lagrangian $\bar\psi G_B[\phi]\psi$ occurs in the effective hadron theory (12) under the meson integral. The fluctuating meson field gives in particular rise to meson exchange correlations between quarks. When we put $\phi = \phi_0$ in the Faddeev equation we discard the meson exchange in the construction of the baryon. The meson exchange is in particular important in the construction of the diquarks. In the actual solution of the Faddeev equation we try to simulate the effect of the omitted meson exchange by renormalizing the coupling constant g_2 of the diquark interaction, which becomes then a parameter independent of the interaction strength g_1 in the meson channel.

In ref. [17]) the relativistic Faddeev equation shown in fig. 5 (a) was solved numerically in the static limit where the exchanged quark (see fig. 4) is considered infinitely heavy, i.e. its kinetic energy is neglected. In this limit the Faddeev equation reduces to a Bethe-Salpeter equation (with ladder approximation) see fig. 5 (b). The parameters of the assumed NJL Lagrangian (1) are the current quark masses m_0^i, the coupling constants g_1 and g_2 and the ultraviolet cut-off

Figure 2: The meson propagator (b) and its inverse (a) in the ladder approximation.

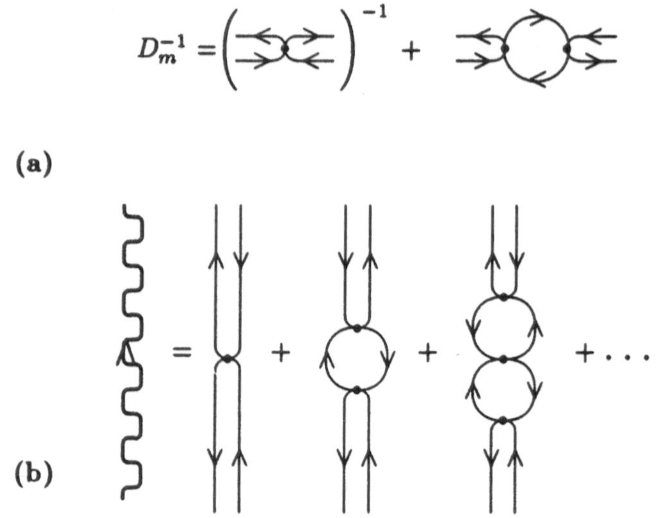

(a)

(b)

Λ, which is necessary since the model (1) is not renormalizable. $m_0^{i=u,d,s}$ and g_1 are fixed in the meson sector from $F_\pi, m_{\pi^\pm}, m_{K^\pm}$ and $m_{K^0} = m_{\bar{K}^0}$ while Λ or equivalently the constituent quark mass $m^{i=u}$ of the u-quark is left as a free parameter. The diquark interaction strength is chosen to reproduce for given m^u the proton mass. The results obtained for the masses of the spin $s = \frac{1}{2}$ baryon octet are shown in table 1 together with the corresponding experimental masses. The predicted masses compare rather well with the experimental data. Table 1 shows also the sum of the relevant constituent quark masses. One notices an appreciable binding in both the diquark and the baryon.

The obtained results are rather encouraging and call for more detailed and systematic studies of baryon properties within the effective hadron theory derived in ref. [11]).

5. The effective hadron action in the large N_c limit: baryons as chiral solitons

The Fierz transformation (2) was performed for $N_c = 3$ colours. If one performs this transformation for an arbitrary number of colours N_c one finds that the diquark interaction $(qq)_{3_c}$ is suppressed as $\frac{1}{N_c}$ compared to the quark–antiquark interaction $(q\bar{q})$. Thus in the limit $N_c \to \infty$ there are no diquark correlations and consequently no diquark and baryon fields. The hadronization approach reduces then to the effective meson theory (13) and baryons show up as solitons of the meson fields. At low energies the mesons will keep their vacuum values

Figure 3: (a) The unpertubated baryon propagator describing the independent propagation of a quark and a diquark. By construction of the diquark propagator this diagram includes already the diagram shown in (b). (c) Exchange diagram to (b).

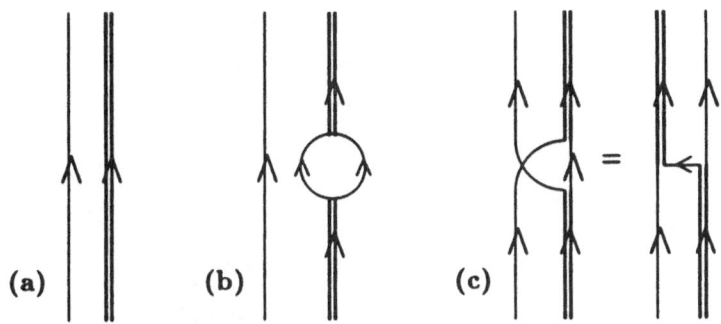

Figure 4: Exchange diagrams to the leading order diagram shown in fig. 3 (a).

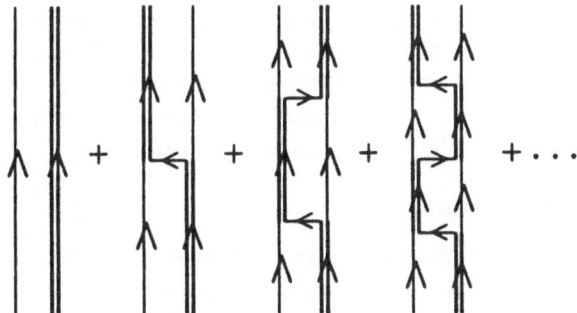

$\phi = \phi_0 = m$, except for the pseudoscalar mesons which are the (would-be for zero current quark masses) Goldstone bosons of spontaneous breakdown of chiral symmetry. If we keep only the pseudoscalar mesons $\pi(x)$ in their non-linear realization, i.e. the chiral field

$$U(x) = exp(i\pi(x)/f_\pi) , \qquad (26)$$

the effective action is uniquely given by the quark loop [3]

$$S[U] = Trlog(i \not{\partial} - mU(x)^{\gamma_5}) . \qquad (27)$$

This action has nontrivial topological soliton solutions $U(x)$ found in ref. [6] and [7] for two flavours. If this classical soliton is quantized in the full $SU_f(3)$ space one finds the mass splittings of the $s = \frac{1}{2}$ and the $s = \frac{3}{2}$ baryons octet shown in table 2 [19] (similar calculations were also performed in ref. [20]). The

Figure 5: Diagrammatic repesentation of the Fadeev equation and its static limit (b).

(a)

(b)

six mass splittings are equally well reproduced as in the quark-diquark picture but the absolute masses come out to large. This can be traced back to the neglect of dynamical meson fluctuations around the classical soliton, which give order $\frac{1}{N_c}$ correction and lower drastically the absolute baryon masses, as similar calculations within the Skyrme model indicate.

6. Concluding remarks

In the actual calculations the three valence quarks of the baryons were considered on top of the mesonic vacuum field configuration $\phi = \phi_0$. One expects however, that the valence quarks of the hadrons will locally disturb the vacuum expectation value of the meson field ϕ_0 as we have seen in the soliton picture. This suggests that one should try to solve for the baryon propagator in an arbitrary meson field configuration ϕ and fix the meson configuration only afterwards by minimizing the action of the valence baryon on top of the vacuum

$$S[\phi, \psi] = S[\phi] + \int d^4x \, \bar{\psi} G_B^{-1}[\phi]\psi \qquad (28)$$

where ψ reprensents now the baryon wave function obtained from the solution of the Faddeev equation. This would lead to non-trivial localized mesonic field configuration in the vicinity of the baryon. In particular the chiral meson field could develop solitonic configurations of the same type as they occur in the chiral soliton picture of the baryons [4,5,6]. The upshot would be a hybrid model which describes baryons as bound states of a quark and a diquark in a chiral soliton field configuration. Such a model would combine the advantages of both the chiral soliton and the valence quark-diquark picture of baryons. Unfortunately such calculations are exceedingly difficult.

Table 1: The masses of the $s = \frac{1}{2}^+$ baryon octet calculated with a constituent quark mass m^u=400 MeV.

		$\sum_i m^i$	m_b(calc.)	m_b(exp.)
p	(uud)	1205	938	938
n	(udd)	1211	938	939
Σ^+	(uus)	1445	1205	1189
Σ^-	(dds)	1455	1214	1197
Ξ^0	(uss)	1690	1292	1315
Ξ^-	(dss)	1695	1300	1321
Σ^0	(uds)	1450	1209	1193
Λ^0	(uds)	1450	1068	1116

Table 2: The mass differences of the $s = \frac{1}{2}^+$ and $s = \frac{3}{2}^+$ baryon states calculated in the chiral soliton picture.

$M_u = 390 MeV$	Λ	Σ	Ξ	Δ	Σ^*	Ξ^*	Ω
calculated	175.	248.	396.	291.	449.	608.	765.
Expt.	177.	254.	379.	293.	446.	591.	733.

Acknowlegdement: Discussions with A. Buck are grately acknowledged.

References

1.) G. 'tHooft, Nucl. Phys. **B72** (1974)461
2.) E. Witten, Nucl. Phys. **B160** (1979)57
3.) D. Ebert and H. Reinhardt,
     Nucl. Phys. **B271** (1986)188; Phys. Lett. **B173** (1986)453
4.) H. Reinhardt and B.V. Dang, Nucl. Phys. **A500** (1989)563
5.) H. Skyrme, Nucl. Phys. **31** (1962)556
6.) H. Reinhardt and R. Wünsch, Phys. Lett. **B215** (1988)577;
     Phys. Lett. **B230** (1989)93;
     R. Alkofer and H. Reinhardt, Phys. Lett. **B244** (1990)46
7.) Th. Meissner, F. Gümmer and K. Goeke, Phys. Lett. **B227** (1989)296
8.) R. Alkofer, Phys. Lett. **B236** (1990)310
9.) H. Reinhardt, Phys. Lett. **B188** (1987)263
10.) J. D. Walecka, Ann. Phys. **83** (1974)491;

B. D. Serot and J. D. Walecka, Adv. Nucl. Phys. vol **16** (1985)

11.) H. Reinhardt, Phys. Lett. **B244** (1990)316

12.) B.-O. Skytt, Diquarks, The Royal Inst. Technology, Stockholm
Report TRITA-TFY-87-04. P. Kroll, this meeting
R.T. Cahill, C.D. Roberts and J. Prashifka, Phys. Rev. **D36** (1987)2804;
B. Stech, Phys. Rev. **D36** (1987)975;
U. Vogl and W. Weise, Progress in Particle and Nuclei, to appear

13.) M. Schaden, H. Reinhardt, P. Amundsen and M. Lavelle,
Nucl. Phys. **B339** (1990)595;
H. Reinhardt, Phys. Lett. **B248** (1990)365; Phys. Lett. **B257** (1991)375

14.) N. Kawamoto and J. Smit, Nucl. Phys. **B192** (1981)100

15.) R. T. Cahill, J. Praschifka and C. J. Burden, Aust. J. Phys. **42** (1989)161

16.) R. Ball, Phys. Lett. **B245** (1990)213

17.) A. Buck, R. Alkofer and H. Reinhardt, Phys. Lett. **B**, in press

18.) H. Reinhardt, Nucl. Phys. **A 503** (1989)825

19.) H. Weigel, R. Alkofer and H. Reinhardt, Phys. Lett. **B284** (1990)296,
subm. to Nucl. Phys. B

20.) A. Blotz, D. Diakonov, K.Goeke, N.W. Park, V. Petrov
and P.V. Probylitsa, Bochum preprint 1992

21.) D. Lichtenstein, this meeting

Strange Quarks in Semibosonized SU(3) Nambu–Jona-Lasinio Model

M. Praszałowicz[1,2], A. Blotz[2], D.I. Diakonov[3],
K. Goeke[2], V.Yu. Petrov[3] and P.V. Pobylitsa[3]

[1] A. von Humboldt Stiftung Fellow, on leave from the Jagellonian University,
    Reymonta 4, 30–059 Kraków, Poland
[2] Institute of Theor. Phys. II, Ruhr-University Bochum, D–4630 Bochum, Germany
[3] St.Petersburg Nucl. Phys. Inst., St.Petersburg-Gatchina 188350, Russia

Abstract. In a semibosonized SU(3) Nambu–Jona-Lasinio model explicit symmetry breaking mass terms are introduced. The soliton solution corresponding to an isospin embedding of the SU(2) *hedgehog* in the SU(3) group is studied. The hyperon splittings and the hadronic part of the isospin mass differences are calculated and a good agreement with experiment is found.

Keywords. Nambu–Jona-Lasinio model, chiral symmetry, soliton, hyperon splittings, isospin splittings.

1   Introduction

It is now widely believed that effective models, like the Nambu–Jona-Lasinio (NJL) model can be derived from QCD [1]. In this talk we will extend the SU(2) semibosonized model, discussed by Meißner [2], in which constituent quarks are coupled to a classical pseudoscalar field, to the 3 flavor case. We will show that not only hyperon splittings, but also isospin breaking effects are accounted for with good accuracy. This work is based on refs[3, 4] and [5] where all details can be found. We therefore do not discuss here neither the details of the mesonic sector, nor the details of the regularization procedure.

In the semibosonised NJL model the energy of a given baryon consists of a *classical* and *quantum* part. The *classical* part, *i.e.* the energy of the soliton, is exactly the same as in the SU(2) model. It gets contributions from the *valence*, as well as from the *sea* quarks. The *quantum* corrections are calculated by adiabatical rotation of the soliton resulting in a hamiltonian analogous to the one of a symmetric top. Then corrections linear in the quark mass matrix are calculated. A novelty is due to the terms linear in current quark mass and in rotational velocities. These terms vanish in an ordinary Skyrme model; in the present model they get main contribution from the *valence* part. The resulting spectrum fits the data with a few percent accuracy.

2   Gell-Mann–Okubo mass formulae

Any dynamical model of low lying hadrons has to reproduce and also explain the Gell-Mann–Okubo [6, 7] mass formulae which are derived assuming that the SU(3) breaking mass operator ΔM transforms like a $Y = 0$, $I = 0$ and $I_3 = 0$ component of the octet tensor operator. Then, due to the Wigner–Eckhart theorem, matrix elements of this operator are given by:

$$\Delta M_B^{(8)} = F \left(\begin{array}{ccc} 8 & 8 & 8_- \\ 000 & B & B \end{array} \right) + D \left(\begin{array}{ccc} 8 & 8 & 8_+ \\ 000 & B & B \end{array} \right) \tag{1}$$

for the octet, and

$$\Delta M_B^{(10)} = C \left(\begin{array}{ccc} 8 & 10 & 10 \\ 000 & B & B \end{array} \right) \tag{2}$$

for the decuplet. The reduced matrix elements F, D and C are free parameters. Symbols in brackets denote SU(3) Clebsch–Gordan coefficients and $B = Y, I, I_3$ for the baryon in question. These Clebsch–Gordan coefficients can be written in terms of the diagonal SU(3) operators [8]:

$$\Delta M_B^{(8)} = -\frac{F}{2}\mathbf{Y} - \frac{D}{\sqrt{5}}\left(1 - \mathbf{I}^2 + \frac{1}{4}\mathbf{Y}^2\right), \qquad \Delta M_B^{(10)} = -\frac{C}{2\sqrt{2}}\mathbf{Y}. \tag{3}$$

The predictive power of the above equations consists in the fact, that the number of free parameters, which could in principle parametrize the mass splittings, is reduced from 3 to 2 for the octet and to 1 for the decuplet.

Equations (3) yield the relations:

$$F = M_\Xi - M_N \tag{4}$$

$$\frac{1}{\sqrt{5}}D = \frac{1}{2}(M_\Sigma - M_\Lambda) = \frac{1}{3}(2M_\Sigma - M_\Xi - M_N) = M_\Xi + M_N - 2M_\Lambda,$$

for the octet, and equal level spacing for the decuplet:

$$\frac{1}{2\sqrt{2}}C^{(10)} = M_{\Sigma^\bullet} - M_\Delta = M_{\Xi^\bullet} - M_{\Sigma^\bullet} = M_\Omega - M_{\Xi^\bullet}. \tag{5}$$

From these relations we can estimate the values of parameters F, D and C:

$$F = 379, \quad D = 79 \pm 17 \text{ and } C = 415 \pm 15 \text{ MeV}. \tag{6}$$

The mass splittings of the baryons belonging to the same isospin multiplet consist of two parts: hadronic and electromagnetic [9], namely:

$$\Delta m_B = (\Delta m_B)_h + (\Delta m_B)_e. \tag{7}$$

If the isospin breaking is assumed to be driven by an octet isovector tensor operator corresponding to $I_3 = 0$ then, in analogy with the Gell-Mann–Okubo mass formulae (1,2), one gets:

$$(\Delta m_B^{(8)})_{\text{h}} = \frac{1}{\sqrt{3}} f \left(\begin{array}{ccc} 8 & 8 & 8_- \\ 010 & B & B \end{array} \right) + \sqrt{\frac{5}{3}} d \left(\begin{array}{ccc} 8 & 8 & 8_+ \\ 010 & B & B \end{array} \right) \qquad (8)$$

for the octet, and

$$(\Delta m_B^{(10)})_{\text{h}} = \sqrt{\frac{2}{3}} c \left(\begin{array}{ccc} 8 & 10 & 10 \\ 010 & B & B \end{array} \right) \qquad (9)$$

for the decuplet (normalization factors in front of the f, d and c coefficients are chosen for future convenience). Evaluating the SU(3) Clebsch-Gordan coefficients gives [8]:

$$(\Delta m_B^{(8)})_{\text{h}} = -\frac{1}{3} f\, \mathbf{I}_3 + d\, \mathbf{Y}\, \mathbf{I}_3, \qquad (\Delta m_B^{(10)})_{\text{h}} = -\frac{1}{3} c\, \mathbf{I}_3. \qquad (10)$$

Electromagnetic part of the isospin splittings was estimated by Gasser and Leutwyler [9] for the octet. Their estimate confirms a reasonable assumption that $\Sigma^- - \Sigma^+$ mass difference has no electromagnetic contribution. This assumption allows us to determine coefficient f, and also c for the decuplet, where no estimate of the electromagnetic part of $\Sigma^{*-} - \Sigma^{*+}$ exists. Coefficient d can be determined from the hadronic part of the n–p mass difference [5]. Altogether we get:

$$f = 12.11 \pm 1.14 , \quad d = 1.73 \pm 0.38 \quad \text{and} \quad c = 6.6 \pm 1.0 \quad \text{MeV.} \qquad (11)$$

Symmetry considerations alone are not able to provide us with any relations between the reduced matrix elements. Dynamical models, like NJL model, make specific predictions for these constants. In the next sections we will calculate coefficients F, D, C, f, d, and c within the framework of the semibosonized SU(3) NJL model.

3 Semibosonized SU(3) Nambu–Jona-Lasinio model

It was already shown in the talk of Meißner [2] that the solitonic solutions of the semibosonized NJL model are studied in terms of an effective action:

$$S_{\text{eff}} = -\text{Sp} \log(i\partial\!\!\!/ - m - M\, U^{\gamma_5}). \qquad (12)$$

This time, however, U is an SU(3) matrix:

$$U = A(t) \left[\begin{array}{cc} \overline{U}_0(\vec{x}) & 0 \\ 0 & 1 \end{array} \right] A^\dagger(t), \qquad (13)$$

where $\overline{U}_0$ is the SU(2) *hedgehog* and M is the constituent quark mass, which is in fact the only free parameter of the model. The bare quark mass matrix can be written in a form:

$$m = \mu_0 \lambda_0 - \mu_8 \lambda_8 - \mu_3 \lambda_3, \qquad (14)$$

where λ_i are Gell-Mann SU(3) matrices $(\lambda_0 = \sqrt{2/3}\,\mathbf{1})$ and

$$\mu_0 = \frac{1}{\sqrt{6}}(m_u + m_d + m_s), \quad \mu_8 = \frac{1}{\sqrt{12}}(2\,m_s - m_u - m_d), \quad \mu_3 = \frac{1}{2}(m_d - m_u).$$

$$(15)$$

The energy of the soliton consists of two parts: the energy of the *continuum*, i.e. the energy corresponding to the effective action (12), and the energy of the *valence* level [2]. In what follows, for simplicity, we confine our discussion to the *continuum* part only, but one always has to remember that the pertinent *valence* contribution has to be added.

The effective action (12) can be rewritten in terms of the Euclidean spectral representation [3, 4]:

$$S_{\text{eff}} = -N_c T \int \frac{d\omega}{2\pi} \text{Tr} \log (i\omega + H) \left[1 + \frac{1}{i\omega + H}(-i\gamma_4 A^\dagger m A + A^\dagger \dot{A}) \right], \quad (16)$$

where H is the hermitean static hamiltonian: $H = \gamma_4 (\gamma_i \partial_i + M U_0)$. The static soliton solution for H reduces to the one found in the SU(2) case. Formula (16) is already written in a form ready to be expanded in a power series in m and in generalized velocities $A^\dagger \dot{A} = i/2\, \Omega_a \lambda_a$. Let us for the moment forget about the mass matrix m and expand (16) in powers of Ω. We get (back in Minkowski metric):

$$L_{\text{rot}} = \frac{1}{2} I_{ab} \Omega_a \Omega_b - \frac{N_c}{2\sqrt{3}} \Omega_8, \qquad (17)$$

where the tensor of inertia I_{ab} is diagonal

$$I_{ab} = \frac{N_c}{4} \int \frac{d\omega}{2\pi} \text{Tr} \left[\frac{1}{i\omega + H} \lambda_a \frac{1}{i\omega + H} \lambda_b \right] = \begin{cases} I_1 \delta_{ab} & \text{for } a, b = 1...3 \\ I_2 \delta_{ab} & \text{for } a, b = 4...7 \\ 0 & \text{for } a, b = 8 \end{cases} \quad (18)$$

but not invertible. (The above expression is assumed to be properly regularized and the full moments of inertia have also a *valence* part.)

The quantization of the rotational lagrangian (17) proceeds exactly as in the case of the Skyrme [10] model. As a result one arrives at the hamiltonian:

$$H_{\text{rot}} = \frac{S(S+1)}{2I_1} + \frac{C_2(\text{SU}(3)) - S(S+1) - \frac{N_c^2}{12}}{2I_2} \qquad (19)$$

whose eigenfunctions are given in terms of Wigner D matrices for SU(3) representation R:

$$\psi(A) = \sqrt{\dim(R)}\, D_{ab}^{(R)}(A) \equiv \sqrt{\dim(R)}\, \langle Y, I, I_3 \mid D^R(A) \mid Y_R, S, -S_3 \rangle. \quad (20)$$

Here S denotes baryon spin and the right hypercharge Y_R is a subject to a constraint:

$$Y_R = \frac{N_c}{3}. \tag{21}$$

The new part consists in the expansion in powers of the rotated matrix m:

$$L_m = -\sigma[\sqrt{6}\mu_0 - \sqrt{3}(\mu_8 D_{88}^{(8)} + \mu_3 D_{38}^{(8)})] - 2(\mu_8 D_{8a}^{(8)} + \mu_3 D_{3a}^{(8)})K_{ab}\Omega_b, \tag{22}$$

where constant σ

$$i\frac{N_c}{4}\int\frac{d\omega}{2\pi}\mathrm{Tr}\left[\frac{1}{i\omega + H}\gamma_4\lambda_a\right] = \begin{cases} \sqrt{6}\,\sigma & \text{for} \quad a = 0 \\ \sqrt{3}\,\sigma & \text{for} \quad a = 8 \\ 0 & \text{for} \quad a = 1...7 \end{cases} \tag{23}$$

is related to the pion-nucleon sigma term $\Sigma = 3/2(m_u + m_d)\sigma$ and *anomalous* tensor K_{ab} is defined as:

$$K_{ab} = i\frac{N_c}{4}\int\frac{d\omega}{2\pi}\mathrm{Tr}\left[\frac{1}{i\omega + H}\gamma_4\lambda_a\frac{1}{i\omega + H}\lambda_b\right] = \begin{cases} K_1\delta_{ab} & \text{for } a,b = 1...3 \\ K_2\delta_{ab} & \text{for } a,b = 4...7 \\ 0 & \text{for } a,b = 8 \end{cases}. \tag{24}$$

We call K_{ab} *anomalous* since it comes from the imaginary part of the effective action, which is related to anomaly, and as such does not require regularization. In fact K_{ab} gets contribution almost entirely from the *valence* level.

The quantized hamiltonian gets two new pieces corresponding to L_m: one which shifts all masses by a constant and another one which splits the spectrum:

$$\begin{aligned} H_{br} &= \sqrt{3}\left(-\sigma + \frac{K_2}{I_2}\right)(\mu_8 D_{88}^{(8)}(A) + \mu_3 D_{38}^{(8)}(A)) \\ &+ 2\left(\frac{K_1}{I_1} - \frac{K_2}{I_2}\right)\sum_{a=1}^{3}(\mu_8 D_{8a}^{(8)}(A) + \mu_3 D_{3a}^{(8)}(A))S_a \\ &- \frac{2K_2}{I_2}\left(\mu_8\frac{\sqrt{3}}{2}Y + \mu_3 I_3\right), \end{aligned} \tag{25}$$

where index 8 corresponds to $Y = 0$, $I = 0$, $I_3 = 0$ and index 3 to $Y = 0$, $I = 1$, $I_3 = 0$.

In the next section the expectation values of the hamiltonian H_{br} are calculated. We adopt the following numerical procedure: first we find the solitonic solution for a range of constituent masses M, then we find the optimal value of M which reproduces the octet-decuplet splitting due to the rotational hamiltonian H_{rot} (19). It turns out [3, 4] that $M = 390$ MeV and the corresponding moments of inertia take the following values: $I_1 = 1.25$, $I_2 = 0.59$, $K_1 = 0.42$ and $K_2 = 0.28$ fm, and $\sigma = 3.07$ [11]. Next we calculate the mass splittings as functions of the current quark masses and compare with experiment.

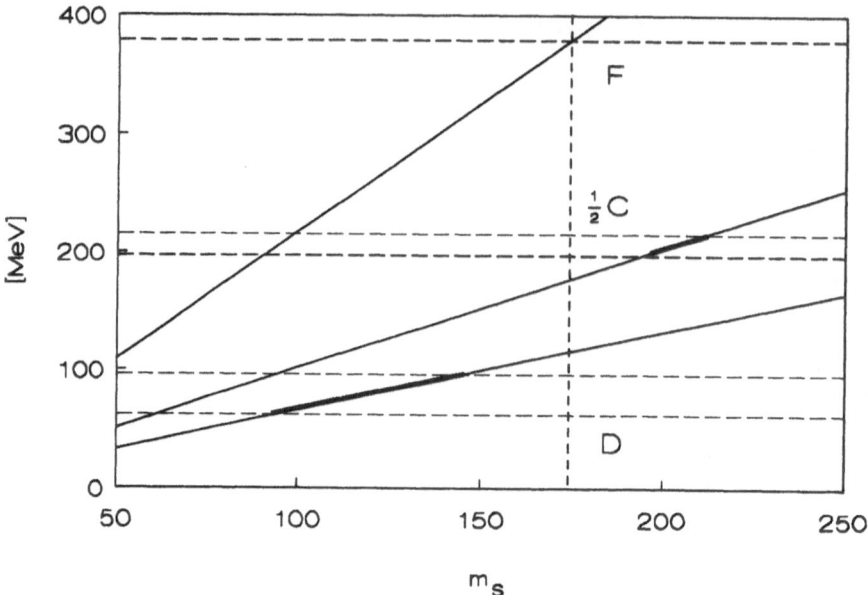

Figure 1: Constants F, D and $1/2\,C$. Solid lines represent model predictions as functions of m_s. Dashed line correspond to the error bars of eq.(6).

4 Mass splittings

The expectation values of H_{br} between the baryonic wave functions (20) are easily expressed as products of two SU(3) Clebsch–Gordan coefficients [8]. One of them coincides with the Clebsch–Gordan coefficients encountered in eqs(1,2,8) and (9). Therefore the Gell-Mann–Okubo mass formulae are naturally reproduced. In order to calculate the reduced matrix elements let us define the following quantities:

$$\varphi = \sigma + 2\frac{I_2}{K_2} + \frac{I_1}{K_1}, \qquad \gamma = \sigma + 2\frac{I_2}{K_2} + 5\frac{I_1}{K_1}, \qquad \delta = \sigma + 2\frac{I_2}{K_2} - 3\frac{I_1}{K_1}. \quad (26)$$

Then we get:

$$F = \frac{1}{2}\,\varphi\,m_{\text{s}}, \qquad D = \frac{1}{2\sqrt{5}}\,\delta\,m_{\text{s}}, \qquad C = \frac{1}{2\sqrt{2}}\,\gamma\,m_{\text{s}}, \quad (27)$$

$$f = \frac{3}{4}\,\varphi\,(m_{\text{d}} - m_{\text{u}}), \qquad d = \frac{3}{20}\,\delta\,(m_{\text{d}} - m_{\text{u}}), \qquad c = \frac{3}{8}\,\gamma\,(m_{\text{d}} - m_{\text{u}}). \quad (28)$$

In figs 1 and 2 we plot the splitting constants of eqs(27,28) as functions of m_{s} and $m_{\text{d}} - m_{\text{u}}$ respectively together with the error bars corresponding to eqs(6,11). For $m_{\text{s}} \approx 175$ MeV the N–Ξ splitting (i.e. constant F) is reproduced. For the same value of m_{s} constant D corresponding to Σ–Λ splitting is overestimated by 35 MeV, whereas constant C is underestimated by 60 MeV. This means that Λ and Σ are displaced by about 20 Mev with respect to their experimental values, whereas splittings in the decuplet are approximately 20 MeV

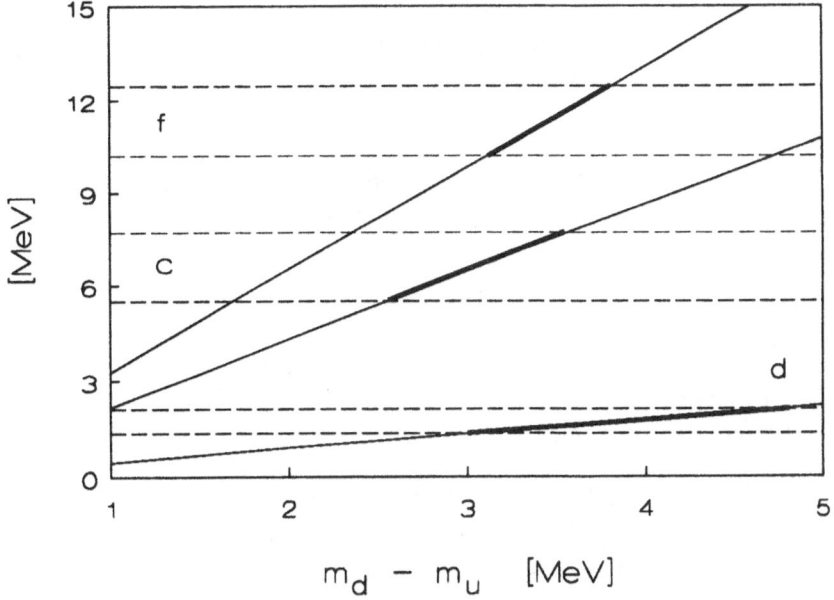

Figure 2: Constants f, d and c. Solid lines represent model predictions as functions of $m_d - m_u$ mass difference. Dashed line correspond to the error bars of eq.(11).

smaller than in nature. On the other hand the isospin breaking constants f, d and c are reproduced within experimental errors for $m_d - m_u \approx 3.5$ MeV.

The present model makes specific prediction for the ratio of hadronic to isospin breaking constants:

$$\frac{f}{F} \ (2.99 \pm 0.3) = \sqrt{5} \frac{d}{D} \ (4.90 \pm 2.15) = \sqrt{2} \frac{c}{C} \ (2.25 \pm 0.42 \quad \times 10^{-2}), \quad (29)$$

where the numbers in brackets correspond to the experimental values of eqs(6,11). Certainly the central values are fairly scattered. We would like to offer the following explanation of this discrepancy. The isospin splittings are proportional to a tiny parameter, namely $m_d - m_u$ and therefore the first order of the perturbation theory is legitimate. On the contrary, for the hyperon splittings which are proportional to the much larger parameter, namely strange quark mass, one may expect some corrections from the higher order of the perturbative expansion in m_s. Indeed, already the second order brings the splittings to their experimental values with an accuracy of a few MeV. A fully consistent incorporation of higher order effects would, however, require to abandon the *hedgehog* Ansatz.

To summarize: we have studied the symmetry breaking effects due to the quark masses in the solitonic sector of the semibosonized NJL model. A satisfactory description of the hadronic mass spectrum including both the hyperon and the isospin splittings was found. The absolute masses are too big, but there exist several mechanisms which may bring them down, namely gluon corrections [12], rotational and translational band subtraction [4] and Casimir energies of quantum fluctuations [13].

This work was partly sponsored by *Polish Research Grant 2.0091.91.01* (MP), by *Graduiertenstipendium des Landes NRW* (AB) and by *Deutsche Forschungsgemainschaft* together with *Russian Academy of Sciences* (D.I.D., V.Yu.P. and P.V.P.).

References

[1] D.I. Diakonov, Diquarks in the instanton picture, talk at this conference and references therein.

[2] Th. Meißner, The Nambu–Jona-Lasino Model..., talk at this conference and references therein.

[3] A. Blotz, D. I. Diakonov, K. Goeke, N. W. Park, V. Petrov, and P. V. Pobylitsa, Strange baryons in the solitonic sector of the Nambu–Jona-Lasinio model, Bochum Univ. preprint RUB-TPII-16/92, 1992, Phys. Lett. B in print.

[4] A. Blotz, D. I. Diakonov, K. Goeke, N. W. Park, V. Petrov, and P. V. Pobylitsa, The SU(3)-Nambu–Jona-Lasinio soliton in the collective quantization formulation, submitted to Nucl. Phys. A, 1992.

[5] M. Praszałowicz, A. Blotz, and K. Goeke, Isospin baryon mass differences in SU(3) Nambu–Jona-Lasinio model, Bochum Univ. preprint RUB-TP2-20/92, 1992.

[6] M. Gell-Mann, *Phys. Rev.*, **125** (1962) 1067.

[7] S. Okubo, *Progr. Theor. Phys.*, **27** (1962) 349.

[8] J. J. de Swart, *Rev. Mod. Phys.*, **35** (1963) 916.

[9] J. Gasser and H. Leutwyler, *Phys. Rep.*, **87** (1982) 77.

[10] E. Guadagnini, *Nucl. Phys.*, **B236** (1984) 35.

[11] Z. Duliński, M. Praszałowicz, and P. Sieber, H dibaryon in chiral quark model, Bochum Univ. preprint RUB-TP2-19/92, 1992.

[12] D. I. Diakonov, J. Jaenicke, and M. V. Polyakov, Gluon exchange corrections to the nucleon mass in the chiral theory, St. Petersburg preprint 1738, 1991.

[13] B. Moussalam and D. Kalafatis, *Phys. Lett.*, **B272** (1991) 196.

The Nambu–Jona-Lasinio Model with Scalar, Pseudoscalar, Vector and Axialvector Mesons: Solitonic Solutions and Nucleon Properties

Th.Meißner[2], C.Schüren[1], F.Döring[1], E.Ruiz Arriola[3], K.Goeke[1]

[1] Institut für Theoretische Physik II, Ruhr-Universität Bochum, D-4630 Bochum 1, Germany
[2] Institut for Nuclear Theory, University of Washington, HN-12, Seattle, WA 98195, USA
[3] Departamento de Fisica Moderna, Universidad de Granada, E-18071 Granada, Spain

Abstract: We solve the NJL model with $SU(2)$-flavour and σ and $\vec{\pi}$- couplings for a baryonic system consisting of $N_c = 3$ valence quarks and the polarized Dirac sea. Solitonic solutions are found, from which various nucleon observables can be calculated. For the first time we present also solitonic solutions of an extended NJL model including ω, $\vec{\rho}$ and $\vec{A}_1$ mesonic couplings.

1. The NJL Model with Scalar and Pseudoscalar Couplings

1.1. LAGRANGEAN AND FIXING OF THE PARAMETERS

We consider the Nambu–Jona-Lasinio (NJL) model [1] for $SU(2)$-flavour (u and d) quarks with a small current mass $m_0 = \frac{1}{2}(m_u + m_d)$ and restrict ourselves first only on scalar- isoscalar (σ) and pseudo scalar-isovector ($\vec{\pi}$) couplings. In the semi bosonized form [2] the Lagrangean reads:

$$\mathcal{L} = \bar{q}i\partial\!\!\!/q - g_\pi \bar{q}(\sigma + i\vec{\tau}\vec{\pi}\gamma_5)\, q - \frac{\mu^2}{2}\left(\sigma^2 + \vec{\pi}^2\right) + \frac{\mu^2}{g}m_0\sigma \qquad (1.1)$$

The meson fields σ and $\vec{\pi}$, which are pure auxiliary fields without explicit kinetic term, are treated classically (stationary phase approximation). Performing the Grassmann integration over the quark fields q and $\bar{q}$ in Euclidean space time one arrives at the effective action consisting of the fermion determinant and the quadratic mesonic mass as well as an explicit chiral symmetry breaking term:

$$S_{eff}[\sigma, \vec{\pi}] = -\mathrm{Sp}_{reg} \ln\left(-i\partial\!\!\!/ + g_\pi(\sigma + i\vec{\tau}\vec{\pi}\gamma_5)\right)$$
$$+ \frac{\mu^2}{2}\int d^4x\left(\sigma^2 + \vec{\pi}^2\right) - f_\pi m_\pi^2 \int d^4x\,\sigma \qquad (1.2)$$

The total trace $\mathrm{Sp}\,\mathcal{A} = N_c \int d^4x\,\mathrm{Tr}_\gamma\mathrm{Tr}_\tau\langle x|\mathcal{A}|x\rangle$ has to be regularized by some UV-cutoff Λ. For our actual numerical calculations we will use for convenience the Schwinger proper time method [3], which consists in regularizing the logarithm:

$$\ln\alpha = -\int_{\frac{1}{\Lambda^2}}^{\infty} \frac{d\tau}{\tau} e^{-\tau\alpha} \qquad (1.3)$$

The equation of motions for the vacuum configurations of the σ field $\frac{\delta S_{eff}[\sigma,\vec{\pi}]}{\delta\sigma} = 0$ (gap equation) can be solved for a finite vacuum value σ_V, which realizes the spontaneous breakdown of chiral symmetry and gives the quarks a finite constituent mass $M = g_\pi\sigma_V$. We now can fix the 4 parameters g_π, Λ, μ^2 and m_0 by adjusting the pion decay constant f_π and the pion mass m_π to their experimental values $f_\pi = 93\text{MeV}$ and $m_\pi = 139\text{MeV}$, respectively, and by demanding that $\sigma_V = f_\pi$.

In order to calculate f_π and m_π one has to expand the effective action $S_{eff}[\sigma,\vec{\pi}]$ up to second in the meson fields around their vacuum values [4]. Because the pion mass is very small one can use for this the gradient or heat kernel expansion corresponding to the soft pion limes.

Fixing the parameters in this way we are left with only one free parameter for the baryonic sector, for which we take for convenience the constituent quark mass $M = g_\pi f_\pi$.

1.2. Systems with Baryon Number $B = 1$ - Solitonic Solutions

For time independent (static) meson field configurations the effective action determines beside a trivial Euclidean time interval T just the total energy of the system: $S_{eff}[\sigma,\vec{\pi}] = TE$. In order to obtain a system with baryon number $B = 1$ one has to add $N_c = 3$ valence quarks [5, 6] which can be formally done by introducing a thermochemical potential μ and using the grand canonical effective action [7]. Explicitly one has:

$$E_{B=1} = E_{mes} + E_{br} + E_{sea} + \eta_{val}\epsilon_{val} \tag{1.4}$$

where E_{mes}, E_{br} and E_{sea} denote the quadratic mesonic, the chiral breaking and the regularized sea parts of the energy, respectively, and

$$\eta_{val} = \begin{cases} 1 & \text{if } \epsilon_{val} \geq 0 \\ 0 & \text{if } \epsilon_{val} < 0 \end{cases} \tag{1.5}$$

This means that we are dealing with a valence quark picture ($\eta_{val} = 1$), if the valence particle is a positive (bound) state. On the other hand, if the valence particle has got part of the negative spectrum ($\eta_{val} = 0$), it is not taken explicitly into account. This corresponds to a fully bosonized picture, where the baryon number of the system is carried by the Dirac sea and equal to the topological winding number of the pion field like it is e.g in the Skyrme model. The important point is that by solving the equations of motion for the $B = 1$ system:

$$\frac{\delta E_{B=1}}{\delta\sigma} = 0 \qquad \frac{\delta E_{B=1}}{\delta\vec{\pi}} = 0 \tag{1.6}$$

for a given constituent quark mass M the system can decide by itself if it wants to adopt $\eta_{val} = 1$ or $\eta_{val} = 0$, unlike other effective chiral models, which work from the beginning with (like e.g. the Gell-Mann–Levi chiral sigma model [8,9]) or without valence quarks (like e.g. the Skyrme model [10,11]). The equations of motions can be solved by a self consistent Hartree like procedure [12,13] or by using a minimization method in a multidimensional space [14].

Using the hedgehog Ansatz for σ and $\vec{\pi}$, which can be shown to be a self consistent symmetry [15] and restricting σ and $\vec{\pi}$ to the chiral circle $\sigma^2 + \vec{\pi} = f_\pi^2$ (non linear model), one finds solitonic solutions, if $M > M_{cr} \approx 350\text{MeV}$ [13, 16]. One has $\epsilon_{val} > 0$ for $M < 700\text{MeV}$ (weak quark-meson coupling g_π) and $\epsilon_{val} < 0$ for $M > 700\text{MeV}$ (strong quark-meson coupling g_π) [13, 16]. Furthermore it turns out that in order to get a reasonable nucleon size (characterized e.g by the isoscalar quadratic radius) the values for M are restricted to a region of $M \approx 400\text{MeV}$, where one obtains a clear valence quark picture. The problems, which arise, if one leaves the restriction to the chiral circle are discussed in detail in refs. [14, 17].

1.3. CRANKING - NUCLEON OBSERVABLES

In order to obtain nucleon and Δ states with good spin and isospin quantum numbers, we use the well known perturbative cranking approach [11]. It consists in performing an adiabatic
(iso-)rotation of the chiral hedgehog field $U_5 = \sigma + i\vec{\tau}\vec{\pi}\gamma_5$ with the cranking frequency $\vec{\Omega}$ by:

$$U_5(\vec{r}) \rightarrow \tilde{U}_5(t, \vec{r}) = e^{it\vec{\tau}\vec{\Omega}} U_5(\vec{r}) e^{-it\vec{\tau}\vec{\Omega}} \qquad , \qquad (1.7)$$

inserting this into the expression for the effective action (1.2) and expanding up to 2nd (more precisely the first non-vanishing) order in Ω. The corresponding inertial parameter is the isorotational moment of inertia Θ, which is given as the properly regularized sum over all particle-hole matrix elements of the isospin operator $\vec{\tau}$ (Inglis formula) [18, 19, 20]. Care has to be taken in applying the regularization in Euclidean space time, because Ω has to be Wick rotated as the time component of a 4-vector [18]. The (iso-) rotational degrees of freedom can now be treated in the standard way by semiclassical collective quantization [11]. One proceeds in the same manner by calculating the expectation value of an arbitrary current $j_a^\mu = \bar{q}\Gamma_a^\mu q$ with the spin-isospin operator Γ [15]:

$$<j_a^\mu> = \frac{d}{d\gamma}\text{Spln}\left[-i\slashed{\partial} + g_\pi(\sigma + i\vec{\tau}\vec{\pi}\gamma_5) - \gamma\Gamma_a^\mu\right]|_{\gamma=0} \qquad (1.8)$$

In order to get absolute values for the masses of the nucleon and the Δ one has to take into account the spurious zero point energies due to the (iso-)rotational and translational symmetries of the model, which are broken for the hedgehog Ansatz. It has to be stated, that up to now it is not clear, how to treat these zero modes consistently within an relativistic, UV-regularized action, but as a first approximation we can use the well known formulas from non relativistic many body theory [21], which consist in subtracting the *band hat energy* $\frac{<T^2>}{2\Theta}$ for the rotational and the *center of mass energy* $\frac{<P^2>}{2M}$ for the translational zero mode [22].

Following this prescription we have calculated: the masses of nucleon and Δ [22]; the nucleon Σ-term [16]; electric form factors and radii of proton and neutron [23]; the axial coupling g_A and axial form factor of the nucleon [15] and the πNN-coupling and form factor [16]. The main results can be summarized as follows:

1. Most of the observables are well reproduced within about $10-15\%$ deviation from their experimental values.

2. The axial vector coupling constant $g_A \approx 0.8$ turns out to be too small in comparison with its experimental value of 1.24 [15]. Nevertheless it should be stated that up to now no effective chiral model is able to reproduce this quantity correctly. The calculated values are either too small (≈ 0.8 in the Skyrme model [11] and the NJL) or too high (≈ 1.7 in the Gell-Mann-Levi chiral sigma model [9]).

3. The neutron electric form factor is off as well, which can be seen manifestly in the value for the neutron mean square electric charge radius $< R^2 >_n = -0.26\mathrm{fm}^2$ (exp.$-0.12\mathrm{fm}^2$) [23].

4. All observables are clearly dominated by the valence quarks, confirming a clear valence quark picture of the nucleon. The only exception is the neutron electric charge distribution, where for large distances only the sea quarks (corresponding to the pion tail) are present [23].

5. The zero point energies are with $\approx 100 MeV$ for the rotational and $\approx 300\mathrm{MeV}$ for the translational motion quite high, which makes the quantization procedure very questionable [22].

2. The NJL Model with ω, $\vec{\rho}$ and $\vec{A}_1$ Mesons

Up to now most calculations in the soliton sector of the NJL model have been restricted to scalar and pseudo scalar couplings. Only very recently calculations adding vectorial degrees of freedom have been performed [24, 25, 26]. We will present now the first investigations within the NJL model including all hedgehog mesons $\sigma, \pi, \omega, \rho$ and A_1. The inclusion of vector mesons allows to implement the concept of vector dominance. In addition experience with effective chiral models suggests that vector mesons change noticeably calculated the baryon properties [27]. It is therefore interesting to investigate the influence of vector mesons in the NJL model.

The bosonized NJL Lagrangean with σ, π, ω, ρ and A_1-meson reads

$$\mathcal{L}_{NJL}(x) = \bar{q} \left(i\partial\!\!\!/ - m_0 + g_\omega \psi\!\!\!/ + g_\rho \frac{\vec{\tau}}{2} \left(\vec{\rho}\!\!\!/ + \vec{A}\!\!\!/ \gamma_5 \right) - g_\pi \left(\sigma + i\gamma_5 \vec{\tau}\vec{\pi} \right) \right) q$$

$$- \frac{\mu^2}{2} \left(\sigma^2 + \pi^2 \right) + \frac{m_\omega^2}{2} \omega_\mu^2 + \frac{m_\rho^2}{2} \left(\vec{\rho}_\mu^2 + \vec{A}_\mu^2 \right)$$

where we have introduced in addition the two independent quark vector-meson coupling constants g_ω and g_ρ. In this context the A field plays the role of the chiral partner of the ρ field. The corresponding effective action in Euclidean space contains both a real and an imaginary part. The imaginary part arises from the Wick rotation of the time component of the vector and the axial vector fields and is UV convergent in contrast to the real part. Therefore we will not regularize it, which has the advantage that the anomalous sector is not affected by a cutoff.

In order to fix the parameters of the extended NJL model we calculate the meson propagators and the corresponding on-shell masses [4]. Introducing the physical values of the vector meson masses ($m_\rho = 770\text{MeV}$, $m_\omega = 783\text{MeV}$) we can in principle relate the two coupling constants g_ω and g_ρ to m_ω and m_ρ as well as to the constituent quark mass $M = g_\pi f_\pi$. Nevertheless it will turn out that in order to get solitonic solutions at all one has to allow g_ω to deviate from the value, which results from the fixing to m_ω and use it as a free parameter. Furthermore it should be stated, that the vector meson masses are high in comparison with the pion mass and therefore the validity of the gradient expansion is questionable.

The method how the real and the imaginary parts of the Euclidean action are calculated and especially their continuation back to Minkowski space is non trivial. It has been treated in detail in ref. [25] and we adopt its notation for the quark eigenvalues here. Like in the scalar sector we use the hedgehog Ansatz [27, 25, 26] for the meson fields, restrict σ and $\vec{\pi}$ to the chiral circle and look for the stationary points of the total energy.

In Tab. 1 we present the results of our calculations for various kinds of couplings. In all cases the constituent quark mass is $M = 340\text{MeV}$. One should mention that as a consequence of the $\pi - A$ mixing the UV-cutoff has a larger value when the A_1–meson is included.

mesons	g_ω	Λ [MeV]	E_{tot} [MeV]	$N_c\epsilon_{val}^{(+)}$ [MeV]	E_{sea} [MeV]	$< R^2 >$ [fm^2]	g_A
σ, π	-	654	1244	818	397	0.85	0.83
$\sigma, \pi, \rho, A, \omega$	0	847	1026	-385	547	0.13	0.10
	1.0	847	1040	290	615	0.55	0.52
	1.7	847	1359	688	912	1.51	0.74

Tab. 1: Properties of the NJL soliton with ω, $\vec{\rho}$ and $\vec{A}_1$ mesons in dependence of g_ω

The size of the soliton grows considerably with increasing g_ω due to the repulsive nature of the ω field. In principle, this effect is desired since the inclusion of the ρ and A mesons leads to a lowering of the soliton size. Unfortunately the attraction increases too strong and for the value of the ω coupling constant $g_\omega = 2.24$, no solitonic solution can be found. The experimental value of the isoscalar mean squared radius is reproduced for an ω coupling constant slightly above $g_\omega = 1.0$. It should be stressed that the sign of $\bar{\epsilon}_{val}$ (which is identical to the well known valence eigenvalue when the ω field is not included [25]) is strongly dependent on the different kinds of couplings which are involved. This implies that the quest for the validity of the valence picture in the vector meson sector can not be answered unambiguously up to now. Finally we present the values for the axial coupling constant g_A calculated directly from the corresponding matrix element. Like in the case discussed in chapter 1 it comes out too small compared to the experimental value.

REFERENCES

[1] J.Nambu, G.Jona-Lasinio *Phys. Rev.* **122** (1961) 345 ; **124** (1961) 246

[2] T.Eguchi *Phys. Rev.* **D 14** (1976) 2755

[3] J.Schwinger *Phys. Rev.* **82** (1951) 664

[4] M.Jaminon, R M.Galain, G.Ripka, P.Stassart *Nucl. Phys.* **A 537** (1992) 418

[5] D.Dyakonov, V.Petrov, P.V.Pobylitsa *Nucl. Phys.* **B 306** (1988) 809

[6] Th.Meißner, E.Ruiz Arriola, F. Grümmer, K. Goeke, H.Mavromatis *Phys. Lett.* **B 214** (1988) 312

[7] Th.Meißner, F. Grümmer, K. Goeke *Ann. of Phys.* **202** (1990) 297

[8] M. Gell-Mann, M. Lévy *Nuovo Cimento* 16 (1960) 705

[9] for a review see e.g. M.C.Birse *Prog. in Part. and Nucl. Phys.* **26** (1990) 1

[10] T.H.R.Skyrme *Nucl. Phys.* **31** (1962) 556

[11] G.S.Adkins, C.R.Nappi, E.Witten *Nucl. Phys.* **B 228** (1983) 552

[12] H.Reinhardt, R.Wünsch *Phys. Lett.* **B 215** (1988) 577

[13] Th.Meißner, F. Grümmer, K. Goeke *Phys. Lett.* **B 227** (1989) 296

[14] P.Sieber, Th.Meißner, F.Grümmer, K.Goeke *Nucl. Phys.* **B** (in press)

[15] Th.Meißner, K.Goeke *Z. Phys.* **A 339** (1991) 513

[16] Th.Meißner, K.Goeke *Nucl. Phys* A **524** (1991) 719

[17] Th.Meißner, G.Ripka, R.Wünsch, P.Sieber, F.Grümmer, K.Goeke preprint RUB-TPII-25/92 (subm. to *Phys. Lett.* **B**)

[18] H.Reinhardt *Nucl. Phys.* **A 503** (1989) 825

[19] K.Goeke, A.Gorski, F.Grümmer, Th.Meißner, H.Reinhard, R.Wünsch *Phys. Lett* **B 256** (1991) 321

[20] M.Wakamatsu, H.Yoshiki *Nucl. Phys.* **A 524** (1991) 561

[21] P.Ring, P.Schuck *The Nuclear Many Body Problem*, Springer 1980

[22] P.Pobylitsa, E.Ruiz Arriola, Th.Meißner, F.Grümmer, K.Goeke, W.Broniowski *Journ. of Phys.* G in press

[23] A.Gorski, F.Grümmer, K.Goeke *Phys. Lett.* **B 278** (1992) 24

[24] R.Alkofer, H.Reinhardt *Phys. Lett.* **B 244** (1990) 461

[25] C.Schüren, E.Ruiz Arriola, K.Goeke *Phys. Lett.* **B** in press

[26] F.Döring, E Ruiz Arriola, K.Goeke *Z. Phys* A in press

[27] For a review see U.G.Meißner *Phys. Rep.* **161** (1988) 213 and references therein.

An Effective Lagrangian Analysis of the U(1) Goldberger-Treiman Relation

Masashi Wakamatsu

Department of Physics, Faculty of Science, Osaka University,
Toyonaka, Osaka 560, JAPAN

Abstract. To get some insight into the role of the gluon field in the recent "proton spin crisis", the $U_A(1)$ Goldberger-Treiman relation is rederived on the basis of some specific effective lagrangians which incorporates the axial anomaly of QCD. This model analysis indicates "quark spin screening" interpretation of the $U_A(1)$ charge due to a nonperturbative effect of axial anomaly.

1. Introduction

Undoubtedly, the so-called "proton spin crisis" aroused by the 2nd EMC measurement is one of the most interesting puzzle in the field of hadron physics [1]. According to their interpretation, the origin of the proton spin is not the quark spin. Later, it was suggested that the introduction of the gluon polarization dictated by the $U_A(1)$ anomaly of QCD might resolve this puzzle [2,3]. However, the large gluon polarization required in this senario, combined with the proton spin sum rule, leads to large and negative orbital angular momentum, which seems to be difficult to understand theoretically.

Since satisfactory treatment of gluon dynamics is far beyond our scope, some indirect methods to attack the problem have been invented. Among them, Cheng and Li's proposal is especially simple [4]. Starting with the anomaly equation

$$\partial_\mu J_5^\mu = \sum_f 2 m_f \bar{q}_f i \gamma_5 q_f + N_f \frac{\alpha_S}{2\pi} \operatorname{tr} G_{\mu\nu} \tilde{G}^{\mu\nu}, \tag{1}$$

and taking its matrix element between the polarized proton states, they are led to a seemingly natural separation of the proton $U_A(1)$ charge into the quark component and the gluon component. Although this separation is gauge invariant, it turned out to have some troubles. That is, the quark and gluon components defined in this way have a considerable amount of isospin violation although their sum is nearly isospin invariant [5]. A peculiar nature of Cheng and Li's identification can also be seen by considering the chiral limit with vanishing current quark masses [6,7]. In this limit, the first term of eq.(1) is absent. This means that, if we still stick to Cheng and Li's definition, it amounts to attributing all the $U_A(1)$ charge to the gluon component. It is quite strange, because irrespective of the quark masses the quark helicity contribution to the $U_A(1)$ charge should be there as far as the physical proton contains the core of three valence quarks [8].

The resolution of this isospin breaking problem was given by several authors. For instance, based on an effective action method of Zumino, Shore and Veneziano derived a generalized Goldberger-Treiman (G-T) relation in the $U_A(1)$ channel [9] (see also the work by Efremov et al. [10]) as :

$$g_A(0) \;=\; \Delta q \;-\; N_f \Delta \Gamma, \tag{2}$$

where

$$\Delta q \;=\; \frac{F}{2\,M_N}\, g_{\eta' NN}, \tag{3}$$

$$\Delta \Gamma \;=\; \frac{1}{4\,N_f^2\,M_N}\, F^2\, m_{\eta'}^2\, g_{QNN}, \tag{4}$$

with M_N the nucleon mass and $F = \sqrt{2N_f}\, F_\pi$. This decomposition does not suffer from the aforementioned isospin breaking problem, since the quark and gluon components here are related to the physical coupling constants : the former is given by the $\eta' NN$ coupling constant and the latter by the coupling constant between the nucleon and gluon composite field Q mocking up the QCD anomaly. Shore and Veneziano further gave a conjecture that these two components may be identified with the quark and gluon helicity contributions in the QCD parton model. In our opinion, however, this last assertion appears highly nontrivial and requires further study.

In this short report, we shall investigate the subtlety of the problem on the basis of a specific effective model of QCD. The point of our study here is to use an effective quark theory not an effective meson theory. This is based on our belief that the physics of "spin contents" can be most transparently understood within the framework of effective theories explicitly containing the quark fields.

2.   Chiral Quark Soliton Model

We start with the model lagrangian of Diakonov and Petrov, which incorporates the most important feature of the low energy QCD, i.e. the spontaneous breaking of the chiral symmetry and the associated appearance of the Goldstone pions [11] :

$$\mathcal{L} \;=\; \bar{\psi}\,(\,i\,\partial\!\!\!/ - M U^{\gamma_5}\,)\,\psi, \tag{5}$$

where

$$U^{\gamma_5}(x) \;=\; e^{\,i\,\gamma_5\, \tau \cdot \pi(x)/f_\pi}. \tag{6}$$

Here $\psi(x)$ and $\pi(x)$ are respectively the quark and pion fields, while M is the "constituent" quark mass which is dynamically generated through the spontaneous chiral symmetry breaking of the QCD vacuum. The absence of the pion kinetic term here implies that it is not an independent field of quarks, but eventually interpreted as a $q\bar{q}$ composite. Under the mean-field treatment, this simple model is known to have a soliton-like localized solution [11-14], which can be identified with the physical nucleon, after performing an appropriate quantization procedure [11,14]. With this simple model, we can calculate any nucleon observables in a nonperturbative manner with full inclusion of the Dirac-sea quark degrees of freedom [14].

Table 1. The spin contents of the proton in dependence of the dynamical quark mass M. The energy of the valence quark level in units M is also shown in the second column for reference.

M(MeV)	E_0/M	contents	valence	vacuum	total
		$< \Sigma_3 >$	0.535	0.002	0.537
350	0.649	$< 2L_3 >$	0.227	0.236	0.463
		$< 2J_3 >$	0.762	0.238	1.000
		$< \Sigma_3 >$	0.357	0.007	0.364
450	0.363	$< 2L_3 >$	0.280	0.356	0.636
		$< 2J_3 >$	0.637	0.363	1.000
		$< \Sigma_3 >$	0.184	0.027	0.211
700	0.038	$< 2L_3 >$	0.309	0.480	0.789
		$< 2J_3 >$	0.493	0.507	1.000
		$< \Sigma_3 >$	-	0.135	0.135
1000	-0.146	$< 2L_3 >$	-	0.865	0.865
		$< 2J_3 >$	-	1.000	1.000

As for the particular problem of the proton spin contents, we can derive the following sum rule,

$$< \Sigma_3 >_{val} + < \Sigma_3 >_{v.p.} + < 2L_3 >_{val} + < 2L_3 >_{v.p.} = 1, \qquad (7)$$

and can calculate each of these contributions nonperturbatively. Here $< \Sigma_3 >_{val}$ and $< \Sigma_3 >_{v.p.}$ respectively stand for the valence and vacuum polarization contributions to the quark spin expectation value, while $< 2L_3 >_{val}$ and $< 2L_3 >_{v.p.}$ are the corresponding quantities for the orbital part of the total proton spin. Shown in table 1 are the contents of the above sum rule given as functions of the dynamical quark mass M [14]. (Although we believe that a reasonable range of the parameter M is more or less restricted by various other physical consideration, we find it amusing to regard it as a free parameter of the model and to see the consequence of its variation. This parameter turns out to play a role of the quark-pion coupling constant and controls the character of the resultant soliton solution.) The energy of the valence quark level (in units M) is also shown for reference in the second column of the table. The detailed contents of the proton spin are shown from the third to the sixth columns. One sees that, for weak coupling case of $M = 350$ MeV, most ($\sim 76\%$) of the proton spin is carried by the valence quark part, of which about 53 % is due to quark spin and about 23 % due to the orbital angular momentum. Only a small fraction ($\sim 24\%$) of it is due to the vacuum quark term. As M increases, however, the vacuum quark contribution grows rapidly. For $M = 700$ MeV, for instance, the vacuum quark part bears about a half of the total proton spin. Moreover, most of it is attributed to the orbital part of the angular momentum. As M is further increased, the valence orbital finally dives into the negative energy region, and in this case there is no valence quark contribution by definition. It is interesting to see that the spin expectation value is only about $0.1 \sim 0.2$ in this case. The proton spin is therefore almost due to the orbital angular momentum. Remembering that in the Skyrme model the nucleon spin totally comes from the collective rotation of the hedgehog mean field, we then realize that the chiral quark soliton model in the strong coupling region gives a microscopic foundation of the Skyrme model. It should be emphasized, however, that for the self-consistent solutions corresponding to a physically reasonable choice of the dynamical quark mass parameter, i.e. $M \simeq (300 \sim 500)$ MeV, the energy of the valence quark level is positive and the quark spin contribution to the proton helicity cannot be extremely small. This illustrates a distinguishable role of the valence quark degrees of freedom in this special observable, even though it is phenomenologically desirable that this quark helicity contribution is canceled by something like the gluonic effect that is not yet incorporated into the above effective model. We emphasize that the chiral quark soliton model with this physically reasonable parameter range predicts very natural nucleon picture, i.e. the core of N_c valence quarks surrounded by the $q\bar{q}$ excitation of pionic nature, which seems to have some direct phenomenological supports [15].

3. Generalized Effective Quark Models

Now we want to generalize the Diakonov-Petrov model so that it effectively incorporates the $U_A(1)$ anomaly of QCD [16]. We first replace the $SU(2)$ matrix $U^{\gamma_5}(x)$ by the $U(2)$ matrix $\tilde{U}^{\gamma_5}(x) \equiv e^{i\gamma_5 (\eta(x) + \tau \cdot \pi(x))/f_\pi}$ by introducing the $U(1)$ meson $\eta(x)$. (Here we confine ourselves to the 2 flavor case, but essential physics would not be altered by the 3 flavor generalization.) The resultant lagrangian is invariant under the global $U_A(1)$ transform as

$$\psi(x) \rightarrow e^{i\gamma_5 \theta} \psi(x), \qquad \eta(x) \rightarrow \eta(x) - 2 f_\pi \theta. \qquad (8)$$

The incorporation of the $U_A(1)$ anomaly can be done as usual by introducing an effective gluon composite Q and its coupling with the $U_A(1)$ meson η as [17]

$$\mathcal{L} = \bar{\psi}(i \not{\partial} - M \tilde{U}^{\gamma_5}) \psi + \frac{1}{8 f_\pi^2 m_\eta^2} Q^2 + \frac{1}{2 f_\pi} Q \cdot \eta. \qquad (9)$$

It is now easy to show that the $U_A(1)$ current obeys the desired anomaly equation with the identification $Q = N_f (\alpha_S/2\pi) \, \text{tr} \, G_{\mu\nu} \tilde{G}^{\mu\nu}$. Since the Q field has no kinetic term, it can be eliminated. In doing so, we get the following lagrangian, which shows that the would-be-Goldstone $U_A(1)$ meson now acquires the anomalously generated mass term :

$$\mathcal{L} = \bar{\psi}(i \not{\partial} - M \tilde{U}^{\gamma_5}) \psi - \frac{1}{2} m_\eta^2 \eta^2. \qquad (10)$$

Now it is convenient to introduce the redefinition of the quark field following Ellwanger and Stech as [18]

$$\psi'(x) = e^{i\gamma_5 \eta(x)/f_\pi} \psi(x). \qquad (11)$$

With this redefinition, the above lagrangian can be cast into the form :

$$\mathcal{L} = \mathcal{L}_0 + \mathcal{L}', \qquad (12)$$

with

$$\mathcal{L}_0 = \bar{\psi}'(i \not{\partial} - M U^{\gamma_5}) \psi', \qquad (13)$$

$$\mathcal{L}' = -\frac{1}{2} m_\eta^2 \eta^2 + \frac{1}{2 f_\pi} \bar{\psi}' \gamma^\mu \gamma_5 \partial_\mu \eta \psi'. \qquad (14)$$

Here we have intentionally divided the total lagrangian into two parts, so that the first part just takes the form of the lagrangian of the chiral quark soliton model, which describes the strongly coupled quark-pion system. Now we assume that the nucleon bound state is generated by this first part of the lagrangian, and the η-quark coupling can be treated as a perturbation. (We recall that a similar assumption is always implicit in the pole saturation derivation of the G-T type relation.) Under this approximation, we can derive the following G-T relation :

$$g_A(0) \;=\; \frac{2 f_\pi}{2 M_N} \, g_{\eta NN}(0) \;=\; \Delta\Sigma. \tag{15}$$

We recall that the left equality was also derived by Fritzsch and also by Ji in more model independent analysis [6,7]. A new feature of the present analysis based on a specific model is that the flavor-singlet axial charge $g_A(0)$ just coincides with the quark spin expectation value which was already calculated in the chiral quark soliton model. Now we clearly recognize that, even though we have started with the anomaly equation without no symmetry-breaking quark mass term, the $U_A(1)$ charge just coincides with the quark helicity component, not the gluon helicity in contradiction with Cheng and Li's prescription.

A further generalization of the above effective quark model may be attained through the introduction of the direct coupling between the quark and gluon composite field Q [19]. Assuming that this coupling is gauge invariant as well as globally $U_A(1)$ invariant, we are led to the following effective lagrangian :

$$\mathcal{L} \;=\; \bar{\psi}[i\gamma^\mu(\partial_\mu + g\,\gamma_5\,\partial_\mu Q) - M\tilde{U}^{\gamma_5}]\psi + \frac{1}{8 f_\pi^2 m_\eta^2}\, Q^2 + \frac{1}{2 f_\pi}\, Q\cdot\eta. \tag{16}$$

Under the similar approximation as before, we can then derive the following relation [16]:

$$g_A(0) \;=\; (1 - 4 f_\pi^2 m_\eta^2\, g)\, \Delta\Sigma. \tag{17}$$

Viewing it perturbatively, the first and the second terms of the r.h.s respectively correspond to the first and the second terms of Shore and Veneziano's formula (2). In the present model, however, $\Delta\Sigma$ appears as a common factor for these two contributions, since both of the η and Q couple to the same quark axial-vector sources in the bound nucleon. We then conclude that the second term of eq.(17) certainly has its origin in the $U_A(1)$ anomaly of QCD, but it has little to do with the helicity contribution of the polarized gluon in the QCD parton model. The effect of the gluon field here is rather to screen (reduce) the quark helicity contribution to the $U_A(1)$ charge.

4. Conclusion

To summarize, Shore and Veneziano have recently derived a Goldberger-Treiman relation in the $U_A(1)$ channel which does not suffer from unphysical isospin breaking problem. However, the identification of these two components with the quark and gluon helicity contributions of the QCD parton model seems nontrivial to us. In fact, our analysis here based on a specific model indicates another possibility that the effect of $U_A(1)$ anomaly is rather to be interpreted as screening the quark helicity contribution to the $U_A(1)$ charge in a nonperturbative manner. This interpretation has an advantage that it necessarily need not large gluon polarization, and consequently large and negative orbital angular momentum inside the nucleon.

References

[1] European Muon Collab., J. Ashman et al., Phys. Lett. B206 (1988) 364 ; Nucl. Phys. B328 (1989) 1.

[2] G. Altarelli and G.G. Ross, Phys. Lett. B212 (1988) 391.

[3] R.D. Carlitz, J.C. Collins and A.H. Mueller, Phys. Lett. B214 (1988) 229.

[4] T.P. Cheng and L.-F. Li, Phys. Rev. Lett. 62 (1989) 1441.

[5] T. Hatsuda, Nucl. Phys. B329 (1990) 376.

[6] H. Fritzsch, Phys. Lett. B242 (1990) 451.

[7] X. Ji, Phys. Rev. Lett. 65 (1990) 408.

[8] S. Forte, Phys. Lett. B224 (1989), 189 ; Nucl. Phys. B331 (1990), 1.

[9] G.M. Shore and G. Veneziano, Phys. Lett. B244 (1990) 75.

[10] A.V. Efremov, J. Soffer and N.A. Törnqvist, Phys. Rev. D44 (1991) 1369.

[11] D.I. Diakonov and V.Yu. Petrov, Nucl. Phys. B272 (1986) 457 ; D.I. Diakonov, V.Yu. Petrov and P.V. Pobylista, Nucl. Phys. B306 (1988) 809.

[12] H. Reinhardt and R. Wünsch, Phys. Lett. B215 (1988) 577.

[13] Th. Meissner, F. Grümmer and K. Goeke, Phys. Lett. B227 (1989) 296.

[14] M. Wakamatsu and H. Yoshiki, Nucl. Phys. A524 (1991) 561.

[15] M. Wakamatsu, Phys. Rev. D44 (1991) R2631 ; Phys. Lett. B269 (1991) 394.

[16] M. Wakamatsu, Phys. Lett. B280 (1992) 97.

[17] C. Rosenzweig, J. Schechter and G. Trahern, Phys. Rev. D21 (1980) 3388 ; P. Di Vecchia and G. Veneziano, Nucl. Phys. B171 (1980) 253 ; K. Kawarabayashi and N. Ohta, Nucl. Phys. B175 (1980) 477.

[18] U. Ellwanger and B. Stech, Z. Phys. C49 (1991) 683.

[19] J. Schechter, V. Soni, A. Subbaraman and H. Weigel, Phys. Rev. Lett. 65 (1990) 2955.

SOFT HIGH ENERGY SCATTERING AND QUARK CLUSTERING

H.G. Dosch

Institut für Theoretische Physik
Universität Heidelberg, Philosophenweg 16, D-6900 Heidelberg, FRG

Abstract: The influence of quark clustering in nucleons on soft high energy scattering is discussed.

In this contribution I shall shortly discuss the influence of quark clustering in nucleons on soft high energy scattering. This is done in the framework of a model [1,2,3] based on the method of QCD vacuum correlators [4,5].

As is well known, the total hadron-hadron cross sections at high energies satisfy the quark counting rule [6], i.e. the ratio of the nucleon-nucleon cross section to that of the meson-nucleon cross section is 9:6, and the prediction for the ratio of the latter one to the meson-meson cross section is 6:4. This favors an interpretation of high energy cross sections as due to an quark-quark interaction, e.g. the exchange of two non-perturbative gluons between quarks [7]. As long as the mass scale of the non-perturbative gluons (introduced by the IR regularization) is large as compared to the inverse hadronic radius, such a model leads to quark additivity in the cross sections. On the other hand, the total cross sections show a distinct flavor dependence. The more strange hadrons are involved, the smaller is the cross section (see Table 1).

Table 1. Total cross sections at $\sqrt{s} \sim 20$ GeV

Reaction	$p\pi$	pK	pp	$p\Sigma$	$p\Xi$
$\sigma_{tot}[mb]$	24	20	39	34	29

This flavour dependence is most naturally explained by a geometrical model, where the total cross section depends on the radius of the hadrons, which in turn depends on the reduced mass of the quarks [8]. In such a model the ratio 9:6 for $\sigma_{NN}/\sigma_{\pi N}$ would be accidental, due to a contingent ratio of the radii. I shall address these problems in the framework of a specific model for non-perturbative effects developed and advocated by Yuri A. Simonov and myself in the last years [4,5]. In its simplest form, it is based on the assumption that the integration over the slowly varying (i.e. non-perturbative) gluon fields can be approximated by a Gaussian stochastic process, with finite correlation length. In order to define a gauge-invariant correlator, we parallel-transport all gluon fields to some reference point z, i.e. we introduce

$$\mathbf{F}_{\mu\nu}(x;z) = \phi^*(x,z)\mathbf{F}_{\mu\nu}(x)\phi(x,z) \tag{1.1}$$

where $\phi(x,z)$ is the parallel-transporter from x to z.

$$\phi = P \ \exp\{ig \int_x^z \mathbf{A}_\mu(y)dy_\mu\} \tag{1.2}$$

and $\mathbf{A}$ and $\mathbf{F}$ are the $su(3)$-valued potentials and fields of the gluons.

For the field correlator we make the simplifying ansatz that it depends only on the difference of the coordinates $h = z - x$, an ansatz which is certainly justified for small and large values of h.

Under these assumptions the most general form of the correlator is given by [5]:

$$\langle\langle g_s^2 F_{\mu\nu}^A(x,z_0) F_{\rho\sigma}^B(y,z_0)\rangle\rangle = \frac{\delta_{AB}}{8} \frac{\langle g_s^2 F_{\mu\nu}^A(0) F_{\mu\nu}^A(0)\rangle}{12} \Big[\{\delta_{\mu\rho}\delta_{\nu\sigma} - \delta_{\mu\sigma}\delta_{\nu\rho}\} \kappa \cdot D(h^2)$$
$$+ \frac{1}{2} \Big[\frac{\partial}{\partial h_\mu}(h_\rho\delta_{\nu\sigma} - h_\sigma\delta_{\nu\rho}) + \frac{\partial}{\partial h_\nu}(h_\sigma\delta_{\mu\rho} - h_\rho\delta_{\mu\sigma}) \Big] (1-\kappa)D_1(h^2) \Big] \tag{1.3}$$

The functions $D(h^2)$ and $D_1(h^2)$ are normalized to unity at $h^2 = 0$. This simple ansatz leads to the following consequences:

1) The correlator proportional D can, due to the Bianchi identities, only occur in a non-Abelian gauge theory, i.e. $\kappa \equiv 0$ in an Abelian gauge theory [5].

2) It yields a linear confining potential (i.e. an area law of the Wegner-Wilson loop) if $\kappa \neq 0$. In particular the slope of the linear potential σ will be given by

$$\sigma = const \cdot \kappa \cdot a^2 \cdot G_2 \tag{1.4}$$

where a is the correlation length of the correlator function D, and

$$G_2 = \langle \frac{\alpha_s}{4\pi} F_{\mu\nu}^A(0) F_{\mu\nu}^A(0)\rangle \tag{1.5}$$

the gluon condensate of [9]. A numerical analysis yields for the correlation length values between 0.3 and 0.5 fm.

The model can be applied to hadron-hadron scattering. If a quark moves with high energy in an external field, its amplitude can be represented by the WKB-approximation [10]. Constructing a hadron from two parallel moving quarks on (essentially) light-like pathes leads thus to a Wegner-Wilson loop for the scattering amplitude of a hadron in an external field (the transversal motion is treated adiabatically). The scattering of two hadrons [1, 2] is thus described by the expectation value of two loops as depicted in fig. 1, where the averaging over the "external field" according to the correlator (1.3) represents the functional integration over the (non-perturbative) gluon fields.

Fig. 1. Light-like Wegner-Wilson loops constituting the hadron-hadron scattering amplitude

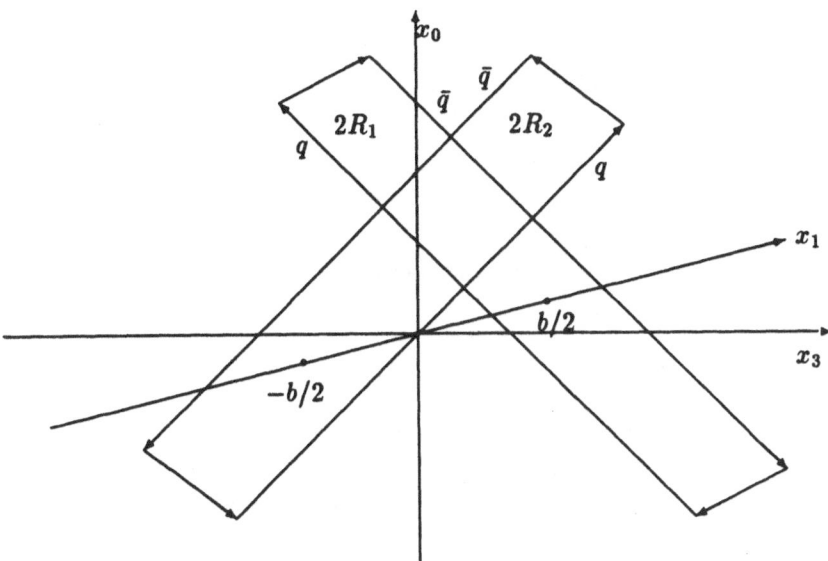

The averaging procedure yields a function $J(b, R_1, R_2, \varphi_1, \varphi_2)$ depending on impact parameter b, i.e. the distance between the light-like middlelines of the loops, their transversal extensions R_i and their azimutal angles φ_i around the middleline . In order to arrive at a hadronic expression, we average over the angles φ_i and over the transversal extensions with a transversal wave function as weight, specifically we use for the mesons

$$\psi_M(R) = \sqrt{\frac{2}{\pi}}\frac{1}{S}e^{-\frac{R^2}{S^2}} \tag{1.6}$$

with an extension parameter S.

For the baryon wave function we make two extreme ansätze:

1) a genuine three-body one, as suggested in a string picture or in strong coupling approximation [11] (see fig. 2a),

2) a quark-diquark ansatz (fig. 2b), leading to a structure like the meson (fig. 2c). For the distances R we assume the wave function (1.6). In such a model the total cross section and the slope parameter $B = -\frac{d}{dt}(\ln \sigma_{el}(t))$ have been evaluated in [2]. It was found [1] that the correlator D, which leads to confinement, also leads to a string-string interaction in soft high energy scattering: Even for hadronic radii large as compared to the correlation length a there is a distinct dependence of the cross section on these radii, i.e. the parameter S in the wave function (1.6). The extension parameter S_N and the correlation length a can be fixed by the $I = 0$ exchange part of the proton-(anti)proton total cross section and slope parameter. This yields

a) for the three-body ansatz (fig. 2a):

$$a = 2.06 \text{ GeV}^{-1}, \quad S_N = 6.42 \text{ GeV}^{-1}$$

and

b) for the diquark ansatz (fig. 2b):

$$a = 2.26 \text{ GeV}^{-1}, \quad S_N = 6.42 \text{ GeV}^{-1}.$$

Fig. 2.   String structure of baryons and mesons

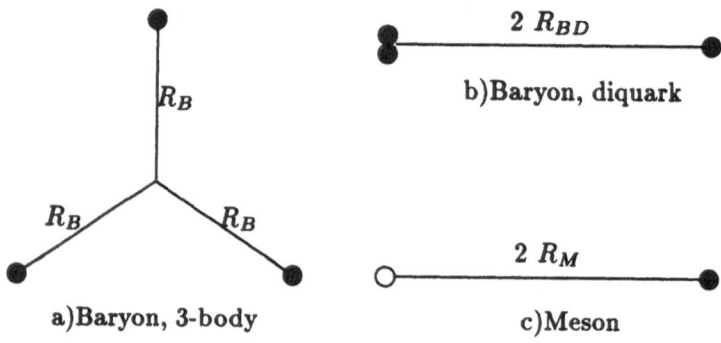

a)Baryon, 3-body   b)Baryon, diquark   c)Meson

The correlation length $a \sim 0.4$ fm is very well compatible with the one determined from the string tension. The expected variation of the radii with the reduced quark mass is able to reproduce the flavour dependence of the cross section. The slope parameter of the $\pi - p$ scattering can be predicted since the extension parameter of the meson wave function can be determined by the total cross section.

Both sets of parameters, i.e. the three-body and the diquark model, yield agreement with experiment within the estimated experimental errors of $\sim 5\%$ (which are hard to assess, since the $I = 0$ exchange part has to be extracted from the data), see table 2.

The two approaches lead to the same prediction for the total cross section, which is in (more or less accidental) agreement with the quark counting rule, but to quite distinct predictions for the slope parameter of high energy $\pi - \pi$-scattering. Experimental data [12] for $\pi^+\pi^-$ scattering are only available at $\sqrt{s} \sim 4$ GeV yielding $\sigma_T \approx 10$ mb and $B \sim 5$ GeV^{-2}, thus are more in favour of the diquark picture.

Table 2.   Total cross sections and the slope parameter for the $I = 0$-exchange contribution

Hadron system	Baryon model	Correl. length [GeV^{-1}] a	Extension [GeV^{-1}] S_1	Extension [GeV^{-1}] S_2	σ_T [mb]	$-B$ [GeV^{-2}] exp.	$-B$ [GeV^{-2}] theor.
pp	3-body	1.94	6.98	6.98	34	13	13*
pp	diquark	2.13	6.98	6.98	34	13	13*
$p\pi$	3-body	1.94	6.98	6.12	20	10.7	11.4
$p\pi$	diquark	2.13	6.98	4.94	20	10.7	9.7
$\pi\pi$		1.94	6.12	6.12	12†		10
$\pi\pi$		2.13	4.94	4.94	12†		6

$\sqrt{s} =$ ca. 15 GeV   *used for fit of a and S_p. † theoretical value.

Acknowledgement: I thank Erasmo Ferreira and Peter V. Landshoff for many valuable discussions. My thanks are also due to the organizers of the workshop for creating such a nice and pleasant atmosphere.

References

[1] A. Krämer, H. G. Dosch, Phys. Lett. **B272** (1991) 114

[2] H. G. Dosch, E. Ferreira, A. Krämer, Soft high-energy hadron-hadron scattering, CERN-TH.6454/92, Phys. Lett. B, to appear

[3] A. Krämer, Dissertation Heidelberg 1991

[4] H. G. Dosch, Phys. Lett. **B190** (1987) 177

[5] H. G. Dosch and Yu. A. Simonov, Z. Phys. **C45** (1989) 147

[6] E. M. Levin, L. L. Frankfurt, JEPT Lett. **2** (1965) 65;
H. J. Lipkin, F. Scheck, Phys. Rev. Lett. **16B** (1966) 71;
H. J. Lipkin, Phys. Rev. Lett. **16B** (1966) 1015;
J. J. J. Kokkedee, L. Van Hove, Nuovo Cimento **42A** (1966) 711

[7] P. V. Landshoff and O. Nachtmann, Z. Phys. **C35** (1987) 405

[8] B. Povh and J. Hüfner, Phys. Rev. Lett. **58** (1987) 1612

[9] M. A. Shifman, A. I. Vainshtein, V. I. Zakharov, Nucl. Phys. **B147** (1979) 385, 448, 519

[10] O. Nachtmann, Ann. Phys. (N.Y.) **209** (1991) 436

[11] H. G. Dosch, V. F. Müller, Nucl. Phys. **B116** (1976) 470

[12] Daum et al., Elastic and total $\pi^+\pi^-$ cross sections from a high statistics measurement of the reaction $\pi^-p \rightarrow \pi^+\pi^-n$ at 63 GeV/c, Int. Conf. on High Energy Physics, Geneva 1979, vol. 2

The Quark Model, the Baryon-Baryon Interaction and Hypernuclei

Amand Faessler[1], G. Lalazissis[2], U.Straub[1], H. Velasquez[3] and Z. Zhang[4]

[1] Institut für theoretische Physik, Universität Tübingen, D-7400 Tübingen, Germany
[2] Departement of Theoretical Physics, University Thessaloniki, GR 54006 Thessaloniki, Greece
[3] Escuela Superior de Fisica y Matematiccs, Instituto Politecnico National, 03910 Mexico, D.F.
[4] Institute of High Energy Physics, Academia Sinica, Beijing, Peoples Rep. China

(talk given by U. Straub)

Abstract. We have extracted from our quark model calculations of the hyperon-nucleon interaction an equivalent non-local ΛN potential at the baryon level. Using this potential the binding energy of the hypernucleus $^5_\Lambda$He is calculated as 2 MeV. The shape of the bound state wave function of this hypernucleus is checked by computing the mesonic decay rates. A nice agreement with experimental data is found.

Keywords. Constituent quark model, resonating group method, hyperon-nucleon interaction, hypernucleus $^5_\Lambda$He, mesonic decay rates

1. Introduction

Although the strong interaction is the basis of nuclear physics it is theoretically not as profoundly understood as e.g. the electromagnetic interaction. We believe that the basic theory underlying the strong interaction is Quantum-Chromo-Dynamics (QCD), but up to now there is no rigorous derivation of the strong interaction between baryons based entirely on the concept of quarks and gluons. Instead several models have been invented which partially take into account some aspects of QCD. One such model is the non-relativistic quark model. Despite its theoretical foundation is not very clear, many calculations based on this model show its usefulness. For example, it is well known that the nucleon-nucleon interaction can be understood in the framework of the non-relativistic quark cluster model [1]. A further test of the quark cluster model is the calculation of the hyperon-nucleon interaction. Again a good agreement of the theoretical calculations with existing experimental data is obtained [2]. However, the experimental data on the free YN interaction are far less precise than the NN data. Therefore one should try to include in the analysis additional information extracted from the study of hypernuclei. Some results of our attempts along this line are presented in this paper.

Supported by the Deutsche Forschungsgemeinschaft

In section 2 we decribe briefly the definition of an equivalent lambda-nucleon potential at the baryon level and the approach to calculate hypernuclei. In section 3 we present the results of the hypernuclear calculations for $^5_\Lambda$He and compare our theoretical results for the mesonic decay rates to experimental data. Due to lack of space we want to give here only the main ideas. The details may be found in ref. [3].

2. Lambda-nucleon potential and hypernuclei

In the non-relativistic quark cluster model [4] one considers the baryons as colorless particles build up by three quarks. The two-baryon system is written as

$$\Psi_{6q} = \mathcal{A}[\Phi_A^{int}\Phi_B^{int}\chi(\mathbf{R}_{AB})] \tag{1}$$

where $\mathcal{A}$ is the six-quark antisymmetrizer which is necessary to fullfil the Pauli principle. $\Phi_{A(B)}^{int}$ are the internal wave functions of the two three-quark clusters A and B, while $\chi(\mathbf{R}_{AB})$ is the relative wave function. The goal of the cluster model is to determine dynamically this relative wave function assuming a reasonable form of the internal wave function. The interaction of two baryons is described by a hamilton operator on the quark level

$$H = \sum_{i=1}^{6} \frac{p_i^2}{2m_i} - T_{cm} + \sum_{i<j} V(i,j) + V^\sigma \tag{2}$$

which consists of the non-relativistic kinetic energy and of two-quark interactions. The potentials on the quark level contain the one-gluon exchange potential, a phenomenological confinement potential and a pseudoscalar-meson (pion, kaon) exchange potential. All parameters of these potentials are fixed from single baryon properties. In addition a phenomenological σ-meson exchange potential is included. The strength $g_\sigma^2/4\pi$ of the σ-meson potential is the only fit parameter.

The unknown relative wave function $\chi(\mathbf{R})$ is computed by solving the Resonating Group equation,

$$\int \left(\mathcal{H}(\mathbf{R},\mathbf{R}') - E\mathcal{N}(\mathbf{R},\mathbf{R}') \right) \chi(\mathbf{R}') d^3R' = 0 \tag{3}$$

where the non-local Hamilton and norm kernels $\mathcal{H}(\mathbf{R},\mathbf{R}')$, $\mathcal{N}(\mathbf{R},\mathbf{R}')$ are defined as expectation values of the Hamilton operator (2) resp. the unit operator $\mathbf{1}$ beween the internal wave functions including the six-quark antisymmetrizer. By a standard transformation the norm kernel can be removed, and subtracting the relative kinetic energy the remaining term is the non-local potential at the baryon level. However this expression is not convenient for further calculations.

A simpler expression can be obtained by the same approach we have used to solve the RG equation. The idea was to expand the relative wave function into a superposition of peaked gaussians. Formally this expansion is written as:

$$|\chi> = \sum_\alpha c_\alpha |\alpha> \tag{4}$$

This expansion is possible as long as the width of the gaussians $|\alpha>$ is larger than typical oscillations of the relative wave function $|\chi>$. Essentially this means the relative kinetic energy should be below 300 MeV in the c.m. system. Under this condition one can consider the set of peaked gaussians $|\alpha>$ as a non-orthogonal basis. Inserting the expansion (4) into the RG eq. (3) one obtains a generalized linear eigen value problem:

$$\sum_{\beta} \left(H_{\alpha\beta} - E N_{\alpha\beta} \right) c_{\beta} = 0$$

where all matrix elements $H_{\alpha\beta}, N_{\alpha\beta}$ can be computed analytically.

Since the relative wave function can be expanded into the non-orthogonal basis it must be possible to expand also the non-local potential into the same basis. Hence we may write the potential as

$$V_{\Lambda N} = \sum_{\alpha\beta} |\alpha> \; V_{\alpha\beta} \; <\beta| \tag{5}$$

The coefficients $V_{\alpha\beta}$ are related to the already calculated matrix elements $H_{\alpha\beta}, N_{\alpha\beta}$ and the direct overlap $<\alpha|\beta>$ (cf. ref. [3]). In the space spanned by the peaked gaussians the expansion (5) is completely equivalent to the original definiton of the potential mentioned above after eq. (3). Since the peaked gaussians form a basis for all "smooth" functions this expansion is the searched expression for the non-local potential at the baryon level. Depending on the matrix elements $H_{\alpha\beta}$ and $N_{\alpha\beta}$ the formula (5) describes either the NN or YN interaction. For the further calculations of the hypernucleus $^{5}_{\Lambda}$He we will need only the lambda-nucleon component since the core part of the $^{5}_{\Lambda}$He wave function is considered to be inert (see below).

The hypernucleus $^{5}_{\Lambda}$He is computed in a RGM approach, i.e. the wave function is considered to be a product of a core wave function describing the nucleon core and a relative motion wave function describing the motion of the lambda particle around the core:

$$\Psi_{^{5}_{\Lambda}He} = \Phi^{int}(u^{int}) \; \chi(\mathbf{r}_{\Lambda} - \mathbf{R}_c) \tag{6}$$

The internal wave function of the ^{4}He core is a gaussian with a width of 1.39 fm adjusted to the experimental rms core radius care taken of the finite nucleon size. We have assumed that the core is not distorted by the presence of the lambda particle. Considering that the nucleons in the ^{4}He core are bound much stronger (7.1 MeV/nucleon) than the lambda particle is bound to the core (3.1 MeV) this assumption should cause no problems. In the full Hamiltonian the nucleon-nucleon potentials of the inert core contribute only to a constant energy shift, whereas the lambda-nucleon potential determines the lambda binding.

For simplicity we take into account only a S-wave component in the relative wave function. In order to be consistent we therefore take into account in the lambda-nucleon potential also only the S-wave component. To describe in this

Λp elastic cross section

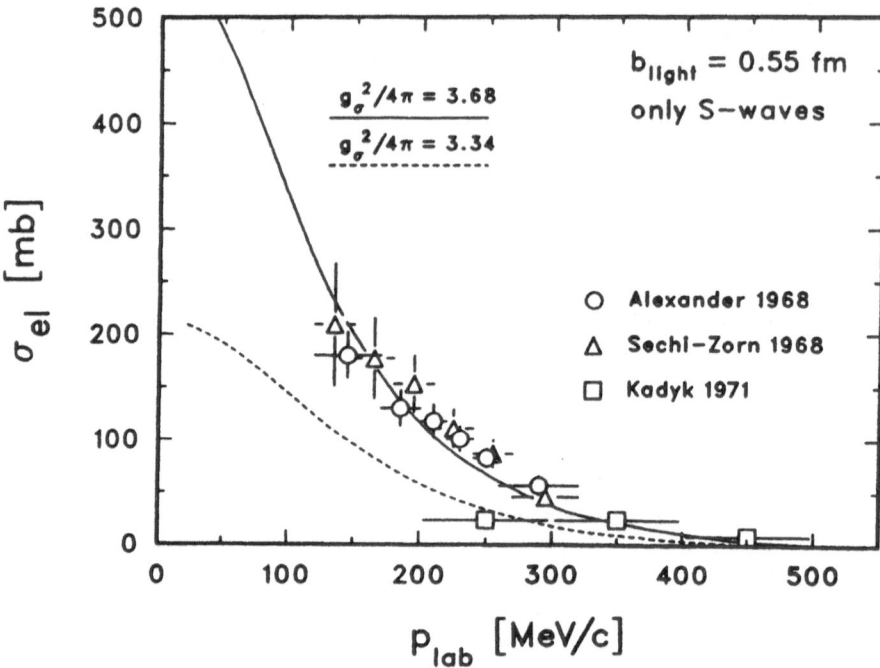

Fig. 1: Λp cross section taking into account only S-waves. The value $g_\sigma^2/4\pi = 3.68$ is determined by fitting the cross sections. For comparison the cross section for $g_\sigma^2/4\pi = 3.34$ is also shown.

simplified approach the free ΛN interaction the strength of the sigma meson exchange has to be adjusted to $g_\sigma^2/4\pi = 3.68$. This is shown in figure 1 where the free ΛN cross section is plotted. It is obvious that only the full curve using this strength fits the experimental data. For comparison also the result for a strength of $g_\sigma^2/4\pi = 3.34$ is shown. This latter value is the strength fitting the ΛN and ΣN data taking into account in addition also tensor forces [2]. However in the present approach no tensor forces are included and therefore the value $g_\sigma^2/4\pi = 3.68$ has to be taken as the strength describing the free ΛN interaction.

The actual calculation of the relative wave function is done similar to the quark model calculations, i.e. the relative wave function is expanded into a set of shifted gaussians with a width parameter adjusted to the size of the ^{4}He core. One derives a linear eigen value problem. From the solution the binding energy and the relative wave function of the lambda particle are obtained.

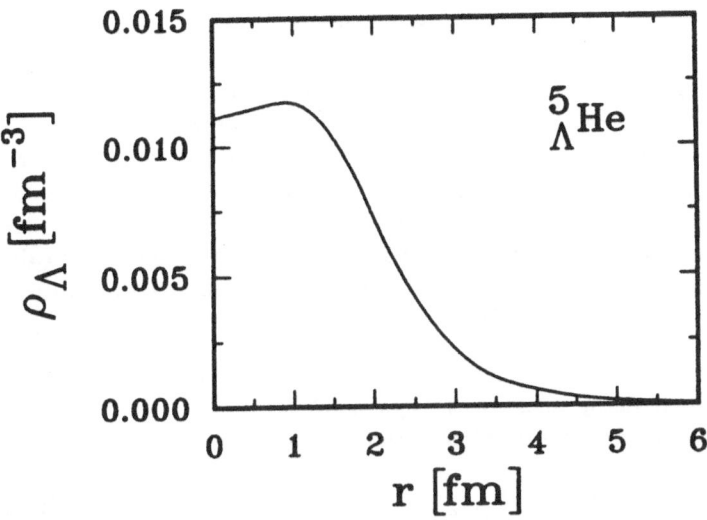

Fig. 2: Density of the relative wave function of $^{5}_{\Lambda}$He describing the motion of the lambda particle relative to the ^{4}He core. Here the lambda particle is bound with 2 MeV.

3. Results for $^{5}_{\Lambda}$He

The binding energy of the lambda particle in $^{5}_{\Lambda}$He is calculated as 2 MeV. The experimental value is 3.1 MeV. It must be noted that the non-local lambda-nucleon potential has been extracted without any fit parameters from the quark model calculations. Therefore we think that the calculated binding energy agrees reasonable well with the experimental value.

The density of the relative wave function is shown in figure 2. It can be seen from the hole at the center that the lambda particle feels a repulsion from the ^{4}He core and thus is pushed to the surface.

The shape of the Λ wave function can be tested by calculating the mesonic decay of this hypernucleus [3]. By the weak interaction the lambda particle decays to a nucleon and a pion. In the free lambda decay the pion is emitted. One has two decay channels: $\Lambda \rightarrow p\pi^{-}$ (64.2 %) and $\Lambda \rightarrow n\pi^{0}$ (35.8 %).

In any hypernucleus the free decay is modified by two effects: First of all, the created nucleon can not occupy states which are occupied already by the nucleons in the core, i.e. there is Pauli blocking and secondly, the created pion also feels the interaction with the core. In heavy nuclei the Pauli blocking leads to a strong suppression of the mesonic decay. However in our case of a light hypernucleus and with the lambda wave function concentrated mainly at the surface, the Pauli blocking is only moderate, while the effect of the distortion of the pions is small since the core is light and the lambda particle is not close to the core. By calculating the mesonic decay rates we can therefore test the lambda wave function.

However it is obvious that on the one hand the binding energy and the mean distance between the lambda particle and the core and on the other hand the mean distance and the decay rates are closely related. Since in our case the calculated binding energy is by 1 MeV too low we have readjusted the sigma meson strength to the experimental binding energy. The needed strength is $g_\sigma^2/4\pi = 3.76$. With this value we have calculated the mesonic decay rates shown in table 1. The rates obtained from the the original lambda wave function with $g_\sigma^2/4\pi = 3.68$ are 10 percent higher.

Table 1: Mesonic decay rates for the hypernucleus $^5_\Lambda$He compared to the total free decay rate. The theoretically calculated ratios and the experimental values are shown.

	Γ^-/Γ_{free}	Γ^0/Γ_{free}	$\Gamma_{tot}/\Gamma_{free}$
Theory	0.431	0.239	0.670
Experiment	0.44 ± 0.11	0.18 ± 0.20	$0.59 ^{+0.44}_{-0.31}$

It can be seen that we obtain an excellent agreement with the experiment even taking the theoretical uncertainties of 10 percent into account. However the experimental errors are still larger.

References

[1] S. Takeuchi, K. Shimizu and K. Yazaki, *Nucl. Phys.* **A504** (1988) 777; K. Bräuer, A. Faessler, F. Fernandez and K. Shimizu, *Nucl. Phys.* **A507** (1990) 599; A. Machavariani, U. Straub and A. Faessler, *Nucl. Phys.* **A** (1992), in press.

[2] U. Straub, Z. Zhang, K. Bräuer, A. Faessler, S.B. Khadkikar and G. Lübeck, *Nucl. Phys.* **A508** (1990) 385c.

[3] U. Straub, A. Faessler, J. Nieves and E. Oset, submitted to *Nucl. Phys.* (1992).

[4] U. Straub, Z. Zhang, K. Bräuer, A. Faessler, S.B. Khadkikar and G. Lübeck, *Nucl. Phys.* **A483** (1988) 686.

Diquark Struscture of Proton and the Problem of Colour Transparency in Wide-Angle Quasielastic PP Scattering on Nuclei

V.V.Anisovich,[1*] L.G.Dakhno,[1] V.A.Nikonov,[1] and M.G.Ryskin[1]

[1] St.Petersburg Nuclear Physics Institute
Gatchina, St.Petersburg district 188350, Russia
* Institut für Theoretische Kernphysik der Universität Bonn, Nussallee 14-16, Bonn-1

Abstract. We suggest a model for the quasielastic wide-angle pp scattering, in which two mechanisms contribute at intermediate energies: a single hard scattering and the Landshoff multiple one. Calculated colour transparency factors, $T(A)$, agree with data. Specific effect, filter mechanism, appears in this model, which can lead to $T > 1$.

The colour transparency phenomenon was suggested to be a direct test of QCD [1-2]. Colour transparency is expected to take place when the hadron participating in the reaction in a nuclear medium is forced to be in a stressed configuration during the time it spends traversing the nucleus. Therefore such an ingredients are needed for colour transparency phenomenon as (i) Preparation of a small and colour neutral state, and (ii) Lorentz time dilation which increases the apparent lifetime of the selected colour neutral state. Hard processes on nuclei can be considered as proper objects for an observation of the colour transparency phenomenon.

In this connection much interest was raised by BNL-experiment [3] where pp quasielastic scattering on Li, C, Al, Cu and Pb at the cm angle 90° was measured at the incident proton lab. momenta P=6,10 and 12 GeV/c.

The transparency factor defined as

$$T = \frac{\frac{d\sigma}{dQ^2}(\text{quasielastic } pp \text{ in nucleus})}{(\text{number of protons in nucleus}) \times \frac{d\sigma}{dQ^2}(\text{elastic } pp)} \tag{1}$$

increases in the momentum range 6-10 GeV/c and falls down at 12 GeV/c. This behaviour of T caused a controversial discussion and the attempts to reanalyse the underlying processes (see, for example, refs.[4,5]).

Two subprocesses control a wide-angle scattering on nucleus: hard proton-proton scattering and soft attenuation of incoming and outgoing waves. The change of the T-behaviour indicates a change of underlying mechanism at least in one of this subprocesses or even in both of them. Interference effects are very important in such cases.

In this paper we consider a realistic model for the quasielastic nucleon- nucleon scattering in a nucleus which allows a consistent quantum-mechanical treatment of hard subprocess and soft nucleon rescatterings in a nuclear medium. The consideration of soft rescatterings is based on a colour screening phenomenon: Gluon structure of pomeron and its "small size" allow one to reformulate the idea about vanishing soft interaction cross sections for stressed neutral-colour configurations in a language of the reggeon exchanges [6]. For example, the sum of the diagrams shown in Fig.1a,b,c for the coupling of pomeron with the $q\bar{q}$ system in the neutral-colour state vanishes when the relative quark-antiquark distance tends to zero. This is possible with the colour screening effects taken into account to describe successfully pp and πp soft scatterings at high energies and, what is important for hadron transparency consideration, to write down the amplitude for hadron-nucleus soft rescatterings [6].

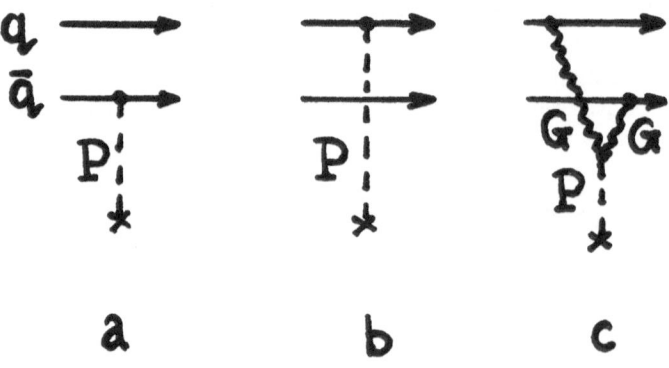

Fig.1. Pomeron coupling to quark-antiquark system: a) to antiquark, b) to quark, c) to both of them through reggeised gluon exchange. The replacement 'antiquark → diquark' leads to the diquark model for proton.

Hard pp-scattering at the discussed energies confronts with two different mechanisms: (i) Purely hard single scattering [7] when colliding hadrons and interaction region are effectively pointlike with typical distance scale $0(1/Q)$, and (ii) Landshoff multiple scattering [8] when transverse distance between constituent quarks is of the order of a size of the colliding hadrons. Both mechanisms give approximetely the same s-dependence of the elastic cross section: $d\sigma/dQ^2(90°) \sim s^n$ with $n_{single} = 10$ and $n_{Landshoff} = 9.6$ [9].

In the intermediate energy region it is difficult to estimate the corresponding amplitudes quantitatively. We choose to fix the relative weight of these processes using the experimental data: The empirical observation of oscillation in the elastic pp cross section $s^{10}d\sigma/dQ^2(90°)$ hints at an importance of both mechanisms in the energy region $s = 10 - 30$ GeV^2 [10].

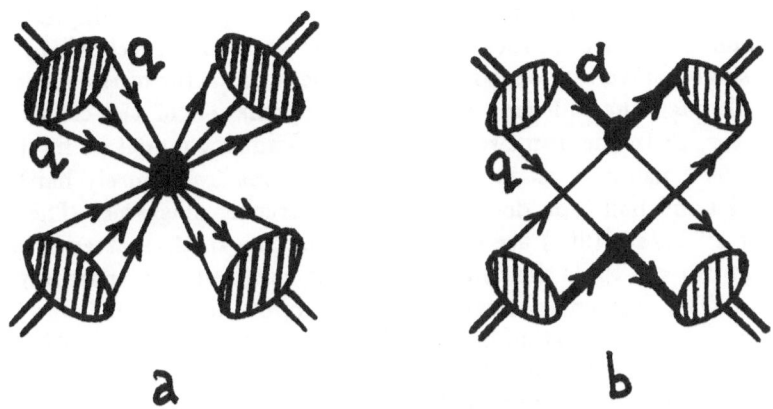

Fig.2. Two types of the mechanism of the wide-angle pp-scattering: a) Purely hard single scattering [7], b)Landshoff multiple scattering [8], c) Multiple scattering in the diquark model.

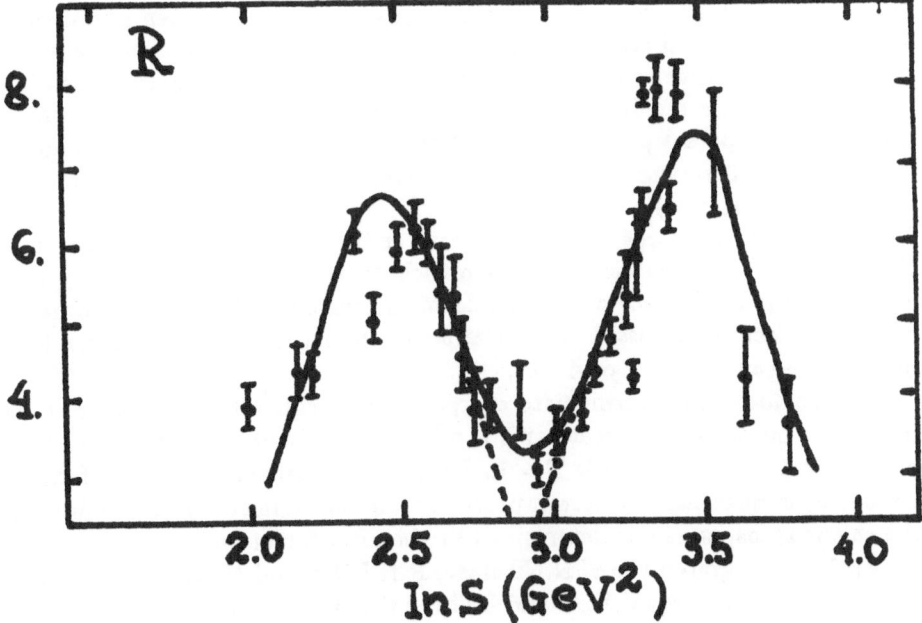

Fig.3. Energy dependence of $R = \text{Const} \cdot s^{10}d\sigma/dQ^2(90°)$ for the pp elastic scattering taken from ref.[10]. Solid line is our fit which includes two contributions: single hard process (dashed line) and Landshoff-type process (dot-dashed line).

For the wide-angle pp-scattering we consider here a simplified picture assuming quark-diquark proton structure. An introduction of diquark for the description of hard processes at moderately large momentum transfer was used for a long time [11,12] and seems to be reasonable approximation (a review of the recent results of the diquark model is given in ref. [13]). In the diquark model the Landshoff mechanism reduces to the double scattering mechanism, which, unlike the single scattering, does not contain the form factor of the nucleon.

We go through the following steps in our calculation of the colour transparency factor, T: In the framework of the quark-diquark model we fit the data for $d\sigma/dQ^2(90°)$ using as an input the contributions from purely hard single (Fig.2a) and Landshoff-type double (Fig.2d) scattering diagrams. The results of our fit of $s^{10}d\sigma/dQ^2(90°)$ are shown in Fig.3. The single scattering process dominates at $\ln s < 3$ while the double scattering contribution is more important at larger $\ln s$. The wide-angle elastic scattering amplitude which is found in such a way is used in the calculation of quasielastic scattering on Li, C, Al and Pb. The results of this calculation, when the colour screening is taken into account for protons traversing the target nucleus, are shown in Fig.4.

In all the versions of our calculation T increases with p in the region 15-30 GeV/c (see Fig.4b, where $T(Al)$ is shown in the region of large p). However this growth of T is not related to the simple colour transparency limit $T = 1$, when hard scattering selects the stressed component of proton wave function, which is traversing the nucleus without interaction. The amplitude of the wide-angle hadron-nucleus scattering, when colour screening effects are taken into account, has more complicated structure. In the intermediate energy region the nuclear medium is not passive spectator of the hard collision: Nucleus, working as a filter because of colour screening factor, selects some stressed proton configuration for which the transparent probability is not too small. (Such an ability of a nucleus to work as a filter was discussed in refs.[14,15] in context of the diffraction dissociation processes).

However at very large Q our treatment of the filter mechanism is not valid. We could expect there either the approach to the colour transparency limit, $T = 1$, or the dominance of the Landshoff mechanism. The latter mechanism does not select the stressed configuration in the hard scattering, so for heavy nucleus one can expect $T \propto A^{-2/3}$.

We conclude: At intermediate energies the specific filter mechanism of the wide-angle quasielastic pp scattering is possible, which allows for the color transparency factor the magnitude $T > 1$. The filter mechanism in its pure form can be studied in quasielastic electron-proton scattering on nuclei: In these processes the colour transparency is determined by the filter factor only.

Authors are grateful to N.N.Nikolaev, R.H.Petry and M.B.Zhalov for useful discussions. Two of us (V.V.A. and L.G.D.) thank M.G.Huber for the hospitality in ITKP, Bonn University, where this work was partly done, and Deutsche Forschungsgemeinschaft for financial support during this visit.

277

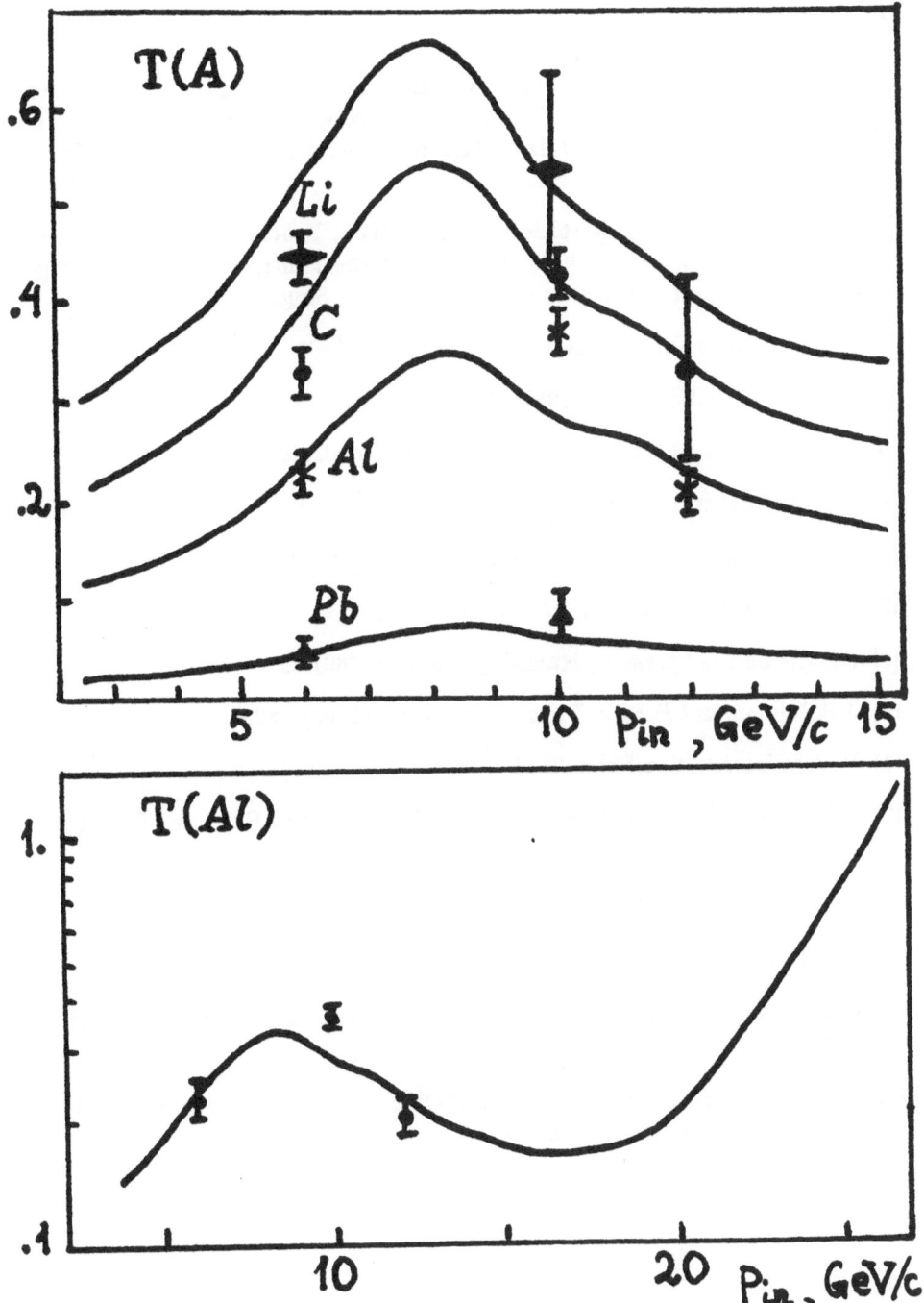

Fig.4 Colour transparency, T, for Li, C, Al and Pb calculated with the use of the colour screening effect.

References

[1] A.H.Mueller, in Proceedings of the 17th Rencontre de Moriond: II.Elementary Hadronic Processes and New Spectroscopy, Les Arcs, France, 1982, edited by J.Tran Thanh Van (Editions Frontierres, Gif-sur-Yvette,1982).

[2] S.J.Brodsky, in Multiparticle Dynamics 1982, Proceedings of the 13th International Symposium, Volendam, The Netherlands, edited by W.Metzger, E.W.Kittel, and A.Stergiou (World Scientific, Singapore, 1983).

[3] A.S.Carroll et al., Phys.Rev.Lett., 61 (1988) 1698.

[4] J.Botts, Phys.Rev.,D44 (1991) 2768.

[5] S.J.Brodsky and G.F.de Teramond, Phys.Rev.Lett.,61 (1988) 1924.

[6] V.V.Anisovich, L.G.Dakhno and V.A.Nikonov, Phys.Rev.,D44 (1991) 1385.

[7] S.J.Brodsky and G.R.Farrar, Phys.Rev.Lett.,31 (1973) 1153.

[8] P.V.Landshoff, Phys.Rev.,D10 (1974) 1024.

[9] J.Botts and G.Sterman, Nucl.Phys.,B325 (1989) 62.

[10] J.P.Ralston and B.Pire, Phys.Rev.Lett.,61 (1988) 1823.

[11] V.V.Anisovich, Sov.Phys.JETP Lett.,21 (1975) 174.

[12] V.V.Anisovich, P.E.Volkovitsky and V.I.Povzun, Sov.Phys.JETP, 43 (1976) 841.

[13] P.Kroll, "New Results from the Diquark Model". Talk given at the 31th Cracow School of Theor.Physics, Zacopane,1991, Wuppertal Univ. Preprint WU-B 91-29(1991).

[14] B.Kopeliovich, L.Lapidus and A.Zamolodchikov, Sov.Phys.JETP Lett., 33 (1981) 612.

[15] E.M.Levin and M.G.Ryskin, 'Ideas of QCD in Nuclei', Proc. of LNPI Winter School, Leningrad, v. 2 (1984) 147.

The Neutron Electric Dipole Moment in Chiral Quark-Meson Models

Judith A. McGovern and Michael C. Birse

Department of Theoretical Physics, University of Manchester
Manchester, M13 9PL, U.K.

Abstract. The neutron electric dipole moment is calculated in the cloudy bag model and the colour dielectric model. In both the $U(1)_A$ anomaly is consistently incorporated. Both models give a similar value for the pion loop contribution, but the cloudy bag model has an equally large direct contribution which is proportional to the anomalous magnetic moment.

Recently interest has re-awakened in an old problem: the calculation of the the strong CP-violating contribution to the neutron electric dipole moment (d_n) in models of nucleon structure. This quantity is proportional to the coefficient θ of the CP-violating term in the QCD Lagrangian,

$$\mathcal{L}_{\text{QCD}}^{\text{CP}} = -\theta \frac{g^2}{32\pi^2} F^{\mu\nu} F_{\mu\nu}^*.$$

It is important as it provides the best experimental limits on θ, and hence places stringent constraints on theories of CP-violation.

The necessity of incorporating both the $U(1)_A$ anomaly and chiral symmetry breaking correctly in models of low-energy QCD was recognised a decade ago.[1] In particular the effects of strong CP-violation should vanish if either the anomaly or any quark mass vanishes. However many calculations of d_n have failed to satisfy this requirement.

We have calculated the neutron electric dipole moment in two widely-used models for the quark structure of baryons, the cloudy bag and colour-dielectric models, using versions of the models which correctly incorporate the axial anomaly. In order to work in the most natural frame, we make an axial rotation to a CP-invariant vacuum. In models based on a non-linear realisation of chiral symmetry such as the cloudy bag model this leads to a CP-violating term in the quark Lagrangian which is similar to that of Baluni,[2] except for a factor which vanishes in the absence of the anomaly. This term leads to a direct quark contribution to d_n, which is the same as that given in ref. 3 apart from the anomaly factor (which is numerically close to one). We have shown that the quark contribution is proportional to the neutron anomalous magnetic moment and so has the same form as the tree-level contribution calculated in baryonic models.[4]

The CP-violating term in the Lagrangian also gives rise to an abnormal-parity πN coupling and hence to a pion-loop contribution to d_n. The magnitude of the loop term we get is similar to that of Morgan and Miller,[3] but has the opposite sign. Hence the quark and pion-loop terms reinforce, as one would expect. The effect of CP-violation on the down quarks is to displace them along the positive z-direction, while the up quark is moved in the opposite direction, so the direct quark contribution to d_n is negative. The pion loop term is due to virtual negative pions. These are emitted from the down quarks in the neutron and so their source is displaced in the positive z-direction. Hence these also make a negative contribution to d_n.

Explicit quark mass terms violate the principle of PCAC which is phenomenologically very successful. We have therefore looked at models with linear realisations of chiral symmetry, in which explicit chiral symmetry breaking is included through terms which change the vacuum expectation values of the scalar fields. In such models there is no direct or tree-level contribution to d_n. Strong CP-violation does lead to an abnormal-parity πN coupling by mixing pions with isovector scalar mesons, and hence there is a loop contribution to d_n. We have calculated this in a chirally-symmetric version of the colour-dielectric model.

Neither of the models we consider leads to a cancellation between two contributions to d_n. Hence the limits we can impose on θ are as stringent as those from most other estimates of d_n. From the most recent experimental limit on the dipole moment, $d_n < 1.2 \times 10^{-25} e$ cm (ref. 5), our calculation of d_n in the colour-dielectric model gives an upper bound of $\theta < 10^{-9}$. This would be about halved for the cloudy bag model.[6]

1. M. A. Shifman, A. I. Vainshtein and V. I. Zakharov, Nucl. Phys. **B166** (1980) 493.
2. V. Baluni, Phys. Rev. **D19** (1979) 2227.
3. M. A. Morgan and G. A. Miller, Phys. Let. **B179** (1986) 379.
4. S. Aoki and T. Hatsuda, Phys. Rev. **D45** (1992) 2427.
5. K. F. Smith, Phys. Let. **B234** (1990) 191.
6. J. A. McGovern and M. C. Birse, Phys. Rev. **D45** (1992) 2437.

The Time-Like E.M. Form Factors of the Neutron: First Measurement at Adone

A. Antonelli[3], R. Baldini[3], M. Bertani[2], M.E.Biagini[3], V. Bidoli[6], C. Bini[5], T. Bressani[7], R.Calabrese[2], R. Cardarelli[6], R. Carlin[4], C. Casari[1], L. Cugusi[1], P. Dalpiaz[2], G. DeZorzi[5], A. Feliciello[7], P.F. Dalpiaz[2], M.L. Ferrer[3], P. Gauzzi[5], P. Gianotti[7], E. Luppi[2], S. Marcello[7], A. Masoni[1], R. Messi[6], M. Morandin[4], L. Paoluzi[6], E. Pasqualucci[6], G. Pauli[8], F. Petrucci[2], M. Posocco[4], M. Preger[3], G. Puddu[1], M. Reale[6], L. Santi[9], R. Santonico[6], P. Sartori[4], M. Savrie[2], S. Serci[1], M. Spinetti[3], S. Tessaro[8], C. Voci [4].

[1] Universitá and INFN, Cagliari, I-09100, Italy
[2] Universitá and INFN, Ferrara, I-44100, Italy
[3] INFN Laboratori Nazionali di Frascati, I-00044, Italy
[4] Universitá and INFN, Padova, I-35131, Italy
[5] Universitá and INFN, Roma I, I-00185, Italy
[6] Universitá and INFN, Roma II, I-00173, Italy
[7] Universitá and INFN, Torino, I-10125, Italy
[8] Universitá and INFN, Trieste, I-34127, Italy
[9] Universitá and INFN, Udine, I-33100, Italy.

Presented by M. Bertani

Abstract. Preliminary results of experiment *FENICE* performed at Adone e^+e^- storage ring in Frascati are presented here. The $J/\psi \to n\bar{n}$ and $J/\psi \to p\bar{p}$ branching ratios have been determined. At $q^2 = 4.4\ GeV^2$ a new measurement of the proton form factor has been performed and at $q^2 = 4.0\ GeV^2$ a first, preliminary value of the neutron form factor $G_n = (0.42 \pm 0.06)$ is obtained, assuming $|G_E| = |G_M|$.

1   Introduction·

One of the fundamental problems in physics is the understanding of the structure of the Nucleon, and this can be achieved through the study of its electromagnetic form factors. In the time-like region ($q^2 > 0$) the Nucleon (N) form factors can be measured through the annihilation process:

$$e^+e^- \to N\bar{N} \tag{1}$$

The differential cross section of this reaction in the 'one photon exchange' approximation, is given by:

$$\left(\frac{d\sigma}{d\Omega}\right)_{c.m.} = \frac{\alpha^2\beta}{4s}\left[|G_M|^2\left(1 + cos^2\theta\right) + \frac{4M^2}{q^2}|G_E|^2 sen^2\theta\right] \tag{2}$$

where: $s = q^2$, θ is the angle between the electron beam and the Nucleon direction, β and M are the velocity and mass of the Nucleon. From eq. 2 it is

Table 1: EVDM predictions for r

EVDM predictions	r
Cabibbo–Gatto (1961)	~ 14
Körner-Kuroda (1977)	~ 2
Voci et al. (1982)	~ 100
Dubnicka (1988)	~ 25

clear that at high q^2 the contribution of the term involving $|G_E|$ becomes very small, while near the Nucleon production threshold the angular distribution is nearly isotropic as expected (S-wave is dominant), and $|G_E| = |G_M|$.

The experimental knowledge of the electromagnetic form factors of the Nucleon is not very satisfactory. Regarding the proton form factors, they have been extensively investigated by high statistics experiments in the space-like region. Recently higher precision data on the proton form factors in the time-like region became available both at low (experiment PS-170 at LEAR [1]) and high q^2 (experiment E760 at Fermilab [1]). On the contrary neutron data are still poor in the space-like region and before this measurement there was no experimental information about the neutron electromagnetic structure in the time-like region.

Perturbative QCD (PQCD), Extended Vector Meson Dominance Models (EVDM) and other theoretical models give very different predictions on the neutron form factors in the time-like region. A suitable quantity to compare different models is the ratio:

$$r = \frac{\sigma(e^+e^- \to n\bar{n})}{\sigma(e^+e^- \to p\bar{p})} \tag{3}$$

According to QCD sum rules there is a leading quark in the Nucleon wave function [2], therefore at high q^2 the ratio r should be equal to the square of the ratio between the electric charges of the leading quarks, namely a value of $r = 0.25$ is expected. EVDM-inspired fits to the space-like form factors in contrast predict $r \simeq 2 \div 100$ depending on the location of the vector meson recurrences, which are not well established. Some EVDM predictions [3] are listed in tab. 1.

From both the theoretical and the experimental situation it is clear that data on $e^+e^- \to n\bar{n}$ are strongly demanded for a better understanding of the Nucleon structure. *FENICE* experiment, at the upgraded Adone e^+e^- storage ring, has been expecially designed for the measurement of the cross section of process 1 and hence for the determination of the neutron form factors in the time-like region. Data was taken in about 6 months periods from 1990 to 1992 at a center of mass energy range from below the $n\bar{n}$ production threshold to 3100 GeV (J/ψ). Data have been analysed for the determination of the branching ratio $J/\psi \to n\bar{n}$, and a preliminary analysis of about 1/3 of the statistics available near threshold has been performed.

2   Experimental Apparatus and Data

Adone is an e^+e^- storage ring operating at a center of mass energy ($\sqrt{s}$) ranging from 1.5 to 3.1 GeV, with a mean luminosity of $\sim 10^{29} cm^{-2} s^{-1}$ at $\sqrt{s} = 2\,GeV$. The *FENICE* experimental apparatus is described elsewhere in details [4]. It is a non magnetic detector, 2.5 m-long, 3 m-diameter, the acceptance for collinear tracks being $0.76 \times 4\pi$. It is a calorimeter consisting of layers of Limited Streamer Tubes (LST) [5], scintillation counters and iron plates in relative proportions as to allow detection of the antineutron in a 80% fraction of events, and of the neutron in a 20% fraction of events. The total thickness is of about 1.5 interaction lengths. The antineutrons are detected by means of the charged products of their annihilation on heavy nuclei, characterized by a typical 'star' topology. Their velocity can be determined by the time of flight (TOF) technique [6]. In order to reduce the interaction of cosmic rays events in the detector, the whole apparatus is surrounded by a 100 cm-thick concrete shield equipped with a double layer of Resistive Plate Counters (RPC) [7] used as veto in the trigger.

The trigger signal is provided by scintillation counters and RPC; many logic triggers are implemented for different final state topologies. The trigger for the detection of the $n\bar{n}$ final state does not make any requirement on the detection of the neutron. The overall trigger rate is about 10 Hz and the dead time due to data acquisition is about 3%.

The luminosity is continuously monitored by single bremmstrahlung in the interaction region with colliding or separated beams. Bhabha events collected by the detector are in agreement with this luminosity within 5%.

3   Data Analysis and Results

The off-line data analysis procedure is mainly based on LST pattern and TOF information. After rejection (97%) of cosmic rays events escaping the veto system and machine background events (produced by interaction between off-momentum beam particles and the beam pipe), we select e^+e^-, $\mu^+\mu^-$ and multihadronic (MH) events for monitoring purposes.

Parallel selections for $p\bar{p}$ and $n\bar{n}$ channels are performed on the remaining events: the $p\bar{p}$ selection is based on the requirement of two collinear (within 10°) tracks in the central detector, one of them ending with the annihilation star of the $\bar{p}$; the $n\bar{n}$ selection is essentially based on TOF cuts and the requirement of the annihilation star of the $\bar{n}$. At this point, for each $p\bar{p}$ candidate event the velocity $\beta = v/c$ for both the proton track and the antiproton annihilation vertex is determined. Analogously for the $n\bar{n}$ candidate events, the $\beta_{\bar{n}}$ of each event is determined from the position and the time of the $\bar{n}$ annihilation vertex. Both the $1/\beta_p$ and the $1/\beta_{\bar{n}}$ distributions are fitted with a gaussian added to a polynomial background, different at each $E_{C.M.}$. Finally from the fit the final number of $p\bar{p}$ and $n\bar{n}$ events is determined.

In the following the preliminary results at $\sqrt{s} = 3100, 2100, 2000\ GeV$ are reported.

3.1 Branching Ratios: $J/\psi \to p\bar{p}$ and $J/\psi \to n\bar{n}$

At the J/ψ resonance energy an integrated luminosity of 32 nb^{-1} has been analysed, corresponding to 43000 MH events, 3700 e^+e^- and 1400 $\mu^+\mu^-$ events. From the total multihadronic cross section integrated on an energy scan performed around the J/ψ a $\Gamma_{ee} = 5.3 \pm 0.3$ MeV is obtained, in good agreement with previous results [8]. The following branching ratios have been normalized to the total multihadronic J/ψ decay, thus allowing cancellation of most systematical errors.

A final sample of 51 events is selected in the $p\bar{p}$ channel. The overall measured efficiency, taking into account trigger, filters and selections, is $\epsilon_{p\bar{p}} = 0.33 \pm 0.03$. The $1/\beta_p$ (β_p being the p and $\bar{p}$ averaged velocities) distribution of the 51 events is reported in fig. 1a. The main sources of background for the channel $J/\psi \to p\bar{p}$ come from the processes: $J/\psi \to p\bar{p}+X$ (where $X = \eta$, π^0 decay in two γ's) and $J/\psi \to \pi^+\pi^-, K^+K^-, \pi^+\pi^-\pi^0$. A relative contamination of 6% is expected from these events. We obtain a branching ratio $Br(J/\psi \to p\bar{p}) = (2.03 \pm 0.32) \times 10^{-3}$, in good agreement with the world average [9] value of $(2.16 \pm 0.11) \times 10^{-3}$.

A final sample of 50 $n\bar{n}$ events is selected, the overall efficiency, including trigger, filters and selections, is $\epsilon_{n\bar{n}} = 0.12 \pm 0.01$. The two main background contributions come from energetic cosmic neutrons annihilating in the fiducial volume and from $J/\psi \to n\bar{n} + X$. In 7 out of the 50 candidates the neutron has also been detected, back-to-back with respect to the antineutron vertex.

The $1/\beta_n$ distribution of the 50 events (fig. 1b) is then fitted to the sum of a gaussian distribution and a two-components background obtained from selected cosmic rays and $n\bar{n} + X$ events. From the fit we obtain $N_{n\bar{n}} = 16 \pm 6$ events, and a branching ratio: $Br(J/\psi \to n\bar{n}) = (1.9 \pm 0.6) \times 10^{-3}$.

This result is consistent with the $p\bar{p}$ branching ratio and in agreement with the only other existing measurement (Bonanza experiment [10]). We should underline that the errors on the quoted branching ratios are only statistical, the systematic errors being negligible.

3.2 The Proton Form Factor Near Threshold

We have analysed 38 nb^{-1} corresponding to about 1/3 of the statistics collected at $\sqrt{s} = 2100\,MeV$. Following the same procedure for the $\bar{p}$ analysis previously described we find $N_{p\bar{p}} = 15 \pm 4$ candidate events without any background coming from e^+e^- interactions. Their $1/\beta_p$ distribution is shown in fig. 2a.

Integrating eq. 2 and correcting for the efficiency $\epsilon_p = 0.42 \pm 0.03$, we obtain a cross section:

$$\sigma(e^+e^- \to p\bar{p}) = (0.82 \pm 0.23)\,nb$$

Near the threshold we can assume $|G_E| = |G_M|$ and therefore it is possible to give a preliminary estimate of the proton form factor:

$$|G_p| = (0.25 \pm 0.04)$$

in agreement with the world average value [11]: $|G_p| = (0.21 \pm 0.02)$.

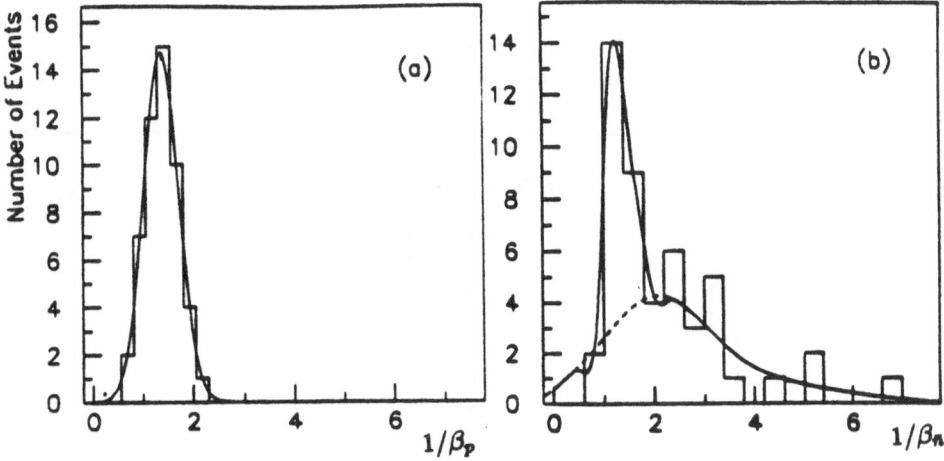

Figure 1: (a) $1/\beta_p$ distribution of the $J/\psi \to p\bar{p}$ candidates and (b) $1/\beta_{\bar{n}}$ distribution of the $J/\psi \to n\bar{n}$ candidates, the superimposed curve is the three-component fit described in the text, the dashed curve shows the background components.

3.3 The Neutron Form Factor Near Threshold

An integrated luminosity of 32 nb^{-1} out of the 100 nb^{-1} collected has been analysed at $\sqrt{s} = 2000~MeV$. Following the same antineutron selection criteria described before, 32 $n\bar{n}$ candidate events are obtained. At this energy the background comes only from neutron cosmic rays annihilating in the detector and simulating an antineutron star. Therefore the $1/\beta_{\bar{n}}$ distribution of these events (fig. 2) is fitted to the sum (continuous line) of a gaussian distribution and a background component (dashed line) obtained from cosmic rays events that passed the same selection criteria. The mean value of the distribution is in agreement with the one expected at this $\sqrt{s}$. The fit yields $N_{n\bar{n}} = 16 \pm 4$ events. Correcting for an overall efficiency $\epsilon_n = 0.27 \pm 0.03$ we measure a cross section

$$\sigma(e^+e^- \to n\bar{n}) = (1.9 \pm 0.5)~nb$$

Under the same assumption that $|G_E| = |G_M|$, it is possible to give a first, preliminary estimate of the neutron form factor:

$$|G_n| = (0.42 \pm 0.06)$$

The analysis of the $p\bar{p}$ events at the same energy is underway. In order to compare the $\sigma(e^+e^- \to n\bar{n})$ with the $\sigma(e^+e^- \to p\bar{p})$ we can use the result from DM2 [11] experiment at this energy ($\sigma(e^+e^- \to p\bar{p}) = 0.66 \pm 0.15nb$) and give an estimate of the ratio r (eq. 3):

$$r = (2.8 \pm 0.9)$$

This value is in agreement with some EVDM predictions, giving the indication of a neutron form factor larger than the proton one. In fig. 3 the *FENICE* preliminary results of the proton form factor at $q^2 = 4.4\ GeV^2$ and of the neutron form factor at $q^2 = 4.0\ GeV^2$ are reported, together with the compilation of the previously measured proton form factors at low q^2.

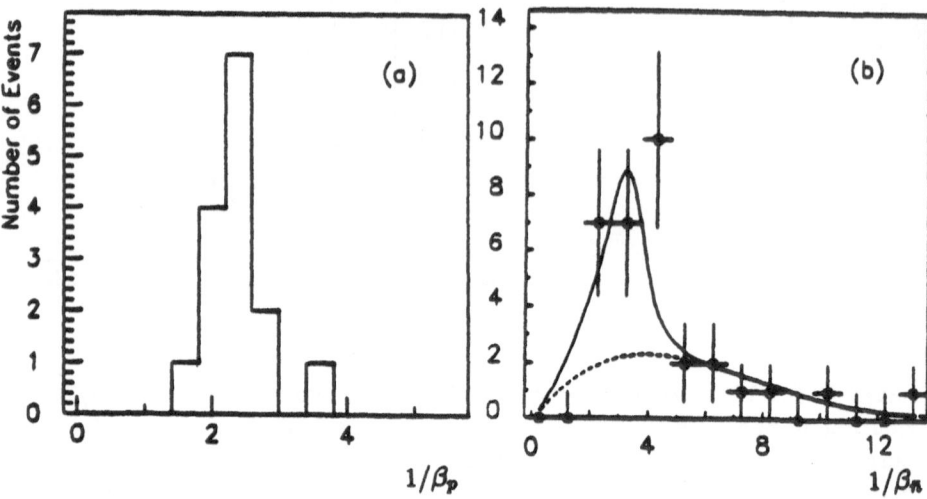

Figure 2: (a) $1/\beta_p$ distribution of the $p\bar{p}$ events at $q^2 = 4.4 GeV^2$ and (b) the $1/\beta_n$ distribution of the $n\bar{n}$ candidate events at $q^2 = 4 GeV^2$, the curve represents the two-components fit described in the text.

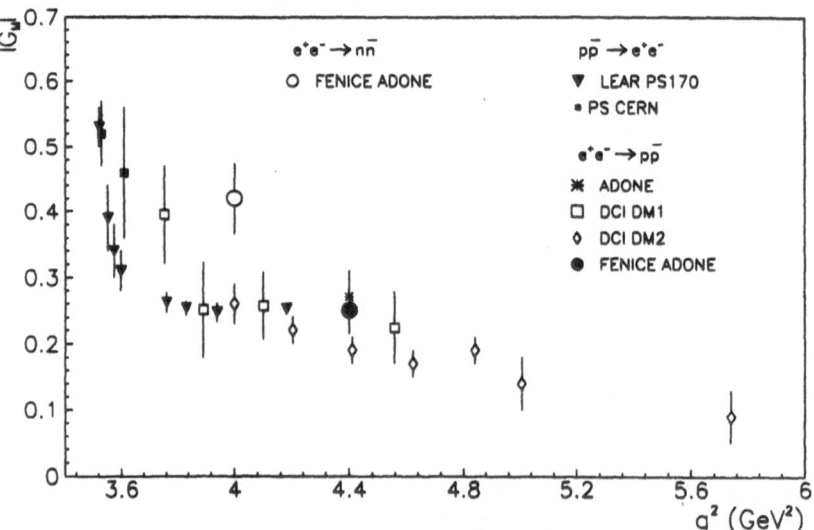

Figure 3: *FENICE* preliminary results on the proton and the neutron form factors together with the compilation of the previously measured proton form factors at low q^2.

4   Summary

We have reported on our preliminary results on the first measurement of the neutron form factor near threshold, and on a new measurement of the proton form factor at $q^2 = 4.4\,GeV^2$. The indication obtained is that the neutron form factor is larger than the proton one. We are analysing the rest of the data in order to reduce the statistical errors as well as to measure the electromagnetic form factors of the neutron also at $q^2 = 3.5$, 3.6, 3.7 and $5.9\,GeV^2$.

References

[1] P.Dalpiaz, invited talk Workshop on Low energy antiproton physics (Moscow, July 1991);
G.Bardin et al., Phys. Lett. **B255** (1991) 149;
G.Bardin et al., Phys. Lett. **B257** (1991) 514.

[2] V.L.Chernyak and I.R.Zhitnitsky, Nucl. Phys. **B246** (1984) 52;
S.J.Brodsky, Nucleon Structure Workshop (Frascati, October 1988).

[3] N.Cabibbo and R.Gatto, Phys. Rev. **124** (1961) 1577;
J.G.Körner and M.Kuroda, Phys. Rev. **D16** (1977) 2165;
P.Cesselli, M.Nigro and C.Voci, Proc. of Workshop on physics at LEAR, Erice (1982);
S.Dubnicka and E.Etim, Nuovo Cimento **100** (1988) 1.

[4] A.Antonelli et al. LNF 87-18(R) (1987);
A.Antonelli et al., submitted to IEEE Trans. on Nucl. Sci. (1992);
A.Antonelli et al., 'LEAP90', Stocholm (1990).

[5] G.Battistoni et al, Nucl. Instr. and Meth. **164** (1979) 57.

[6] T.Bressani et al., Nucl. Inst. and Meth. **A292** (1990) 563.

[7] R.Santonico and R.Cardarelli, Nucl. Instr. and Meth. **187** (1981) 377.

[8] J.P.Alexander et al. Phys. Rev **D37** (1988) 56; and references therein.

[9] I.Peruzzi et al., Phys Rev. **D17** (1978) 2901;
R.Brandelik et al., Z. Phys. **C1** (1979) 233;
M.W.Eaton et al., Phys. Rev. **D29** (1984) 804;
D.Pallin et al., Nucl. Phys. **B292** (1987) 653.

[10] H.J.Besh et al., Phys. Lett. **B78** (1978) 374.

[11] D.Bisello et al., Nucl. Phys. **B224** (1983) 379.

Diquarks in the Instanton Picture

Dmitri Diakonov and Victor Petrov

St.Petersburg Nuclear Physics Institute, Gatchina, St.Petersburg 188350, Russia

Abstract. We recall the 't Hooft-like many-quark interactions induced by instantons. By introducing auxiliary boson fields these interactions can be bosonized in two different ways, "standard" and "non-standard". The "standard" bosonization involving only scalar and pseudoscalar meson fields is justified in the limit $N_c \to \infty$. We consider in some detail a special case of QCD with $N_c = N_f = 2$ whose global symmetry is not $U(2)_L \times U(2)_R$ but actually $U(4)$. The "standard" bosonization violates this symmetry while the "non-standard" preserves it. The "non-standard" bosonization stresses the diquark degrees of freedom. The real world with $N_c = N_f = 3$ lies perhaps in between the imaginary worlds with $N_c = N_f = 2$ and $N_c = \infty$. Therefore, one would expect that the diquark degrees of freedom could play some role in the real world as well.

Keywords. Chiral symmetry breaking, instantons, 't Hooft interactions, diquarks

1 't Hooft interactions

It is well known that to have chiral symmetry breaking one needs a finite spectral density $\nu(\lambda)$ of the Dirac operator at zero eigenvalues, since the chiral condensate is proportional exactly to this quantity: $<\bar{\psi}\psi> = -\pi\nu(0)/V^{(4)}$[1]. A natural way to get $\nu(0) \neq 0$ is to have a finite density of instantons and antiinstantons in the 4-dimensional space-time $N/V^{(4)}$. Indeed, in the presence of the topologically non-trivial gluon fluctuations fermions necessarily have an exact zero mode, as it follows from the Atiah – Singer index theorem. Inside the ensemble of instantons and anti-instantons (I's and $\bar{I}$'s for short) the would-be zero fermion modes in the background field of individual pseudoparticles are smeared into a band with a finite spectral density at zero eigenvalues [2]. The instanton vacuum provides thus a beautiful mechanism of chiral symmetry breaking [3].

There are two mathematically equivalent ways to treat quark observables in the instanton vacuum. One is to calculate an observable in a given instanton background and then to average over the collective coordinates of I's and $\bar{I}$'s. This method was developed in ref.[3]. An equivalent though probably more convenient method was developed in ref.[4, 5]. The idea was to average first over the instanton ensemble. This averaging induces many-quark interactions whose simplified version was first suggested by 't Hooft[6]. Using this effective quark interaction theory one can then calculate various observables. According to the derivation of ref.[4, 5] averaging over the instanton ensemble leads to the following form of the QCD partition function (N_f is the number of light fermion flavours while N_c denotes the number of colours):

$$\mathcal{Z} = \int D\psi D\bar{\psi} \, \exp \left(i \int d^4x \sum_{f=1}^{N_f} \bar{\psi}_f i\partial\psi_f + iY_{N_f}^{(+)} + iY_{N_f}^{(-)} \right) \qquad (1)$$

where $Y_{N_f}^{(\pm)}$ are 't-Hooft-like $2N_f$-fermion vertices generated by I's and $\bar{I}$'s respectively. It will be convenient to decompose the 4-component Dirac spinors describing quark fields into left- and right-handed 2-component Weyl spinors which we denote as

$$\psi_{L(R)}^{f\alpha i}$$

where $f = 1 \ldots N_f$ stand for flavour, $\alpha = 1 \ldots N_c$ stand for colour and $i = 1, 2$ stand for spin indices. The vertices can be conveniently written in the momentum space and using the above Weyl spinors for quarks:

$$Y_{N_f}^{(+)} = \frac{N_+}{V} \left(\frac{N_c V}{N_+} \right)^{N_f} \int \frac{d^4k_1}{(2\pi^4)} \cdots \int \frac{d^4k_{N_f}}{(2\pi^4)} \int \frac{d^4l_1}{(2\pi^4)} \cdots \int \frac{d^4l_{N_f}}{(2\pi^4)}$$

$$\times (2\pi)^4 \delta(l_1 + \ldots + l_{N_f} - k_1 - \ldots - k_{N_f}) \prod_{n=1}^{N_f} \sqrt{M(k_n)M(l_n)}$$

$$\times \varepsilon^{i_n \gamma_n} \varepsilon_{j_n \beta_n} \int dU U_{\gamma_n}^{\alpha_n} U_{\delta_n}^{\dagger \beta_n} \left(-\bar{\psi}_{Rf_n \alpha_n i_n}(k_n) \psi_L^{g_n \delta_n j_n}(l_n) \right). \qquad (2)$$

where N_+/V is the 4-dimensional density of I's. For the vertices $Y^{(-)}$ generated by $\bar{I}$'s one has to replace all left-handed Weyl spinors by right-handed ones and vice versa and the number of I's N_+ by the number of $\bar{I}$'s N_-; in the CP symmetrical vacuum $N_+ = N_- = N/2$. The integral $\int dU$ means averaging over orientations of I's ($\bar{I}$'s) in the colour space. In particular, one has

$$\int dU = 1,$$

$$\int dU U_\gamma^\alpha U_\delta^{\dagger \beta} = \frac{1}{N_c} \delta_\delta^\alpha \delta_\gamma^\beta, \text{ etc.} \qquad (3)$$

Note that in eq. (2) there is a formfactor function $\sqrt{M(k)}$ attributed to every fermion entering the vertex. This function is related to the Fourier-transformation of the zero fermion mode, to be precise,

$$\sqrt{\frac{M(k)}{M(0)}} = 2z[I_0(z)K_1(z) - I_1(z)K_0(z) - \frac{1}{z}I_1(z)K_1(z)] \xrightarrow{z \gg 1} \frac{3}{4z^3}$$

$$z = k\rho/2, \qquad (4)$$

where ρ is the average size of instantons in the vacuum.

To get the $2N_f$-fermion vertices in a closed form one has to perform explicitly the integration over the instanton orientations. We cite here the results for $N_f = 1, 2$ and 3.

$N_f = 1$. In this case the 't Hooft "vertex" is just a mass term,

$$Y_1^{(+)} = -\int \frac{d^4k}{(2\pi^4)} \bar{\psi}_R(k)\psi_L(-k)M(k) \tag{5}$$

with a momentum-dependent dynamical fermion mass whose value at zero momentum is found from a "self-consistency" condition[3, 4]:

$$\frac{4VN_c}{N} \int \frac{d^4k}{(2\pi^4)} \frac{M^2(k)}{M^2(k) + k^2} = 1. \tag{6}$$

Therefore, if the properties of the instanton vacuum are known, such as the average size ρ and the density N/V, the dynamical mass of quarks is immediately found from eqs.(4,6). For example, we usually choose the values $\rho = (600 \ MeV)^{-1}$, $N/V = (200 \ MeV)^{-4}$, corresponding to the correct value of the gluon condensate and of the topological susceptibility of the vacuum[7]. We then find $M(0) \approx 350 \ MeV$.

$N_f = 2$. In this case one gets a nontrivial 4-fermion interaction:

$$Y_2^{(+)} = \frac{2N_c^2 V}{N} \int \frac{d^4k_1 d^4k_2 d^4l_1 d^4l_2}{(2\pi)^{12}} \delta(l_1 + l_2 - k_1 - k_2)\sqrt{M(k_1)M(k_2)M(l_1)M(l_2)}$$

$$\times \frac{1}{2!(N_c^2 - 1)} \varepsilon^{f_1 f_2} \varepsilon_{g_1 g_2} \left[\left(1 - \frac{1}{2N_c}\right) (\bar{\psi}_{Rf_1}(k_1)\psi_L^{g_1}(l_1)) \, (\bar{\psi}_{Rf_2}(k_2)\psi_L^{g_2}(l_2)) \right.$$

$$\left. + \frac{1}{8N_c} (\bar{\psi}_{Rf_1}(k_1)\sigma_{\mu\nu}\psi_L^{g_1}(l_1)) \, (\bar{\psi}_{Rf_2}(k_2)\sigma_{\mu\nu}\psi_L^{g_2}(l_2)) \right] \tag{7}$$

Note that the second (tensor) term here is negligible at $N_c \to \infty$. Using the identity

$$2\varepsilon^{f_1 f_2}\varepsilon_{g_1 g_2} = \delta_{g_1}^{f_1}\delta_{g_2}^{f_2} - (\tau^A)_{g_1}^{f_1}(\tau^A)_{g_2}^{f_2} \tag{8}$$

one can rewrite the leading (first) term of eq. (7) as

$$(\bar{\psi}\psi)^2 + (\bar{\psi}\gamma_5\psi)^2 - (\bar{\psi}\tau^A\psi)^2 - (\bar{\psi}\tau^A\gamma_5\psi)^2 \tag{9}$$

which resembles closely the Nambu—Jona-Lasinio model. We would like to stress that in contrast to that *ad hoc* model the vertex of eq. (7) i) violates explicitly the $U_A(1)$ symmetry, ii) has a fixed interaction strength and iii) contains an intrinsic ultra-violet cut-off due to the formfactor function $M(k)$.

$N_f = 3$. In this case one has a 6-fermion vertex of the following structure:

$$Y_3^{(+)} = \frac{N}{2V} \left(\frac{2N_c V}{N}\right)^3 \sqrt{M^6(k)} \frac{1}{3!N_c(N_c^2 - 1)} \varepsilon^{f_1 f_2 f_3} \varepsilon_{g_1 g_2 g_3}$$

$$\times \left[(1 - \frac{3}{2(N_c + 2)})\, (\bar{\psi}_{Rf_1} \psi_L^{g_1})\, (\bar{\psi}_{Rf_2} \psi_L^{g_2})\, (\bar{\psi}_{Rf_3} \psi_L^{g_3}) \right.$$

$$\left. + \frac{3}{8(N_c + 2)}\, (\bar{\psi}_{Rf_1} \psi_L^{g_1})\, (\bar{\psi}_{Rf_2} \sigma_{\mu\nu} \psi_L^{g_2})\, (\bar{\psi}_{Rf_3} \sigma_{\mu\nu} \psi_L^{g_3}) \right]. \tag{10}$$

Here the second term is again suppressed at $N_c \to \infty$.

Any N_f. Let us note that for *any* N_f the *leading* term at $N_c \to \infty$ can be written as a determinant,

$$Y_{N_f}^{(\pm)} \stackrel{N_c \to \infty}{=} \left(\frac{2V}{N}\right)^{N_f - 1} \int d^4x \, \det J^{(\pm)}, \tag{11}$$

where $J^{(\pm)}$ are $N_f \times N_f$ matrices made of quark bilinears:

$$J_{fg}^{(\pm)}(x) = \int \frac{d^4k \, d^4l}{(2\pi)^8} e^{i(k-l,x)} \sqrt{M(k)M(l)} \bar{\psi}^f(k) \frac{1 \pm \gamma_5}{2} \psi^g(l). \tag{12}$$

2 Standard bosonization

One can linearize the many-fermion vertices by introducing integration over auxiliary boson fields. This formal procedure is called "bosonization" of the theory. Roughly speaking, when one has a theory with a four-fermion interaction, it can be viewed as a limit of a one-boson-exchange when the mass of intermediate boson tends to infinity. This is the meaning of the bosonization.

In case $N_f = 2$ the 't Hooft interactions are 4-fermion ones (see eqs.(7,9)) and it is very simple to bosonize them by introducing scalar and pseudoscalar fields. Note that the non-leading second term in eq. (7) needs additional tensor fields for the bosonization. In case $N_f \geq 3$ the 't Hooft interactions become 6-fermion and so on, and the bosonization becomes less trivial. Let us show how it can be performed in the limit of $N_c \to \infty$[4, 5].

We note first that at $N_c \to \infty$ the 't Hooft interactions can be written in the determinant form (see eq. (11)). Second, we note that if one introduces $N_f \times N_f$ matrices $\mathcal{M}$ the following equation becomes true in the saddle-point approximation:

$$\int d\mathcal{M} \exp\left(-ia(\det \mathcal{M})^{\frac{1}{N_f-1}} + i\text{Tr}(\mathcal{M}J)\right) = \exp\left[i \det J \left(\frac{(N_f - 1)}{a}\right)^{(N_f - 1)}\right]. \tag{13}$$

This remarkable formula enables one to bosonize the many-fermion interactions of the determinant type. It should be stressed however that the procedure is justified only at large N_c: otherwise i) the 't Hooft interactions have not the simple determinant form and ii) the saddle-point evaluation of the integral in eq. (13) is not justified. We notice that at $N_f = 2$ eq. (13) becomes exact since in this case it deals with a Gaussian integral.

Using eqs.(13,11) we can rewrite our theory (eq. (1)) in the following bosonized form:

$$\mathcal{Z} = \int D\mathcal{M}_R \int D\mathcal{M}_L \int D\psi \int D\bar{\psi} \exp\left[i \int d^4x\, \bar{\psi} i\partial \psi\right]$$

$$\times \exp\left[-i\frac{N(N_f - 1)}{2V} \int d^4x \left((\det\mathcal{M}_R)^{\frac{1}{N_f-1}} + (\det\mathcal{M}_L)^{\frac{1}{N_f-1}}\right)\right]$$

$$\times \exp\left[-i \int d^4x \int \frac{d^4k\, d^4l}{(2\pi)^8} \sqrt{M(k)M(l)}\, e^{i(k-l,x)} \bar{\psi}(k) \left(\mathcal{M}_L(x)\frac{1+\gamma_5}{2}\right.\right.$$

$$\left.\left. +\mathcal{M}_R(x)\frac{1-\gamma_5}{2}\right) \psi(l)\right]. \tag{14}$$

It is the theory of massless quarks interacting with two meson fields $\mathcal{M}_L(x)$ and $\mathcal{M}_R(x)$ (being $N_f \times N_f$ matrices) which have no kinetic energy term. A useful parametrization of the meson fields is given by

$$\begin{aligned}\mathcal{M}_L(x) &= [\sigma(x) + \eta(x)]U(x)V(x),\\ \mathcal{M}_R(x) &= [\sigma(x) - \eta(x)]V(x)U^\dagger(x)\end{aligned} \tag{15}$$

(U and V are unitary $N_f \times N_f$ matrices). It introduces scalar-isoscalar field $\sigma(x)$, pseudoscalar-isoscalar field $\eta(x)$ and two more fields: scalar $\sigma^A(x)$ and pseudoscalar $\pi^A(x)$ which belong to the adjoint representation of the $SU(N_f)$ flavour group:

$$\begin{aligned}V(x) &= \exp i\sigma^A(x)\lambda^A,\\ U(x) &= \exp i\pi^A(x)\lambda^A\end{aligned} \tag{16}$$

where λ^A are the Hermitian generators of $SU(N_f)$.

The vacuum state of the theory given by eq. (14) corresponds to a nonzero value of the σ field. It leads to the spontaneous breakdown of chiral symmetry, and the π^A fields become Goldstone particles.

To reveal the nonzero value of the σ field let us calculate the effective potential for it. Putting

$$\sigma = const, \eta = 0, U = V = 1 \tag{17}$$

and integrating out the quark degrees of freedom we obtain the following formula for the effective potential $V_{eff}(\sigma)$:

$$V_{eff}(\sigma) = \frac{N}{V(N_f - 1)}\sigma^{\frac{N_f}{N_f-1}} - 2N_f N_c \int \frac{d^4k}{(2\pi^4)} \log\left(1 + \frac{\sigma^2 M^2(k)}{k^2}\right). \tag{18}$$

The minimum of this potential in σ is found from the equation $\partial V/\partial\sigma = 0$:

$$\sigma^{\frac{N_f}{N_f-1}} = \frac{4V N_c}{N} \int \frac{d^4k}{(2\pi^4)} \frac{\sigma^2 M^2(k)}{\sigma^2 M^2(k) + k^2} \tag{19}$$

which has a solution

$$\sigma = 1 \tag{20}$$

if we recall the "self-consistency" equation (see eq. (6)).

Let us note that constant fields π^A correspond to chiral rotations of the meson fields. They be can be eliminated from the quark determinant by changing the integration variables in eq. (14) according to

$$\psi \longrightarrow \exp[\frac{i}{2}\pi^A \lambda^A \gamma_5]\psi. \tag{21}$$

It means that the effective Lagrangian for the π^A fields contain only derivatives, i.e. pions are massless—in accordance with the Goldstone theorem. As to the singlet η, the quark determinant does not depend on a constant $U(1)$ chiral phase as well. However $\det \mathcal{M}_L$ and $\det \mathcal{M}_R$ depend explicitly on this phase. Therefore, the η meson is not a Goldstone particle, so that the $U(1)$ problem is solved by the 't Hooft interactions[6] .

3   Imaginary world with $N_c = N_f = 2$.

Our world corresponds to 3 colours (N_c =3) and to 3 light fermion species (N_f=3). Therefore, one may wonder whether the logic of "large N_c" should work numerically well for the real world.

In this section we investigate in a sense an opposite case: QCD with $N_c = N_f = 2$. It is quite a remarkable theory where certain exact statements can be made owing to an additional global symmetry of QCD in that case.

First of all we notice that baryons, i.e. colour-antisymmetric quark states of the type $\varepsilon_{\alpha\beta}\psi^\alpha\psi^\beta$ are bosons since the number of colours is even. Therefore, not only the usual chiral condensate $< \bar\psi\psi >$, but also that of the type $< \psi\psi >$ or $< \bar\psi\bar\psi >$ are in principle possible, and one may wonder which of them is more favourable. The answer is: it is in fact the *same* condensate: one can be obtained from another by certain global rotations. The point is, the global symmetry of QCD with $N_c = N_f = 2$ is not $U_L(2) \times U_R(2)$ as one would naively think but actually much larger— it is the $U(4)$ symmetry. To see it explicitly, let us use the spinor basis,

$$\gamma_\mu = \begin{pmatrix} 0 & \sigma_\mu^- \\ \sigma_\mu^+ & 0 \end{pmatrix}, \; \sigma_\mu^\pm = (\pm\vec\sigma, 1) \tag{22}$$

and let us decompose both Dirac fermions $\psi^{1,2}$ of the theory into 2-component Weyl spinors ϕ. One needs four Weyl spinors to write down two Dirac bi-spinors:

$$\psi^{1\alpha} = \begin{pmatrix} \varepsilon^{\alpha\beta}\varepsilon^{ij}\phi^+_{3\beta j} \\ \phi^{1\alpha i} \end{pmatrix} \boxed{\begin{matrix} R \\ L \end{matrix}}, \; \psi^{2\alpha} = \begin{pmatrix} \varepsilon^{\alpha\beta}\varepsilon^{ij}\phi^+_{4\beta j} \\ \phi^{2\alpha i} \end{pmatrix} \boxed{\begin{matrix} R \\ L \end{matrix}},$$

$$\psi^+_{1\alpha} = (\phi^+_{1\alpha i}, \; \varepsilon^{\alpha\beta}\varepsilon^{ij}\phi_{3\beta j}), \; \psi^+_{2\alpha} = (\phi^+_{2\alpha i}, \; \varepsilon^{\alpha\beta}\varepsilon^{ij}\phi_{4\beta j}). \tag{23}$$

The fermion Lagrangian written in terms of the two 4-component Dirac spinors,

$$\mathcal{L} = \sum_{f=1,2} \psi^+_{f\alpha}\gamma_\mu \left(i\partial_\mu 1 + g A^A_\mu t^A\right)^\alpha_\beta \psi^{f\beta} \tag{24}$$

can be identically rewritten through four 2-component Weyl spinors,

$$\mathcal{L} = \sum_{p=1}^{4} \phi_{p\alpha i}^{+}(\sigma_{\mu}^{-})_{j}^{i} \, (i\partial_{\mu}\, 1 + g A_{\mu}^{A} t^{A})_{\beta}^{\alpha} \, \phi^{p\beta j}, \tag{25}$$

the latter being evidently invariant under the $U(4)$ transformations mixing these four Weyl spinors,

$$\phi^{p\alpha i} \longrightarrow U_{q}^{p} \phi^{q\alpha i},$$
$$\phi_{p\alpha i}^{+} \longrightarrow \phi_{q\alpha i}^{+}(U^{+})_{p}^{q} \tag{26}$$

where $U \in U(4)$; $p, q = 1, 2, 3, 4$. The chiral $U_{L}(2) \times U_{R}(2)$ transformations form a subgroup of this $U(4)$ symmetry, corresponding to separate rotations of the first two and of the second two Weyl spinors.[1]

It is easy to check that under a transformation of a special form,

$$U = \begin{pmatrix} 1 & 0 & 0 & 0 \\ 0 & 0 & i & 0 \\ 0 & i & 0 & 0 \\ 0 & 0 & 0 & 1 \end{pmatrix} \tag{27}$$

the chiral condensate $i < \psi_{f\alpha i}^{+} \psi^{f\alpha i} >$ is transformed into a combination of "baryon" and "antibaryon" condensates:

$$i < \psi_{f\alpha i}^{+} \psi^{f\alpha i} > \xrightarrow{U} \frac{1}{2} \left(\varepsilon_{fg} \varepsilon_{\alpha\beta} \varepsilon_{ij} < \psi^{f\alpha i} \psi^{g\beta j} > + \varepsilon^{fg} \varepsilon^{\alpha\beta} \varepsilon^{ij} < \psi_{f\alpha i}^{+} \psi^{g\beta j} > \right). \tag{28}$$

It means that in a particular case of $N_{c} = N_{f} = 2$ there are more "flat" directions and hence more Goldstone modes in case of the spontaneous breaking of the underlying global symmetry.

Let us now discuss the instanton-induced 't Hooft interactions in the case of $N_{c} = N_{f} = 2$. The general form of these interactions is given by eq. (7) where one has to put now $N_{c} = 2$. However the $U(4)$ symmetry of eq. (7) at $N_{c} = 2$ is not what strikes the eye. To make it more explicit let us rewrite eq. (7) using the four 2-component Weyl spinors ϕ^{p} introduced by eq. (23). After some work with eq. (7) we get:

$$Y_{2} \sim \varepsilon_{\alpha\gamma} \varepsilon_{ik} \varepsilon_{\beta\delta} \varepsilon_{jl} \varepsilon_{pqrs} \phi^{p\alpha i} \phi^{q\beta j} \phi^{r\gamma k} \phi^{s\delta l},$$

$$\alpha, \beta, \gamma, \delta = 1, 2; \ i, j, k, l = 1, 2; \ p, q, r, s = 1, 2, 3, 4 \tag{29}$$

(a Hermitian-conjugate vertex with ϕ replaced by ϕ^{+} should be added to eq. (29) as due to the antiinstantons). The vertex given by eq. (29) is apparently invariant

---

[1] It is not easy to point out the author of the $U(4)$ symmetry. In general, symmetry involving fermions and antifermions has been discussed by Pauli and GÈrsey in the 50's[8, 9]. A concrete statement about the $U(4)$ symmetry of QCD with $N_{c} = N_{f} = 2$ was first heard by one of us from A.I.Vainshtein in the end of 70's. However, the earliest published paper with that statement which we have found is by Dimopoulos in 1980[10], see also [11]

under the $SU(4)$ (not $U(4)$!) transformations of eq. (26); as usually, the $U(1)$ symmetry is explicitly broken by instantons.

It is possible to bosonize exactly the 4-fermion interactions given by eq. (29). To this end we introduce 6 meson or, better to say, boson fields, as part of them will have the meaning of baryons or diquarks. The quark content of these bosons is

$$\mathcal{M}_{[pq]} \sim \varepsilon_{pqrs}\varepsilon_{\alpha\beta}\varepsilon_{ij}\phi^{r\alpha i}\phi^{s\beta j}; \tag{30}$$

6 other boson fields are obtained by Hermitian conjugation. All in all we need 12 auxiliary boson fields to bosonize the theory. They can be enumerated also in more familiar terms: 8 fields are the usual σ, η, σ^A, π^A ($A = 1, 2, 3$) scalar and pseudoscalar meson fields of the "standard" bosonization (see the previous section) and the extra 4 are the diquark fields,

$$\varepsilon^{\alpha\beta}\varepsilon^{ij}\psi^+_{L1\alpha i}\psi^+_{L2\beta j},\ \varepsilon_{\alpha\beta}\varepsilon_{ij}\psi^{1\alpha i}_L\psi^{2\beta j}_L,\ \text{and}\ (\text{L}) \longrightarrow (\text{R}). \tag{31}$$

As seen from eq. (30), the necessary boson fields form the 6-dimensional antisymmetric tensor representation of the $SU(4)$ group or, equivalently, a vector representation of the $SO(6)$ group.

Let us introduce the quark bilinear currents,

$$J^{pq} = \varepsilon_{\alpha\beta}\varepsilon_{ij}\phi^{p\alpha i}\phi^{q\beta j}. \tag{32}$$

The 't Hooft interactions (eq. (29) can be written as $\varepsilon_{pqrs}J^{pq}J^{rs}$+h.c., which can be bosonized by integrating over 12 auxiliary boson fields:

$$\exp \int d^4x\,(\varepsilon_{pqrs}J^{pq}J^{rs} + h.c.)$$

$$= \int D\mathcal{M}D\mathcal{M}^+ \exp \int d^4x\left[b(\varepsilon_{pqrs}J^{pq}\mathcal{M}^{rs} + h.c.) - \frac{b^2}{4a}(\varepsilon_{pqrs}\mathcal{M}^{pq}\mathcal{M}^{rs} + h.c.)\right]. \tag{33}$$

As in the case of the "standard" bosonization one can integrate first over the fermions. The fermion Lagrangian (in Euclidean space) has the form

$$\mathcal{L} = \phi^+_{p\alpha i}i\partial_\mu(\sigma^-_\mu)^i_j\phi_{p\alpha j} + \phi^{p\alpha i}\phi^{q\beta j}\mathcal{M}_{pq}\varepsilon_{\alpha\beta}\varepsilon_{ij} + \phi^+_{p\alpha i}\phi^+_{q\beta j}\mathcal{M}^{+pq}\varepsilon^{\alpha\beta}\varepsilon^{ij}, \tag{34}$$

as its determinant for a constant boson field $\mathcal{M}$ is

$$V_{eff} = 4\int \frac{d^4k}{(2\pi^4)}\log\left(1 + \frac{4\lambda_+}{k^2}\right)\left(1 + \frac{4\lambda_-}{k^2}\right) \tag{35}$$

where $\lambda_\pm$ are the (twice degenerate) eigenvalues of the 4×4 matrix

$$\mathcal{N}^\tau_t = \mathcal{M}_{pq}\mathcal{M}^{+p'q'}\varepsilon^{pqrs}\varepsilon_{p'q'ts}. \tag{36}$$

These eigenvalues can be expressed through the $SU(4)$ invariants,

$$\lambda_\pm = \frac{1}{2}\left(I \pm \sqrt{I^2 - KK^+}\right) \tag{37}$$

where

$$I = \mathcal{M}_{pq}\mathcal{M}^{+pq}, \ K = \mathcal{M}_{pq}\mathcal{M}_{rs}\epsilon_{pqrs}, \ K^+ = \mathcal{M}^{+pq}\mathcal{M}^{+rs}\epsilon_{pqrs}. \quad (38)$$

We remind the reader that one has to add to the V_{eff} of eq. (35) the quadratic mass terms (see eq. (33)). In notations of eq. (37) they are proportional to $K + K^+$. It means that V_{eff} depends in fact on 3 $SU(4)$ invariants. We conclude that out of 12 boson field necessary to bosonize the theory, $12 - 3 = 9$ fields are massless Goldstone particles. Out of these 9 three are the usual pseudoscalar pions and 6 others are certain combinations of "meson" and "diquark" fields.

It is remarkable that the existence of 6 additional massless exitations in this particular theory follows solely from symmetry considerations. Let us note that had we treated the theory from the "large N_c" point of view and applied the "standard" bosonization procedure, the actual $U(4)$ symmetry would be lost, and no extra massless bosons in addition to pions would be ever found. Of course, with $N_c = 2$ it looks risky to use the "large N_c" logic, but is the real case of $N_c = 3$ far better in that respect? It seems plausible that the real world with $N_c = N_f = 3$ lies in between the two idealizations — $N_c = N_f = 2$ and $N_c = \infty$. Therefore, one would expect that certain diquark degrees of freedom could be light ones and thus could play an important role in the real world as well.

Finally, let us note that the meson-diquark bosonization of a theory with colour has been first considered in ref.[12], see [13] and references therein. More recently the diquark bosonization has been studied in refs.[14, 15, 16], see also the talk by D.Ebert and M.Lutz at this conference.

Acknowledgement D.D would like to thank warmly the organizers of the conference — Professor Klaus Goeke, Professor Peter Kroll and Professor Herbert Petry for their generous hospitality at Bad Honnef and for the friendly and creative atmosphere throughout the conference.

References

[1] T.Banks and A.Casher, *Nucl. Phys.* **B169** (1980) 103

[2] D.Dyakonov and V.Petrov, *Phys. Lett.* **147B** (1984) 351

[3] D.Dyakonov and V.Petrov, *Sov.Phys.JETP* **62** (1985) 204,431; *Nucl. Phys.* **B272** (1986) 457

[4] D.Dyakonov and V.Petrov, preprint LNPI-1153 (1986); published (in Russian) in: *Hadron matter under extreme conditions*, Kiew (1986) p.192

[5] D.Dyakonov, Dr.Hab.Thesis, LNPI (1986) (unpublished)

[6] G.'t Hooft, *Phys. Rev.* **D** (1976) 8

[7] D.Dyakonov and V.Petrov, *Nucl. Phys.* **B245** (1984) 259

[8] W.Pauli, *Nuovo Cim.* **6** (1957) 206

[9] F.GÈrsey, *Nuovo Cim.* **7** (1958) 411

[10] S.Dimopoulos, *Nucl. Phys.* **B168** (1980) 69

[11] M.E.Peskin, *Nucl. Phys.* **B175** (1980) 197

[12] D.Ebert and V.N.Pervushin, *Theor. Math. Phys.* **36** (1978) 759

[13] D.Ebert and L.Kaschluhn, *Nucl. Phys* **B355** (1991) 123

[14] R.Cahill, J.Praschifka and C.J.Burden, *Austr. J. Phys* **42** (1989) 161

[15] U.Vogl, *Z. Phys.* **A337** (1990) 191

[16] D.Kahana and U.Vogl, *Phys. Lett.* **244B** (1990) 10

Lecture Notes in Physics

For information about Vols. 1–379
please contact your bookseller or Springer-Verlag

New Series m: Monographs